This book is about how to understand quantum mechanics by means of a modal interpretation. Modal interpretations provide a general framework within which quantum mechanics can be considered as a theory that describes reality in terms of physical systems possessing definite properties. The text surveys results obtained using modal interpretations, and is intended as both an accessible survey that can be read from cover to cover, and a systematic reference book.

Quantum mechanics is standardly understood to be a theory about probabilities with which measurements have outcomes. Modal interpretations are relatively new attempts, first proposed in the 1970s and 1980s, to present quantum mechanics as a theory which, like other physical theories, describes an observer-independent reality. In the 1990s much work has been carried out to develop fully these interpretations. In this book, Pieter Vermaas summarises the results of this work. A basic acquaintance with quantum mechanics is assumed.

This book will be of great value to undergraduates, graduate students and researchers in philosophy of science and physics departments with an interest in learning about modal interpretations of quantum mechanics.

PIETER VERMAAS studied philosophy and theoretical physics in his home town at the University of Amsterdam. He obtained his PhD with research on modal interpretations of quantum mechanics at Utrecht University with Dennis Dieks. He published several papers on especially the modal interpretation in the version proposed by Simon Kochen, Dennis Dieks and Richard Healey, in physics and philosophy journals ranging from *Physical Review Letters* to *Minnesota Studies of Philosophy of Science*. Together with Dennis Dieks he proposed a generalised modal interpretation. This generalisation has since formed the basis of much further research on modal interpretations. He has worked at the University of Cambridge with a British Council Fellowship. Currently he is a Research Fellow at the Delft University of Technology, where he is involved in developing the new field of philosophy of technology.

A Philosopher's Understanding of Quantum Mechanics

Possibilities and Impossibilities of a Modal Interpretation

Pieter E. Vermaas

Delft University of Technology

CAMBRIDGE UNIVERSITY PRESS

CAMBRIDGE UNIVERSITY PRESS
Cambridge, New York, Melbourne, Madrid, Cape Town, Singapore, São Paulo

Cambridge University Press
The Edinburgh Building, Cambridge CB2 2RU, UK

Published in the United States of America by Cambridge University Press, New York

www.cambridge.org
Information on this title: www.cambridge.org/9780521651080

First published 1999
This digitally printed first paperback version 2005

A catalogue record for this publication is available from the British Library

Library of Congress Cataloguing in Publication data
Vermaas, Pieter E.
A philosopher's understanding of quantum mechanics: possibilities
and impossibilities of a modal interpretation / Pieter E. Vermaas.
p.   cm.
Includes bibliographical references and index.
ISBN 0 521 65108 5
1. Quantum theory–Mathematical models.   I. Title
QC174.12.V46   1999
530.12–dc21   99-14416   CIP

ISBN-13 978-0-521-65108-0 hardback
ISBN-10 0-521-65108-5 hardback

ISBN-13 978-0-521-67567-3 paperback
ISBN-10 0-521-67567-7 paperback

Contents

Preface

When I decided to enter research on modal interpretations of quantum mechanics, I barely knew what it was about. I had attended a talk on the subject and read bits about them, but the ideas behind these interpretations didn't stick in my mind. Modal interpretations were at that time (1993) not widely known, and their approach to quantum mechanics was not common knowledge in the philosophy of physics. So my decision was a step in the dark. But what I did know was that I was beginning research on one of the most irritating and challenging problems of contemporary physics. Namely, the problem that quantum theories, unlike all other fundamental theories in physics, cannot be understood as descriptions of an outside world consisting of systems with definite physical properties.

Your decision to read this book may be a step in the dark as well, because modal interpretations are presently, especially among physicists, still rather unknown. The reason for this may lie in their somewhat isolated and slow development. The first modal interpretation was formulated in 1972 by Van Fraassen. Then, in the 1980s, Kochen, Dieks and Healey put forward similar proposals which, later on, were united under Van Fraassen's heading as modal interpretations. But these proposals were not immediately developed to fully elaborated accounts of quantum mechanics. Moreover, modal interpretations were proposed and discussed in journals and at conferences which were mainly directed towards philosophers of physics, rather than towards general physicists. Modal interpretations are in that sense true philosophers' understandings of quantum mechanics. But, as a possible down-side of that, the discussion of the possibilities and the impossibilities of the modal account remained slightly formal and therefore maybe not that appealing to the general physicist.

In the 1990s, however, the development of modal interpretations gained momentum and took a turn which made them much more accessible and interesting to a wider audience. A group of researchers started to work on modal interpretations and took up the challenge to systematically answer physical and theoretical questions about the way these interpretations describe our world. This has led to a burst of results about, for instance, the algebraic structure of the properties ascribed by modal interpretations, the correlations and the dynamics of those properties, the way in which modal interpretations describe measurements, and how one can philosophically and physically motivate modal interpretations.

These efforts have meant that nowadays many of the more important issues for modal interpretations have been resolved or have been proved to be unresolvable. Modal interpretations have thus matured into what can be taken as a well-developed and general framework to convert quantum mechanics into a description of a world of physical systems with definite properties. This general framework is of interest to anyone who aims at understanding quantum mechanics. Presently, one can therefore witness a second burst of activity, namely a burst of publications which present modal interpretations to the wider communities of physicists and philosophers and to those interested in philosophy and physics. This book introduces the reader to modal interpretations and guides him or her through many of their results. It may also be used as a reference book which can be consulted without the need to read it from cover to cover. The text is accessible to those who have a basic understanding of the quantum mechanical formalism. For experts I have added proofs of the various results in separate subsections.

This book is the result of five years of research at the *Institute for History and Foundations of Mathematics and the Natural Sciences* of Utrecht University. This research has started as a PhD project, supervised by Dennis Dieks and financially supported by the Foundation for Fundamental Research on Matter (FOM) and by the Foundation for Research in the Field of Philosophy and Theology (SFT) which is subsidised by the Netherlands Organisation for Scientific Research (NWO).

I thank Dennis Dieks for his invitation to work on modal interpretations. I feel indebted for the way in which he, one of the modal pioneers, supported my work and enabled me to develop my own views on the subject. I am also grateful to Tim Budden, Fred Muller and Jos Uffink, for their helpful discussions and advice, and for their friendship during my time at Utrecht University.

In addition to Dennis Dieks, I acknowledge the fruitfulness and importance for my work of discussions and joint projects with Guido Bacciagaluppi and Rob Clifton as well as with Michael Dickson, Matthew Donald and Meir Hemmo. I also thank the British Council for providing a fellowship to visit Cambridge University, and I thank Jeremy Butterfield for his friendly support.

Finally a word of dedication: It has become a tradition to dedicate academic books to those who are important to the author. However, to be honest, I have not written this book to honour my family, my friends or the one I love. Instead I have written it for those who wish to read it and, possibly, in dedication to the academic adventure to get to the heart of the matter. (And adding the names of the ones I am close to on one of the first pages of this book seems to me an academic variation of tattooing them on one of my arms, which, incidently, I haven't done either.) However, to meet tradition halfway, I heartily thank my parents, send sincere apologies to my friends for being absent during the period that I have worked on this book, and express my deep affection to Florentien Vaillant.

Delft University of Technology   Pieter Vermaas

1

Introduction

Imagine this strange island you have just set foot on. The travel agencies had advertised it as the latest and most exciting place to visit, an absolute must for those who still want to explore the unknown. So, of course, you decided to visit this island and booked with your friends a three-week stay. And now you've arrived and are sitting in a cab taking you from the airport into town. The landscape looks beautiful but strange. For some reason you can't take it in at one glance. You clearly see the part right in front of you, but, possibly because of the tiring flight, everything in the corners of your eyes appears more blurred than usual.

In town you buy a map. They don't sell one single map of the island but offer instead a booklet containing on each page a little map which covers only a small patch of the town or of the surrounding countryside. 'How convenient,' your friends say and off they go to explore this new and exciting place. But you approach things differently. You want to figure out where the places of interest are. So you buy the booklet, seek the nearest café, take out the pages and try to join the little maps together to make a single big one. Unfortunately you don't succeed; the little maps seem not to match at the edges. You start to suspect that the little maps are in some way incorrect. However, your friends, when they come around to see what's keeping you, tell you that the maps are fine: you just use the map containing your present position and when you reach the edge, you simply take the next map, look up your new position and continue.

You then try to convince your friends that something funny is going on. This island is patchwise accurately described by little maps but it is impossible to construct a map that depicts the island as a whole. Your friends agree, they have had some strange experiences themselves. For instance, in a bar they have found that if you order a drink, then, when you are absorbed in some discussion, you sometimes find that most of your drink is gone, even though

1

you can't remember having taken a single sip. When this happened the first time, they complained about it, but the bartender didn't look impressed and mentioned something about tunneling times.

During such a holiday I would definitely try to solve the puzzle of the map. I would try to draw a map myself, firstly only of, say, the coastline and then also of the main roads and streets. And I would test all kinds of hypotheses: maybe the little maps depict the island on different scales, maybe roads which are straight lines on the maps are actually curved, or maybe it is the other way round. But what if it really is impossible to make a general map? All the islands and towns we know of can be described by single maps, so what is wrong with this one? Is it some fantastic amusement park full of *trompes l'oeil* constructed by those travel agencies? Or does this island perhaps not exist at all? Have you perhaps landed in a huge *fata morgana*?

The possibility that there really doesn't exist a general map of an island would be challenging. All the islands we know can be described by single maps and we therefore can understand that these islands can be described by booklets of little maps as well: the little maps are just fragments of the general map. The question is now whether we can also understand a description of an island in terms of little maps if these maps are not fragments of a general description. In addition to this epistemic question, the challenge has an ontological twist. We usually assume that the islands on which we set foot are part of a physical world which exists outside of us and independently of us. And given that islands form sufficiently smooth spatial surfaces in this world, it is clear that there exist single maps of islands. But, conversely, if it is impossible to draw a single map of an island, then that island can't exist in the assumed way. The ontological part of the challenge is thus whether you are forced to conclude that an island on our planet doesn't exist if it can only be described patchwise by little maps. And if you indeed are forced to this conclusion, can you then still understand that this non-existent island is describable by maps at all?

There are at least three ways to handle this challenge. The first is the boldest one, namely to deny that there is a challenge in the first place: maps are only meant to find your way, and any requirement on maps over and above the requirement that they are effective for finding your way, is philosophically unfounded and superfluous. Hence, if there exists an island which can only be described by little maps, then this is simply a fact of life; there is no need to understand this patchwise description in terms of a single map, nor is it meaningful to draw on the basis of these little map any conclusion about the existence or non-existence of the island because maps of islands do not conceal information about the (ontological) existence of islands.

The second way is the pragmatic one, namely to just ignore the challenge, to join your friends and to enjoy the rest of your stay. Tomorrow your friends will hire a car — a vw-Golf — and drive through one of the town gates. This gate has two passages and if you drive through it with your eyes open, you will pass it in a fairly straight line, taking one of the passages. But if you drive through it with your eyes closed, you'll feel that your car is making the most peculiar manoeuvres. You certainly don't pass through the gate in a straight line and at the end it's more or less impossible to determine through which of the passages you went.[1]

The third way to approach the challenge is a more introspective one, namely to question your notions about islands. If all the known islands allow a description in terms of a single map, it is natural to assume that the island you're presently on can be described by a single map as well. So, if your notions are telling you that such a map doesn't exist, it seems sensible to assume that there is something wrong with your notions. Or, from an ontological point of view, the island you're on does exist in some sort of way: you can see it, live on it and find your way about on it. So, if your notions about islands tell you that it can't exist, then there must again be something wrong with your notions. If you take this third route, your stay will start to resemble the adventures of Raphaël Hythlodaeus in *Utopia*, of Mr Higgs in *Erewhon* and of Captain Gulliver during his travels,[2] or, for those who prefer more contemporary science fiction, of Captain Kirk in *Star Trek*. All these stories have in common that what appears to be an exploration of the unknown is also an investigation of our own presuppositions. In the three novels, the challenged presuppositions mainly concern our views about society, but in modern science fiction our ideas about physical reality are also questioned. Captain Kirk is thus not only exploring strange new worlds; our own is investigated as well.

Quantum mechanics is, of course, not the theory with which one describes exotic islands. Instead, it is a theory about light and about elementary particles such as electrons and protons. However, quantum mechanics does confront you with questions which are similar to the ones presented by the island. On the one hand, quantum mechanics gives descriptions of the behaviour of light and elementary particles which conform with our observations. For instance, according to quantum mechanics light is diffracted by slits in walls in specific ways and using photographic plates we can

---

[1] Some even whisper that you pass through both passages simultaneously.
[2] See More (1516), Butler (1872) and Swift (1726).

indeed register the resulting diffraction patterns. Also quantum mechanics predicts that the electrons emitted by radioactive atoms, are emitted at specific rates and leave trails of bubbles when they fly through chambers filled with supersaturated water vapour, and we can check those rates and observe those trails. The predictions by quantum mechanics are even so reliable that they can be put to work: the laser which scans the discs in your CD-player, for instance, functions according to quantum mechanical principles. Moreover, quantum mechanics is the first in a series of more sophisticated quantum theories, such as quantum field theories, which are generally seen as physically fundamental and universally valid. Hence, if there exists a description of light and elementary particles, there are good reasons to assume that this description is consistent with the predictions given by quantum mechanics.

On the other hand, quantum mechanics doesn't provide a full description of light and elementary particles. In its standard formulation quantum mechanics assigns a quantum mechanical state to a system and that state has a meaning only in terms of outcomes of measurements performed on that system. Imagine, for instance, an experiment in which you shoot a particle, say an electron, at a distant screen. Quantum mechanics then tells you that given that the electron is shot, you can assign a certain state to the electron, and from this state you can calculate that the electron hits the screen with certain probabilities at specific spots. However, quantum mechanics is silent about how the electron flies from the source to the spot where it finally hits the screen: it doesn't give a trajectory through space which the electron follows, nor does it give values for magnitudes like the velocity and the energy of the electron. Now, in this particular example it seems easy to supplement quantum mechanics and fill in the details of how the electron flew (along a straight line with constant velocity, isn't it?) but in general it is much harder to determine what happens. That is, there have been many attempts to describe the behaviour of light and elementary particles when no measurements are performed, but up to now all these attempts haven't lead to a generally accepted picture.

Quantum mechanics in its standard formulation is thus as challenging to our view of the world as the non-existence of a general map of an island would be: quantum mechanics gives partial descriptions of the behaviour of light and elementary particles. However, attempts to fix a general description of light and particles which includes these partial descriptions have not yet been fully successful. This makes quantum mechanics the first fundamental and universally valid theory in physics which cannot be straightforwardly understood in terms of a general description of nature, which seems to rule out that the systems it describes exist in a usual way in the outside world.

Some physicists and philosophers have concluded from this that quantum mechanics should be taken solely as an instrumentalistic theory about our observations of light, electrons, protons, etc. Such instrumentalists thus reject the idea that there is an all-encompassing description of quantum reality or that light and elementary particles exist 'out there.' And even though instrumentalists also tend to explain quantum mechanics by giving descriptions which exceed the quantum mechanical predictions about measurements (when they explain the setup of some experiment, instrumentalists also draw pictures of the unobservable trajectories through space along slits and beam-splitters, etc., which the particles in the experiment are supposed to follow), they hold that one can understand quantum mechanics without ever giving such descriptions. Physicists in general, however, simply ignore the challenge and continue to explore the partial descriptions that quantum mechanics does provide.

This is a book about *modal interpretations* of quantum mechanics and can be seen as an attempt to take the third approach to the challenge of quantum mechanics. That is, in this book it is assumed that there does exist a general description of light and elementary particles. And the questions which are addressed are questions about how this description looks according to modal interpretations, and about which of our standard notions about the description of light and particles can still be upheld and which of these notions have to be abandoned.

In general the aim of an interpretation of quantum mechanics is defined as to provide a description of what reality would be like if quantum mechanics were true.[3] As I said before, quantum mechanics itself does not yield such a description because in its standard formulation it is a theory which assigns states to systems which only describe the outcomes of measurements performed on those systems. Modal interpretations now modify the standard formulation by giving the quantum mechanical state of a system at all times a meaning in terms of properties possessed by that system. With this modification quantum mechanics does provide a description of reality because now systems always have properties regardless of whether or not measurements are performed.

Modal interpretations aim furthermore to provide a description of what reality would be like in the case that measurements are treated as ordinary physical interactions. The reason for this is that in the standard formula-

---

[3] See page 6 of Healey (1989) and Section 8.1 of Van Fraassen (1991).

tion of quantum mechanics interactions between systems and measurement devices have a special status as compared to other interactions between systems. According to the standard formulation, the evolution of the states of systems is governed by the Schrödinger equation except if a measurement is performed; if a measurement is performed, states evolve according to the so-called projection postulate. It is, however, felt that a description of reality, or a physical theory in general, should be formulated without giving such a special rôle to measurements: measurement interactions are in physics only instances of interactions between two or more systems and a measurement interaction should therefore affect the dynamics of states in the same way as any other interaction affects this dynamics. This requirement is implemented in modal interpretations by assuming that quantum mechanical states always evolve according to the Schrödinger equation, even if measurements are performed. Modal interpretations thus reject the projection postulate.

In this book I explore the possibilities and impossibilities to understand quantum mechanics in terms of a general description of a world. This book thus mainly deals with the epistemic side of the challenge of quantum mechanics, and not with the ontological side. I am therefore not entering the ongoing debate about scientific realism (the position that scientific theories aim at giving a literally true description about what the outside world is like). However, the results presented do have a bearing on this debate. For if it can be proved that there does not exist an (acceptable) general description of the world which is consistent with the partial descriptions provided by quantum mechanics, then it becomes quite difficult to still maintain that light and elementary particles exist in the sense in which we usually assume that physical systems exist. This would be a fantastic ontological conclusion, as it would be a fantastic conclusion if it could be proved that there are islands which do not exist in the usual sense.

In this book I focus specifically on the version of the modal interpretation proposed by Kochen (1985) and Dieks (1988), as well as on two generalisations of this version. The first generalisation is the one presented in Vermaas and Dieks (1995) and the second is the one proposed by Bacciagaluppi and Dickson (1997) and Dieks (1998b). I develop these three modal interpretations to full descriptions of reality, to determine whether these interpretations are able to give empirically adequate descriptions of measurements and to consider the question of whether they can be taken as metaphysically tenable interpretations of quantum mechanics.

In addition to these three interpretations, there exist other versions of the modal interpretation, notably the very first one by Van Fraassen (1972,

1973),[4] the modal interpretation by Healey (1989) and the interpretation by Bub (1992). These further modal interpretations are not the subject of this book, although many of the results presented also apply to them.[5]

The contents of this book are organised such that it can be accessed in at least three ways. Firstly, if the reader wishes to be introduced to modal interpretations and to follow their development step by step, the book can be read linearly. In Chapter 2 I start by giving a brief survey of quantum mechanics and by discussing the problems one encounters if one tries to interpret this theory. In Chapter 3 I introduce modal interpretations in general by giving their common characteristics and by defining the way in which they describe reality. The remainder of the book is then organised around the three tasks which I mentioned above. In Part one the different modal interpretations are defined and their descriptions of reality are developed as far as possible. In Part two the empirical adequacy of modal interpretations is assessed by determining how they describe measurements. In Part three the metaphysical tenability of modal interpretations is discussed, and in Chapters 14 and 15 I collect the more important results about modal interpretations and draw general conclusions.

Secondly, if the reader is not interested in yet another introduction to the conceptual problems of quantum mechanics, he or she can decide to have a quick look at the last paragraph of Section 3.2 (page 29) and then to go directly to Part one, which starts with Chapter 4 in which the different versions of the modal interpretation are introduced. Chapter 5 fixes the full set of properties ascribed to a single system, and deals with the question of how this property ascription induces a value assignment to the magnitudes pertaining to that system. In Chapter 6 I consider the joint ascription of properties to different systems and discuss the possibility of correlating these properties. In this chapter a no-go theorem is derived which substantially limits the existence of such correlations.

Chapters 7 and 8 are concerned with the dynamics of the ascribed properties. Chapter 7 gives the proof that the dynamics of the set of properties which a system *possibly* possesses is discontinuous and highly unstable. Chapter 8 discusses the dynamics of the properties which a system *actually* possesses, and shows that this dynamics is not uniquely fixed by the dynamics of the states of systems. In Chapter 9 it is proved that this loose relation between the dynamics of the actually possessed properties and the dynamics of the

---

[4] If authors refer to the modal interpretation by Van Fraassen, they are referring to the one given in Van Fraassen (1973).

[5] In Section 4.2 I briefly discuss Van Fraassen's modal interpretation and in Section 4.6 I briefly discuss Bub's. See footnote 27 for references to Healey's modal interpretation.

states of systems allows the description of reality by modal interpretations to be non-local in a quite explicit way.

In Part two it is determined whether modal interpretations are empirically adequate when applied to measurement situations. In Chapter 10 I consider the question of whether modal interpretations solve the so-called measurement problem by ascribing outcomes to (pointers of) measurement devices at the end of measurement interactions. In Chapter 11 I prove that if modal interpretations solve this measurement problem, then they ascribe and correlate outcomes of measurements with the empirically correct Born probabilities.

In Part three modal interpretations are analysed from a more philosophical point of view. In Chapter 12 I motivate the criteria I impose on metaphysically tenable interpretations of quantum mechanics. Then I analyse the relations between properties, states and outcomes of measurements in modal interpretations and I discuss how modal interpretations, when restricted to the description of measurement outcomes, recover the standard formulation of quantum mechanics. Chapter 13 concerns the relations between the properties ascribed to composite systems and subsystems. I show that the modal interpretation by Kochen (1985) and Dieks (1988) as well as the one by Vermaas and Dieks (1995) can be characterised as holistic and non-reductionistic. The interpretation proposed by Bacciagaluppi and Dickson (1997) and Dieks (1998b) is, on the other hand, non-holistic and, to a certain extent, reductionistic. I argue that notwithstanding the lack of reductionism or holism, the description of reality by these interpretations can still be taken as tenable. Finally, as I said above, Chapters 14 and 15 are used to collect the more important results and to reach general conclusions.

The third way to use this book is to not read it at all but only to consult it as a reference book. For this third way, I have included an index at the end of the book.

Note, finally, that many of the proofs of the different results are put in separate subsections, which are called 'MATHEMATICS' and which appear, when necessary, at the end of sections. Please do not read these as parts of the running text but consult them when desired. The proofs are intended to be rigorous with regard to quantum systems defined on finite-dimensional Hilbert spaces; the modifications necessary to also include the infinite-dimensional case are not always discussed.

2

Quantum mechanics

I start by briefly overviewing quantum mechanics as it is standardly formulated and by discussing the question of whether this standard formulation needs to be supplemented by an interpretation. The overview is based mainly on Von Neumann (1932, 1955) and its aim is not to give the reader a crash course in quantum mechanics, but to present those parts of the standard formalism which I use in this book. For a more complete treatment, the reader may consult the standard textbooks on quantum mechanics or, for instance, Sudbery (1986) or Redhead (1987).

2.1 The standard formulation

The standard formulation of quantum mechanics can be introduced in four steps. The first step is that in quantum mechanics one describes the physics of a system by means of a Hilbert space $\mathcal{H}$. This Hilbert space is a complex linear vector space on which an inner product is defined.[6] Let's adopt the convention (please consult the Glossary at the end of the book for notational conventions) that α refers to a system and that $\mathcal{H}^{\alpha}$ is the Hilbert space that is associated with this system. Let $|\psi^{\alpha}\rangle$ denote a vector in $\mathcal{H}^{\alpha}$ and let $\langle\psi^{\alpha}|\phi^{\alpha}\rangle$ be the inner product between the vectors $|\psi^{\alpha}\rangle$ and $|\phi^{\alpha}\rangle$. With this notation I can give a few definitions: a normalised vector $|\psi^{\alpha}\rangle$ is a vector for which it holds that the Hilbert space norm $\||\psi^{\alpha}\rangle\| := \sqrt{\langle\psi^{\alpha}|\psi^{\alpha}\rangle}$ is equal to 1 (all vectors considered in this book are assumed to be normalised except in a few explicitly stated cases); two vectors $|\psi^{\alpha}\rangle$ and $|\phi^{\alpha}\rangle$ are orthogonal if their inner product $\langle\psi^{\alpha}|\phi^{\alpha}\rangle$ is equal to 0; and an orthonormal basis for

---

[6] To be more precise, a Hilbert space $\mathcal{H}$ is a complex linear vector space (A) on which an inner product is defined, (B) which is separable (that is, there exists a denumerably infinite sequence of vectors in $\mathcal{H}$ which lies dense in $\mathcal{H}$) and (c) which is complete (that is, every Cauchy sequence of vectors in $\mathcal{H}$ converges to a vector in $\mathcal{H}$). A finite-dimensional complex vector space with an inner product is automatically a Hilbert space (Redhead 1987, App. II).

Discrete spectral resolution

If a self-adjoint operator A is defined on a finite-dimensional Hilbert space or if it is trace class (see footnote 7), then this operator is compact. A self-adjoint compact operator allows a discrete decomposition[a]

$$A = \sum_j a_j \sum_k |a_{jk}\rangle\langle a_{jk}|. \tag{2.1}$$

Here, $\{a_j\}_j$ is a set of real and distinct values which are the eigenvalues of A. The set of vectors $\{|a_{jk}\rangle\}_{j,k}$ are the eigenvectors of A and form an orthonormal basis for $\mathcal{H}$. The set of projections $\{\sum_k |a_{jk}\rangle\langle a_{jk}|\}_j$ are the pair-wise orthogonal eigenprojections of A that correspond one-to-one to the eigenvalues $\{a_j\}_j$. This decomposition (2.1) is called the discrete spectral resolution of A and is unique in the sense that the set of eigenvalues and the corresponding eigenprojections are uniquely fixed by A.

[a] See Reed and Simon (1972, Sect. VI.5).

$\mathcal{H}^\alpha$ is given by a set of normalised and pair-wise orthogonal vectors $\{|e_j^\alpha\rangle\}_j$ (the vectors thus satisfy $\langle e_j^\alpha|e_k^\alpha\rangle = \delta_{jk}$) with which one can decompose every vector $|\psi^\alpha\rangle$ in $\mathcal{H}^\alpha$ as $|\psi^\alpha\rangle = \sum_j c_j |e_j^\alpha\rangle$, where c_j is equal to $\langle e_j^\alpha|\psi^\alpha\rangle$. For every Hilbert space there exist such orthonormal bases. If a Hilbert space $\mathcal{H}^\alpha$ is N-dimensional (with N either finite or equal to ∞), then any basis $\{|e_j^\alpha\rangle\}_j$ of $\mathcal{H}^\alpha$ contains exactly N elements.

The second step is that quantum mechanics speaks about observables pertaining to a system α. Examples are the position, the spin and the energy of α. These observables are all represented by self-adjoint linear operators defined on $\mathcal{H}^\alpha$. Let A^α denote such an operator. Self-adjoint linear operators allow in a number of cases a so-called **discrete spectral resolution**, for instance, if they are defined on finite-dimensional Hilbert spaces or, more generally, if they are trace class.[7,8] This discrete spectral resolution has the form of a discrete sum:

$$A^\alpha = \sum_j a_j \sum_k |a_{jk}^\alpha\rangle\langle a_{jk}^\alpha|. \tag{2.2}$$

---

[7] A self-adjoint linear operator A is trace class if its trace norm $\|A\|_1 := \text{Tr}|A| = \sum_j \langle e_j|\sqrt{A^\dagger A}|e_j\rangle$, with $\{|e_j\rangle\}_j$ an orthonormal basis for $\mathcal{H}$, is finite (Reed and Simon 1972, Sect. VI.6).

[8] A self-adjoint (hypermaximal) linear operator A in general has not a discrete but a continuous spectral resolution. A continuous spectral resolution has the form of a Stieltjes integral $A = \int_{\lambda=-\infty}^\infty \lambda \, dE_A((-\infty, \lambda])$. The operator $E_A(\Gamma)$ is a projection on $\mathcal{H}$ and is a member of the so-called spectral family of A and Γ is a (Borel) set of values. See formula (5.18) for the properties of the spectral family $\{E_A(\Gamma)\}_\Gamma$ and see Von Neumann (1955, Sect. II) and, in a more accessible form, Jauch (1968, especially Sect. 4.3) for the general theory of spectral resolutions.

In this sum the values $\{a_j\}_j$ are the different eigenvalues of A^α and are all real-valued. The vectors $\{|a_{jk}^\alpha\rangle\}_{j,k}$ are the eigenvectors of A^α and form an orthonormal basis for $\mathscr{H}^\alpha$ (so $\langle a_{jk}^\alpha | a_{j'k'}^\alpha\rangle = \delta_{jj'}\delta_{kk'}$). An eigenvector $|a_{jk}^\alpha\rangle$ corresponds to an eigenvalue a_j and if two or more different eigenvectors correspond to the same eigenvalue, one calls the spectral resolution (2.2) as well as A^α itself *degenerate* (the second sum in (2.2) with the label k runs over possible degeneracies). One can define for every eigenvalue a_j an eigenprojection $\sum_k |a_{jk}^\alpha\rangle\langle a_{jk}^\alpha|$ of A^α. The spectral resolution (2.2) is unique in the sense that A^α uniquely fixes the set of eigenvalues $\{a_j\}_j$ and the corresponding eigenprojections $\{\sum_k |a_{jk}^\alpha\rangle\langle a_{jk}^\alpha|\}_j$.

A special class of self-adjoint linear operators on $\mathscr{H}^\alpha$ is given by the idempotent projections. Let Q^α denote such a projection. It satisfies $[Q^\alpha]^2 = Q^\alpha = [Q^\alpha]^\dagger$, where $[Q^\alpha]^\dagger$ is the adjoint of Q^α. An example of such a projection is given by $Q^\alpha = \sum_{k=1}^n |e_k^\alpha\rangle\langle e_k^\alpha|$ with $\{|e_k^\alpha\rangle\}_{k=1}^n$ a set of normalised and pairwise orthogonal vectors. This projection has a discrete spectral resolution with only the two eigenvalues 0 and 1. This projection is called an *n*-dimensional projection because it projects vectors in $\mathscr{H}^\alpha$ onto the *n*-dimensional subspace of $\mathscr{H}^\alpha$ spanned by the vectors $\{|e_k^\alpha\rangle\}_{k=1}^n$. Two projections Q^α and $\widehat{Q}^\alpha$ are called mutually orthogonal if $Q^\alpha\widehat{Q}^\alpha = \widehat{Q}^\alpha Q^\alpha = 0$.

With these definitions it follows that the eigenprojections $\{\sum_k |a_{jk}^\alpha\rangle\langle a_{jk}^\alpha|\}_j$ of an operator A^α with a discrete spectral resolution (2.2) are pair-wise orthogonal projections. If this resolution is non-degenerate, then all the eigenprojections are one-dimensional, whereas if it is degenerate, some eigenprojections are multi-dimensional.

A first prediction of quantum mechanics is now that the possible outcomes of a measurement of an observable represented by an operator A^α with a discrete spectral resolution (2.2) correspond one-to-one to the eigenvalues $\{a_j\}_j$ of A^α. That is, a measurement of such an observable always has an eigenvalue of A^α as an outcome.

The third step is to assign **states** to systems. In quantum mechanics the state of a system α is represented by a density operator W^α defined on $\mathscr{H}^\alpha$ and a special case of such a density operator is given by a one-dimensional projection $|\psi^\alpha\rangle\langle\psi^\alpha|$, with $|\psi^\alpha\rangle$ a normalised vector in $\mathscr{H}^\alpha$. In this case one says that the state of α is pure and one can speak about the state vector $|\psi^\alpha\rangle$ of α.

The states of composite systems and subsystems are related. Consider, for instance, two disjoint systems α and β.[9] The Hilbert space associated with the

---

[9] Two systems α and β are disjoint if they have no subsystems in common. Loosely speaking α and β are disjoint if you can simultaneously put α in one box and β in another. Two different electrons are thus disjoint but a chair and a leg of that chair are not.

Quantum mechanical state

The state of a system α is represented by a density operator W^α defined on $\mathcal{H}^\alpha$ which by definition satisfies

$$\langle \psi^\alpha | W^\alpha | \psi^\alpha \rangle \geq 0 \quad \forall | \psi^\alpha \rangle \in \mathcal{H}^\alpha, \quad [W^\alpha]^\dagger = W^\alpha, \quad \mathrm{Tr}^\alpha(W^\alpha) = 1. \qquad (2.3)$$

The set of density operators is convex: if W_1^α and W_2^α are density operators, then so is $w_1 W_1^\alpha + w_2 W_2^\alpha$, provided that w_1 and w_2 are both positive and sum to 1. The pure states represented by one-dimensional projections $|\psi^\alpha\rangle\langle\psi^\alpha|$ with $|\psi^\alpha\rangle \in \mathcal{H}^\alpha$ are the extreme elements of this convex set.

---

composite system consisting of both α and β, which is denoted by $\alpha\beta$, is then the tensor product of the Hilbert spaces $\mathcal{H}^\alpha$ and $\mathcal{H}^\beta$, so $\mathcal{H}^{\alpha\beta} = \mathcal{H}^\alpha \otimes \mathcal{H}^\beta$.[10] The states of α, β and $\alpha\beta$ are related by means of so-called partial traces:

$$\left. \begin{aligned} W^\alpha &= \mathrm{Tr}^\beta(W^{\alpha\beta}) := \sum_b \langle e_b^\beta | W^{\alpha\beta} | e_b^\beta \rangle, \\ W^\beta &= \mathrm{Tr}^\alpha(W^{\alpha\beta}) := \sum_a \langle e_a^\alpha | W^{\alpha\beta} | e_a^\alpha \rangle, \end{aligned} \right\} \qquad (2.4)$$

with $\{|e_a^\alpha\rangle\}_a$ and $\{|e_b^\beta\rangle\}_b$ arbitrary orthonormal bases for, respectively, $\mathcal{H}^\alpha$ and $\mathcal{H}^\beta$.[11] The partial traces W^α and W^β are called the reduced states.

A second prediction of quantum mechanics is that if a system α has the state W^α and one performs a perfect measurement[12] of an observable A^α with a discrete spectral resolution (2.2), then one obtains with probability

$$p_{\mathrm{Born}}(a_j) = \mathrm{Tr}^\alpha\left(W^\alpha \sum_k |a_{jk}^\alpha\rangle\langle a_{jk}^\alpha|\right) \qquad (2.5)$$

an outcome corresponding to the eigenvalue a_j. This rule is called the Born rule and the probability $p_{\mathrm{Born}}(a_j)$ is called the Born probability.

---

[10] The tensor product $\mathcal{H}^{\alpha\beta} := \mathcal{H}^\alpha \otimes \mathcal{H}^\beta$ of two Hilbert spaces $\mathcal{H}^\alpha$ and $\mathcal{H}^\beta$ is the Hilbert space which contains all the linear combinations of the tensor product vectors $|\psi^\alpha\rangle \otimes |\phi^\beta\rangle$ with $|\psi^\alpha\rangle \in \mathcal{H}^\alpha$ and $|\phi^\beta\rangle \in \mathcal{H}^\beta$. If $\{|e_j^\alpha\rangle\}_j$ and $\{|f_k^\beta\rangle\}_k$ are orthonormal bases for $\mathcal{H}^\alpha$ and $\mathcal{H}^\beta$, respectively, then $\{|e_j^\alpha\rangle \otimes |f_k^\beta\rangle\}_{j,k}$ is an orthonormal basis for $\mathcal{H}^{\alpha\beta}$. And if $|\Psi^{\alpha\beta}\rangle = \sum_{jk} c_{jk} |e_j^\alpha\rangle \otimes |f_k^\beta\rangle$ and if $|\Phi^{\alpha\beta}\rangle = \sum_{jk} d_{jk} |e_j^\alpha\rangle \otimes |f_k^\beta\rangle$, then the inner product $\langle \Psi^{\alpha\beta} | \Phi^{\alpha\beta} \rangle$ is equal to $\sum_{jk,j'k'} \bar{c}_{jk} d_{j'k'} \langle e_j^\alpha | e_{j'}^\alpha \rangle \langle f_k^\beta | f_{k'}^\beta \rangle = \sum_{jk} \bar{c}_{jk} d_{jk}$. Finally, the tensor product of the operators A^α and B^β defines an operator on $\mathcal{H}^{\alpha\beta}$ by $A^\alpha \otimes B^\beta |\Psi^{\alpha\beta}\rangle = \sum_{jk} c_{jk} A^\alpha |e_j^\alpha\rangle \otimes B^\beta |f_k^\beta\rangle$, where $|\Psi^{\alpha\beta}\rangle = \sum_{jk} c_{jk} |e_j^\alpha\rangle \otimes |f_k^\beta\rangle$.

[11] It follows from the relations (2.4) that $W^{\alpha\beta}$ uniquely fixes the states W^α and W^β. Conversely, the states W^α and W^β uniquely determine $W^{\alpha\beta}$ if and only if W^α or W^β is pure: if W^α or W^β is pure, then $W^{\alpha\beta}$ is equal to $W^\alpha \otimes W^\beta$ (for a proof see Von Neumann (1955, Sect. VI.2)) and if neither W^α nor W^β is pure, then there exist in addition to $W^{\alpha\beta} = W^\alpha \otimes W^\beta$ other states $W^{\alpha\beta}$ which have W^α and W^β as partial traces.

[12] A perfect measurement of an observable A^α with a discrete spectral resolution (2.2) is defined as a measurement which yields with probability 1 an outcome corresponding to the eigenvalue a_j if the state of α is given by $W^\alpha = |a_{jk}^\alpha\rangle\langle a_{jk}^\alpha|$ (with k arbitrary).

The fourth step lays down the dynamics of the states of systems. In the standard formulation of quantum mechanics there are two ways in which this dynamics comes about. In general, that is, if no measurements are performed, states evolve according to the Schrödinger equation. The most simple case of such so-called Schrödinger evolution is given by a system α which does not interact with other systems and which has at time t a pure state, so $W^\alpha(t) = |\psi^\alpha(t)\rangle\langle\psi^\alpha(t)|$. The evolution of the state vector $|\psi^\alpha(t)\rangle$ is then given by $i\hbar\,(d/dt)|\psi^\alpha(t)\rangle = H^\alpha\,|\psi^\alpha(t)\rangle$, where H^α is the Hamiltonian of α.[13] This Hamiltonian is a self-adjoint operator and represents the energy of α. A slightly more general case is given by a system α which does not have a pure state but which still evolves without interacting with other systems. The dynamics of the state of α is in this case governed by the generalised Schrödinger equation

$$\frac{d}{dt}W^\alpha(t) = \frac{1}{i\hbar}[H^\alpha, W^\alpha(t)], \tag{2.6}$$

where $[H^\alpha, W^\alpha(t)]$ is the commutator $H^\alpha W^\alpha(t) - W^\alpha(t)H^\alpha$. The solution to this equation is

$$W^\alpha(t) = U^\alpha(t,s)\,W^\alpha(s)\,U^\alpha(s,t), \tag{2.7}$$

with s some initial instant and with $U^\alpha(x,y)$ a unitary operator equal to $\exp([(x-y)/i\hbar]\,H^\alpha)$.

The most general case is given by as system α with a pure or non-pure state which interacts with other systems. In this case the dynamics is determined as follows. Firstly, one takes a composite system, call it ω, which contains α as a subsystem and which itself does not interact with systems disjoint from ω (in the most extreme case one may take the whole universe as ω). The state $W^\omega(t)$ of ω then evolves as given in (2.7). Secondly, one calculates the evolution of the state of α by taking at all times the partial trace of $W^\omega(t)$. So, for interacting systems the Schrödinger evolution is

$$W^\alpha(t) = \mathrm{Tr}^{\omega/\alpha}(U^\omega(t,s)\,W^\omega(s)\,U^\omega(s,t)), \tag{2.8}$$

where ω/α is shorthand for the system given by ω 'minus' α.

This Schrödinger evolution (2.7) or (2.8) of states is now interrupted at the end of a measurement. The states of systems then change according to a second type of dynamics which is governed by the so-called projection postulate. In order to properly formulate this postulate, I continue with some measurement theory.

---

[13] In this book I always use the Schrödinger representation.

The simplest model of a measurement has been given by Von Neumann (1955, Sect. VI.3). Let α be the system on which the measurement is performed and let $A^\alpha = \sum_j a_j |a_j^\alpha\rangle\langle a_j^\alpha|$ represent the (non-degenerate) observable which is measured. Let μ be a measurement device and let M^μ be an observable (pertaining to the device) which we can directly observe. This observable M^μ is traditionally called the pointer reading observable. If its spectral resolution is given by $M^\mu = \sum_j m_j |R_j^\mu\rangle\langle R_j^\mu|$, then the projections $\{|R_j^\mu\rangle\langle R_j^\mu|\}_j$ represent the individual readings of the pointer. In a Von Neumann measurement it is assumed that, before the measurement, the system α is in some pure state $|\psi^\alpha\rangle$ (which can be written as a superposition $|\psi^\alpha\rangle = \sum_j c_j |a_j^\alpha\rangle$ of the eigenvectors of A^α) and that the measurement device is in a pure 'ready to measure' state $|R_0^\mu\rangle$. The interaction between α and μ during the measurement is then (for every possible set of coefficients $\{c_j\}_j$) supposed to be such that the state of the composite $\alpha\mu$ evolves as

$$|\Psi^{\alpha\mu}\rangle = \left[\sum_j c_j |a_j^\alpha\rangle\right] \otimes |R_0^\mu\rangle \longmapsto |\widetilde{\Psi}^{\alpha\mu}\rangle = \sum_j c_j |a_j^\alpha\rangle \otimes |R_j^\mu\rangle. \qquad (2.9)$$

(There indeed exist Hamiltonians $H^{\alpha\mu}$ such that this evolution is obtained through the Schrödinger evolution (2.7) or (2.8).)

In order to see that this interaction (2.9) models a perfect measurement, one can apply the Born rule (2.5) to α. Initially the state of α is $|\psi^\alpha\rangle\langle\psi^\alpha|$. Hence the Born rule yields that any perfect measurement of A^α produces an outcome corresponding to the eigenvalue a_j with a probability $p_{\text{Born}}(a_j)$ equal to $\text{Tr}^\alpha(|\psi^\alpha\rangle\langle\psi^\alpha| \, |a_j^\alpha\rangle\langle a_j^\alpha|) = |c_j|^2$. Let's see whether the interaction (2.9) satisfies this description. After the interaction the state of the device is $\widetilde{W}^\mu = \sum_j |c_j|^2 |R_j^\mu\rangle\langle R_j^\mu|$. The Born rule, this time applied to μ, yields that a measurement of the observable M^μ gives with a probability $p_{\text{Born}}(m_j)$ equal to $\text{Tr}^\mu(\widetilde{W}^\mu |R_j^\mu\rangle\langle R_j^\mu|) = |c_j|^2$ an outcome corresponding to m_j. Since it is assumed that one can directly observe the readings of M^μ, that is, that a look at the device μ counts as a measurement of M^μ, this implies that a direct observation of μ yields with probability $p_{\text{Born}}(m_j) = |c_j|^2$ the reading m_j of the pointer observable M^μ. So, the interaction (2.9) means that the Born probabilities for outcomes of a measurement of A^α are exactly transferred to the probabilities with which we see that the pointer displays one of its readings (that is, $p_{\text{Born}}(a_j) = p_{\text{Born}}(m_j)$ for all j). Hence, if one identifies the observation that the device displays outcome m_j after the interaction (2.9) with the outcome that corresponds to a_j, it follows that this interaction is a proper model for a measurement of A^α. Moreover, the interaction (2.9) models a perfect measurement in the sense of footnote 12: if the initial state of α is given by an eigenvector of A^α, say $|\psi^\alpha\rangle = |a_j^\alpha\rangle$, then the final state of the device is $|\mathrm{R}_j^\mu\rangle\langle\mathrm{R}_j^\mu|$ such that an observation yields with probability 1 the outcome m_j corresponding to the eigenvalue a_j.

There exist many other interactions between a system α and a measurement device μ which count as measurements of the observable A^α.[14] They all have in common that after the interaction, μ is in some final state $\widetilde{W}^\mu$, that the possible outcomes are represented by a pointer observable $M^\mu = \sum_j m_j R_j^\mu$ (the projections $\{R_j^\mu\}_j$ are the eigenprojections of M^μ) and that $\mathrm{Tr}^\mu(\widetilde{W}^\mu R_j^\mu)$ is the probability that the measurement yields the outcome corresponding to the eigenvalue a_j. For a Von Neumann measurement this probability $\mathrm{Tr}^\mu(\widetilde{W}^\mu R_j^\mu)$ is equal to the Born probability $\mathrm{Tr}^\alpha(W^\alpha |a_j^\alpha\rangle\langle a_j^\alpha|)$ with W^α the state before the measurement. In general, however, this need not be the case (there can be errors in a measurement, for instance).

Let's return to the projection postulate. As I said before, this postulate governs the dynamics of states at the end of a measurement. Let ω denote the universe and assume that a measurement performed on some subsystem of ω by means of a device μ ends at time t. Then the projection postulate states that at t the state of the universe makes the following instantaneous transition with probability $\mathrm{Tr}^\mu(\widetilde{W}^\mu R_j^\mu)$:

$$W^\omega(t) \longmapsto \widehat{W}^\omega(t) = \frac{[R_j^\mu \otimes \mathbb{I}^{\omega/\mu}]\, W^\omega(t)\, [R_j^\mu \otimes \mathbb{I}^{\omega/\mu}]}{\mathrm{Tr}^\omega(W^\omega(t)\, [R_j^\mu \otimes \mathbb{I}^{\omega/\mu}])}. \tag{2.10}$$

Here $\mathbb{I}^{\omega/\mu}$ is the unit operator defined on the Hilbert space $\mathcal{H}^{\omega/\mu}$ associated with the universe ω minus the measurement device μ. Note that this state transition is fundamentally different from the Schrödinger evolution. By definition the universe does not interact with systems disjoint from the universe, so if the state of the universe evolves according to the Schrödinger equation, then this evolution is given by (2.7). However, there exist no unitary operators $U^\omega(x, y)$ which reproduce the transition (2.10).

If one applies the projection postulate after the Von Neumann measurement (2.9), one obtains that the state $|\widetilde{\Psi}^{\alpha\mu}\rangle$ of $\alpha\mu$ at the end of the measurement becomes equal to $|a_j^\alpha\rangle \otimes |\mathrm{R}_j^\mu\rangle$ with probability $|c_j|^2$. Then, if one takes the partial trace of this evolution of the state of $\alpha\mu$, the result is that the state of α evolves with probability $|c_j|^2$ as

$$|\psi^\alpha\rangle = \sum_j c_j |a_j^\alpha\rangle \longmapsto \widetilde{W}^\alpha = \sum_j |c_j|^2 |a_j^\alpha\rangle\langle a_j^\alpha| \longmapsto |\widehat{\psi}^\alpha\rangle = |a_j^\alpha\rangle. \tag{2.11}$$

The state of α before the measurement is thus projected onto one of the

---

[14] I discuss measurements more generally in Chapter 10 but see Busch, Lahti and Mittelsteadt (1991) for a rigorous treatment of quantum mechanical measurements.

eigenvectors of the measured observable A^α (which explains the name 'projection postulate'). The state of α collapses to one of the eigenvectors, so to say.

One can also consider series of measurements (joint measurements as well as sequential measurements) and calculate with the standard formulation conditional and joint probabilities for the outcomes of such measurements (such calculations are given in Sections 8.2, 11.2, 11.3 and A.2). The resulting predictions are all in good agreement with our observations of the outcomes of measurements.

2.2 The need for an interpretation

There are two reasons to be dissatisfied with the standard formulation of quantum mechanics. The first is that the standard formulation does not say much about the quantum mechanical systems themselves. The second concerns the exceptional status of measurements in this formulation. I start by discussing the first reason.

If one settles for a strict instrumental approach to physical theories, one has reached the end of the story. Quantum mechanics in the standard formulation yields predictions about outcomes of measurements on elementary particles. These predictions are in good agreement with our observations and that is all an instrumentalist desires from a theory. If, instead, one adopts a more realist attitude towards quantum mechanics and assumes that it is a theory about electrons, protons, etc., which exist independently of us and independently of the performance of measurements, then the standard formulation can only be a beginning. In the realist conception, a true physical theory about elementary particles, aims at (literally) describing the properties of those particles as they exist out there. And the fact that quantum mechanics is in good agreement with observation adds support to the assumption that quantum mechanics is such a true theory. However, in its standard formulation, quantum mechanics clearly does not give a description of elementary particles; it only says something about measurements on these particles. Hence, from a realist point of view one arrives at the need for an interpretation, that is, the need to provide a description of what reality would be like if quantum mechanics were true. But, as I pointed out in the introduction, even if one rejects this realist conception and takes a more agnostic view towards ontological claims about the existence of elementary particles, then one can still be interested in finding an interpretation of quantum mechanics from an epistemic point of view.[15] Our common understanding of scientific theories is that they describe

---

[15] Consider, for instance, Van Fraassen (1980, 1991) who took such an agnostic point of view about the existence of quantum mechanical systems and proposed the first modal interpretation.

entities that exist independently of us. Newtonian mechanics, for instance, is usually understood and explained as a theory about the properties that objects like apples, billiard balls and planets have independently of us. An ontological agnostic can now be interested in an interpretation of quantum mechanics in order to also understand and explain quantum mechanics in this way: namely, in terms of a description of the properties of hypothetical objects named photons, electrons, etc.

If one accepts the need for an interpretation of quantum mechanics, the question arises of what such an interpretation should provide. A first demand is that an interpretation should give a *well-developed* description of reality. That is, an interpretation should take things like position, spin and energy not as merely observables of systems (things which can be observed on those systems). Instead, an interpretation should take such things as normal physical magnitudes which pertain to systems and which exist independently of the notion of observation or measurement (things like position, spin and energy should be *be-ables* in the words of Bell (1987, page 52); things which can exist). An interpretation should, moreover, ascribe properties to systems, that is, it should yield that the physical magnitudes of those systems have definite values and an interpretation should yield a fully-fledged theory of these properties. A second demand is that the description of reality given by an interpretation should be *empirically adequate*. This means that an interpretation should reproduce the predictions of the standard formulation of quantum mechanics with regard to the outcomes of measurements. A third demand is that an interpretation should give a *metaphysically tenable* description of the magnitudes and properties of systems.

In the next chapter I make these demands more explicit, but let's adopt for the moment the following starting points about how an interpretation describes reality.[16] Firstly, all the observables defined by the standard formulation are taken as physical magnitudes in an interpretation. And the magnitude that corresponds to an observable represented by an operator A^α in the standard formulation is in an interpretation represented by that same operator. Secondly, a magnitude represented by an operator A^α with a discrete spectral resolution (2.2) may assume one of the eigenvalues $\{a_j\}_j$ of A^α as a definite value. And the property that this magnitude has value a_j is represented by the eigenprojection $\sum_k |a_{jk}^\alpha\rangle\langle a_{jk}^\alpha|$ corresponding to a_j. Hence, the magnitude A^α has value a_j if and only if the property $\sum_k |a_{jk}^\alpha\rangle\langle a_{jk}^\alpha|$ is possessed.

Now let $[A^\alpha] = a_j$ denote that A^α has the value a_j and let $[\sum_k |a_{jk}^\alpha\rangle\langle a_{jk}^\alpha|] =$

---

[16] These starting points are consistent with the ones I adopt in Section 3.2.

1 denote that α possesses the property represented by $\sum_k |a^\alpha_{jk}\rangle\langle a^\alpha_{jk}|$. The relation between the value assignment to magnitudes and the ascription of properties is then captured by

$$[A^\alpha] = a_j \quad \textit{if and only if} \quad [\sum_k |a^\alpha_{jk}\rangle\langle a^\alpha_{jk}|] = 1. \tag{2.12}$$

An example of an interpretation of quantum mechanics which satisfies these starting points is the so-called *orthodox interpretation*. This interpretation is based on an assumption that can be found in Von Neumann (1955, e.g. Sect. III.3), namely, that a magnitude A^α has the eigenvalue a_j as definite value if and only if the state of α is given by an eigenvector $|a^\alpha_{jk}\rangle$ corresponding to that eigenvalue a_j. A motivation for this assumption can be that if the state of α is indeed an eigenvector $|a^\alpha_{jk}\rangle$ of A^α, then a Von Neumann measurement (2.9) of A^α yields with certainty (with probability 1) the outcome a_j. A good explanation for this certain outcome is that A^α actually has value a_j. This assumption has become known as the (generalised) eigenvalue–eigenstate link:[17] a magnitude A^α has value a_j if and only if the state of α is such that a_j has Born probability 1, or, in terms of properties:

$$[\sum_k |a^\alpha_{jk}\rangle\langle a^\alpha_{jk}|] = 1 \quad \textit{if and only if} \quad \mathrm{Tr}^\alpha(W^\alpha \sum_k |a^\alpha_{jk}\rangle\langle a^\alpha_{jk}|) = 1. \tag{2.13}$$

The orthodox interpretation is now the interpretation obtained by adding this eigenvalue–eigenstate link to the standard formulation of quantum mechanics.

This orthodox interpretation is in complete harmony with the standard formulation. That is, the orthodox interpretation reproduces the prediction of the standard formulation that a measurement by means of a device with a pointer reading magnitude $M^\mu = \sum_j m_j R^\mu_j$ yields with probability $\mathrm{Tr}^\mu(\widetilde{W}^\mu R^\mu_j)$ the outcome m_j that corresponds to the eigenvalue a_j of the measured magnitude A^α. To see this note that by the projection postulate, the state of the measurement device collapses with probability $\mathrm{Tr}^\mu(\widetilde{W}^\mu R^\mu_j)$ to the state

$$\widehat{W}^\mu = \frac{R^\mu_j \widetilde{W}^\mu R^\mu_j}{\mathrm{Tr}^\mu(\widetilde{W}^\mu R^\mu_j)}. \tag{2.14}$$

Given this collapsed state $\widehat{W}^\mu$, the Born probability $\mathrm{Tr}^\mu(\widehat{W}^\mu R^\mu_j)$ is 1. Hence, the orthodox interpretation yields that $[R^\mu_j] = 1$ and by the relation (2.12) it then follows that the pointer reading magnitude M^μ has with probability

---

[17] The name 'eigenvalue–eigenstate link' was introduced by Fine (1973).

$\mathrm{Tr}^{\mu}(\widetilde{W}^{\mu} R_j^{\mu})$ the value m_j after the measurement. This is exactly the prediction of the standard formulation.

One may think that with the orthodox interpretation, one has again reached the end of the story: the orthodox interpretation assigns values to magnitudes, ascribes properties to systems, reproduces the predictions of the standard formulation and thus yields an acceptable description of reality if quantum mechanics were true. However, there still is the question of whether the orthodox interpretation gives a tenable theory about these magnitudes and properties. This question leads us to the second reason to be dissatisfied with the standard formulation.

One of the striking characteristics of quantum mechanics in its standard formulation is that measurement interactions have an exceptional status as compared to other interactions between systems. This status hangs together with the projection postulate: if an interaction between a system and a measurement device counts as a measurement, one should apply this postulate and the state of the universe makes the transition (2.10); and if this interaction does not count as a measurement, one should not apply the projection postulate and the state of the universe evolves according to the Schrödinger equation as in (2.7). As I said in the previous section, these two types of state dynamics are fundamentally different. From a methodological point of view this exceptional status of measurements is, however, quite strange. In physical theories a measurement is considered as a special instance of an interaction between systems. It seems therefore that a measurement interaction should affect the dynamics of states in the same way as any other interaction affects this dynamics. Hence, from a methodological point of view it seems preferable if one could remove the exceptional status of measurements in quantum mechanics. The same remarks hold *mutatis mutandis* for interpretations of quantum mechanics. So, within interpretations it also seems methodologically more sound if measurement interactions could be taken as ordinary physical interactions.

The orthodox interpretation does not meet this desideratum. Consider, for instance, three systems α, β and μ. Let the initial state of the composite of these systems be

$$|\Psi^{\alpha\beta\mu}\rangle = \Big[\sum_j c_j |a_j^{\alpha}\rangle \otimes |b_j^{\beta}\rangle\Big] \otimes |\mathrm{R}_0^{\mu}\rangle, \tag{2.15}$$

where the vectors $\{|a_j^{\alpha}\rangle\}_j$ are the pair-wise orthogonal eigenvectors of an operator A^{α} and the vectors $\{|b_j^{\beta}\rangle\}_j$ are pair-wise orthogonal vectors in $\mathcal{H}^{\beta}$. Assume that the state of β remains constant (say, β is very remote from α and μ and does not interact with these two systems) and assume that the states of α and μ evolve according to the interaction (2.9) which can model a Von Neumann measurement of the magnitude A^α. The state of the composite $\alpha\beta\mu$ after the interaction is then

$$|\widetilde{\Psi}^{\alpha\beta\mu}\rangle = \sum_j c_j |a_j^\alpha\rangle \otimes |b_j^\beta\rangle \otimes |\mathrm{R}_j^\mu\rangle. \tag{2.16}$$

If now the interaction between α and μ is an ordinary interaction, that is, if it is not a measurement, the state of β before and after the interaction is equal to $\sum_j |c_j|^2 |b_j^\beta\rangle\langle b_j^\beta|$. The orthodox interpretation then constantly ascribes the property $\sum_j |b_j^\beta\rangle\langle b_j^\beta|$ to β. If, on the other hand, the interaction between α and μ counts as a measurement, the projection postulate yields that the state of β collapses at the conclusion of the measurement from $\widetilde{W}^\beta = \sum_j |c_j|^2 |b_j^\beta\rangle\langle b_j^\beta|$ to, say, $\widehat{W}^\beta = |b_k^\beta\rangle\langle b_k^\beta|$. So, the property ascribed to β by the orthodox interpretation then changes from $\sum_j |b_j^\beta\rangle\langle b_j^\beta|$ to $|b_k^\beta\rangle\langle b_k^\beta|$. Hence, the properties ascribed to β depend on whether the interaction (2.9) between α and μ is an ordinary interaction or is modelling a measurement. Measurement interactions thus have an exceptional status and cannot therefore be taken as ordinary interactions in the orthodox interpretation.

So, one is still empty-handed if one wants to give an interpretation of quantum mechanics which does not grant an exceptional status to measurements. There have now been many attempts to give an interpretation of quantum mechanics in which measurements are taken as ordinary interactions. Examples are Bohmian mechanics, the consistent histories approach, many worlds and many minds interpretations and, more recently, modal interpretations. This book is about three versions of these modal interpretations.

An obstacle one can encounter when trying to define an interpretation which takes measurements as ordinary interactions is given by what I will call the *measurement problem* for interpretations. Consider the interpretation which accepts the eigenvalue–eigenstate link and which says that the states of systems always evolve according to the Schrödinger equation (measurements thus lose their exceptional status in this interpretation because the projection postulate, which distinguishes measurements from ordinary interactions, is rejected). Such an interpretation is unable to reproduce the predictions of the standard formulation because it is unable to ascribe outcomes after measurements. Take again the Von Neumann measurement (2.9). After this measurement, the state of the device μ is still $\widetilde{W}^\mu = \sum_j |c_j|^2 |\mathrm{R}_j^\mu\rangle\langle \mathrm{R}_j^\mu|$ because one has rejected the projection postulate. Application of the eigenvalue–eigenstate link now yields in general that μ possesses the property $\sum_j |\mathrm{R}_j^\mu\rangle\langle \mathrm{R}_j^\mu|$ and not one of the individual pointer readings $\{|\mathrm{R}_j^\mu\rangle\langle \mathrm{R}_j^\mu|\}_j$. Hence, our interpretation does not reproduce the prediction of the standard formulation that the measurement (2.9) has with probability $|c_j|^2$ the outcome $|R_j^\mu\rangle\langle R_j^\mu|$ that corresponds to the eigenvalue a_j. This problem that interpretations sometimes fail to ascribe outcomes to measurement devices is the measurement problem for interpretations.

So, to sum up, from a realist point of view and from an epistemic point of view one may want to add an interpretation to quantum mechanics, and from a methodological point of view one may want to remove the exceptional status of measurements. However, if one tries to do so, one can be confronted with the measurement problem for interpretations. In order to formulate this measurement problem more generally, consider the state $\widetilde{W}^\mu$ of a measurement device after a measurement of A^α. Let $M^\mu = \sum_j m_j R_j^\mu$ be the pointer reading magnitude. Then the standard formulation predicts that the measurement yields with probability $\text{Tr}^\mu(\widetilde{W}^\mu R_j^\mu)$ the outcome corresponding to the eigenvalue a_j of A^α. This prediction implies that the measurement device possesses the property R_j^μ with probability $\text{Tr}^\mu(\widetilde{W}^\mu R_j^\mu)$. So, if an interpretation of quantum mechanics should reproduce this prediction, it should ascribe with probability $\text{Tr}^\mu(\widetilde{W}^\mu R_j^\mu)$ the pointer reading R_j^μ to the measurement device. The orthodox interpretation achieves this but still grants measurements their exceptional status. An interpretation that accepts the eigenvalue–eigenstate link and rejects the projection postulate is not able to ascribe the readings. The question is now whether modal interpretations can avoid the measurement problem as successfully as the orthodox interpretation while, at the same time, taking measurements as ordinary physical interactions.

3

Modal interpretations

Before considering the particulars of the different versions of the modal interpretation in the next chapter, I first present their common characteristics. Then I list the starting points from which I develop these modal interpretations to fully-fledged descriptions of reality. Finally, I give the criteria I think interpretations should meet and present a number of desiderata I hope they meet.

3.1 General characteristics

The name 'modal interpretation' originates with Van Fraassen (1972) who, in order to interpret quantum mechanics, transposed the semantic analysis of modal logics to the analysis of quantum logic. The resulting interpretation was for obvious reasons called the modal interpretation of quantum logic. Since then, the term modal interpretation has acquired a much more general meaning and lost its direct kinship with modal logics. In particular new interpretations of quantum mechanics developed in the 1980s by Kochen (1985), Krips (1987), Dieks (1988), Healey (1989) and Bub (1992) became known as modal interpretations and also older traditions like Bohmian mechanics (Bohm 1952; Bohm and Hiley 1993) were identified as modal ones. But why are all these interpretations still called modal? And what is the present-day meaning of this term?

I think part of the answer to the first question has to do with public relations. The name 'modal' is short, sounds nice and is rather intriguing. Furthermore, I guess that also Van Fraassen's prestige as a philosopher of science adds a special gloss to the term. But, apart from all this, I believe the name 'modal interpretation' is quite suited. This name pin-points a feature all modal interpretations share, and that brings me to the second question.

In order to delineate this feature, I first present the common characteristics of modal interpretations and then propose a general definition.

A first characteristic of modal interpretations is that they keep close to the standard formulation of quantum mechanics. That is, they all accept that the quantum mechanical description of a system α is defined on a Hilbert space $\mathcal{H}^\alpha$. So, magnitudes of α are represented by self-adjoint operators A^α and the state of α is given by a density operator W^α. In modal interpretations it is thus not assumed that there exists a more precise state for a system represented by something different to a density operator. In this sense modal interpretations are not so-called hidden-variable theories because in such theories one does assume that there exist more precise states.

Secondly, states of systems evolve in modal interpretations only according to the Schrödinger equation; the projection postulate is rejected.

Thirdly, modal interpretations take quantum mechanics as a universal theory of nature. Quantum mechanics thus applies not only to elementary particles, but also to macroscopic systems like measurement devices, planets, cats and elephants.

Fourthly, modal interpretations give rules to ascribe properties to systems at all times. This property ascription depends on the states of systems and applies regardless of whether or not measurements are performed. States of systems thus have a meaning in terms of properties possessed by systems and not merely in terms of outcomes of measurements.

Fifthly, these rules by which properties are ascribed are stochastic. So, a system α is not simply ascribed one set of properties (as was the case with the eigenvalue–eigenstate link (2.13)) but is ascribed a number of sets of properties with corresponding probabilities. Each set contains properties possibly possessed by α and the corresponding probability gives the probability that these properties are actually possessed by α.

A final common characteristic is that the probabilities with which modal interpretations ascribe properties to a system α are taken as representing ignorance about the actual properties of α only. These probabilities thus do not represent ignorance about the state of α. To make this point clear, I briefly discuss the so-called *ignorance interpretation* of quantum mechanics.

Consider for a moment an ensemble of N similar systems α, say N electrons. Assume that all these systems are in pure states but that the ensemble is inhomogeneous with regard to these states. Say, $N_1 \leq N$ systems are in state $|\psi^\alpha\rangle\langle\psi^\alpha|$ and $N_2 = N - N_1$ systems are in $|\phi^\alpha\rangle\langle\phi^\alpha|$. If one now wants to give predictions about measurements on a system randomly chosen from this ensemble, one can proceed as follows. The state of such a random system is with probability N_1/N equal to $|\psi^\alpha\rangle\langle\psi^\alpha|$ and with probability N_2/N equal to $|\phi^\alpha\rangle\langle\phi^\alpha|$. Hence, by the Born rule (2.5), there exists a weighted probability

$$p_{\text{Born}}(a_j) = \frac{N_1}{N}\,\text{Tr}^\alpha(|\psi^\alpha\rangle\langle\psi^\alpha|\,|a_j^\alpha\rangle\langle a_j^\alpha|) + \frac{N_2}{N}\,\text{Tr}^\alpha(|\phi^\alpha\rangle\langle\phi^\alpha|\,|a_j^\alpha\rangle\langle a_j^\alpha|) \qquad (3.1)$$

that a measurement of $A^\alpha = \sum_j a_j\,|a_j^\alpha\rangle\langle a_j^\alpha|$ yields the outcome corresponding to a_j. This prediction can now be reproduced by assigning a statistical or *mixed state* W_{mix}^α to the ensemble which is equal to the weighted sum of the states of the systems in the ensemble, so, $W_{\text{mix}}^\alpha = (N_1/N)\,|\psi^\alpha\rangle\langle\psi^\alpha| + (N_2/N)\,|\phi^\alpha\rangle\langle\phi^\alpha|$.[18] For, if one applies the Born rule to this mixed state, one directly obtains the Born probability (3.1).

The ignorance interpretation is now the interpretation one obtains if one accepts the eigenvalue–eigenstate link and makes the assumption that every non-pure state W^α assigned by quantum mechanics should be taken as a mixed state that describes an inhomogeneous ensemble of systems in pure states. A prevalent idea among physicists is that this ignorance interpretation solves the measurement problem. Consider, for instance, the Von Neumann measurement (2.9). The state of the measurement device μ after the measurement is $\widetilde{W}^\mu = \sum_j |c_j|^2\,|\text{R}_j^\mu\rangle\langle\text{R}_j^\mu|$. According to our assumption, one can take this state as describing an ensemble consisting of $|c_1|^2 N$ devices with state $|\text{R}_1^\mu\rangle\langle\text{R}_1^\mu|$, of $|c_2|^2 N$ devices with state $|\text{R}_2^\mu\rangle\langle\text{R}_2^\mu|$, etc.[19] If one then applies the eigenvalue–eigenstate link to this ensemble, one obtains that $|c_1|^2 N$ devices possess the reading $|\text{R}_1^\mu\rangle\langle\text{R}_1^\mu|$, that $|c_2|^2 N$ devices possess the reading $|\text{R}_2^\mu\rangle\langle\text{R}_2^\mu|$, etc. It thus follows that a device chosen randomly from this ensemble possesses reading $|\text{R}_j^\mu\rangle\langle\text{R}_j^\mu|$ with probability $|c_j|^2$, which is exactly the prediction of the standard formulation. Now, apart from the fact that this solution to the measurement problem doesn't work,[20] one understands within the

---

[18] Note that mixed states satisfy the definition of a quantum mechanical state (see the box on page 12) because mixed states are convex sums of pure states.

[19] Given the assumption that the state $\widetilde{W}^\mu$ describes an inhomogeneous ensemble of systems with pure states, it is not yet fixed that this state describes an ensemble with $|c_j|^2 N$ devices in state $|\text{R}_j^\mu\rangle\langle\text{R}_j^\mu|$. The decomposition $\widetilde{W}^\mu = \sum_j |c_j|^2\,|\text{R}_j^\mu\rangle\langle\text{R}_j^\mu|$ is not the only possible decomposition of $\widetilde{W}^\mu$ in terms of pure states, so $\widetilde{W}^\mu$ may equally describe another ensemble. However, let us, for the sake of argument, put this worry about uniqueness aside.

[20] The ignorance interpretation does not solve the measurement problem because the assumption that $\widetilde{W}^\mu$ describes an inhomogeneous ensemble, leads to inconsistencies. Proof: Consider the states generated by a Von Neumann measurement (2.9). The state of the device is $\widetilde{W}^\mu = \sum_j |c_j|^2\,|\text{R}_j^\mu\rangle\langle\text{R}_j^\mu|$. In order to solve the measurement problem, the ignorance interpretation takes this state as describing an inhomogeneous ensemble of N devices with $|c_j|^2 N$ devices in the state $|\text{R}_j^\mu\rangle\langle\text{R}_j^\mu|$ $(j = 1, 2, \dots)$. The state of the object system is $\widetilde{W}^\alpha = \sum_j |c_j|^2\,|a_j^\alpha\rangle\langle a_j^\alpha|$ and in the ignorance interpretation also describes an inhomogeneous ensemble, say, an ensemble of N' object systems with $\lambda_i' N'$ systems in the state $|\phi_i^\alpha\rangle\langle\phi_i^\alpha|$ $(i = 1, 2, \dots)$, where $\sum_i \lambda_i' = 1$ and $\widetilde{W}^\alpha = \sum_i \lambda_i'\,|\phi_i^\alpha\rangle\langle\phi_i^\alpha|$. The state of the composite $\alpha\mu$ is equal to $|\Psi^{\alpha\mu}\rangle = \sum_j c_j\,|a_j^\alpha\rangle \otimes |\text{R}_j^\mu\rangle$ and there are now two routes to apply the ignorance interpretation to this composite state. Firstly, one can construct an ensemble of composites $\alpha\mu$ which

ignorance interpretation a non-pure state as a description of a system which leaves one ignorant not only about the precise properties of the system, but also about its precise state. For it follows that, for instance, the final state $\widetilde{W}^\mu$ assigned to the measurement device is not the real state of the device. Instead, in the ignorance interpretation the real state is with probability $|c_j|^2$ equal to $|{\rm R}_j^\mu\rangle\langle{\rm R}_j^\mu|$.

The final characteristic of modal interpretations is now that this igno-rance with regard to states is rejected. Within modal interpretations the state assigned to a system is the state of an individual system and not a description of an ensemble of systems. The probabilities with which modal interpretations ascribe properties represent ignorance only with regard to the actual properties of a system and not with regard to the actual state of the system. Consequently, if at the end of a measurement the state of the device is $\widetilde{W}^\mu = \sum_j |c_j|^2 \, |{\rm R}_j^\mu\rangle\langle{\rm R}_j^\mu|$ and one observes that it possesses the reading $|{\rm R}_j^\mu\rangle\langle{\rm R}_j^\mu|$, one does not conclude that the state of the device is actually $|{\rm R}_j^\mu\rangle\langle{\rm R}_j^\mu|$. Or, put differently, one never uses the actually possessed properties of a system to 'update' the state of that system.

I believe that this last characteristic is the common feature that dis-tinguishes modal interpretations from other interpretations of quantum me-chanics. Take the orthodox interpretation, for instance. After a Von Neumann measurement the uncollapsed state $\widetilde{W}^\mu = \sum_j |c_j|^2 \, |{\rm R}_j^\mu\rangle\langle{\rm R}_j^\mu|$ of the measure-ment device is taken as that it actually possesses the outcome $|{\rm R}_j^\mu\rangle\langle{\rm R}_j^\mu|$ with Born probability $p_{\rm Born}(a_j) = |c_j|^2$. And the device possesses this outcome if and only if its state has collapsed to $\widehat{W}^\mu = |{\rm R}_j^\mu\rangle\langle{\rm R}_j^\mu|$. The probability $p_{\rm Born}(a_j)$ is thus also the probability that the device actually has the state $|{\rm R}_j^\mu\rangle\langle{\rm R}_j^\mu|$. Hence, in the orthodox interpretation the Born probabilities rep-

---

is consistent with the ensembles described by $\widetilde{W}^\alpha$ and by $\widetilde{W}^\mu$. Let this ensemble contain N'' systems and consider one of its elements $\alpha\mu$. It has already been established that the subsystem α of this element has one of the pure states $\{|\phi_i^\alpha\rangle\langle\phi_i^\alpha|\}_i$, say, $|\phi_a^\alpha\rangle\langle\phi_a^\alpha|$. And the subsystem μ of this element has one of the pure states $\{|{\rm R}_j^\mu\rangle\langle{\rm R}_j^\mu|\}_j$, say, $|{\rm R}_b^\mu\rangle\langle{\rm R}_b^\mu|$. It then follows (see footnote 11) that the state of the element $\alpha\mu$ itself is uniquely $|\phi_a^\alpha\rangle\langle\phi_a^\alpha| \otimes |{\rm R}_b^\mu\rangle\langle{\rm R}_b^\mu|$. This result holds for every element of the ensemble of composites, hence, every element has one of the product states $\{|\phi_i^\alpha\rangle\langle\phi_i^\alpha| \otimes |{\rm R}_j^\mu\rangle\langle{\rm R}_j^\mu|\}_{i,j}$. In order to get the distributions right, the ensemble of composites contains $\lambda''_{ij}N''$ elements with the state $|\phi_i^\alpha\rangle\langle\phi_i^\alpha| \otimes |{\rm R}_j^\mu\rangle\langle{\rm R}_j^\mu|$, where λ''_{ij} satisfies $\sum_j \lambda''_{ij} = \lambda'_i$ and $\sum_i \lambda''_{ij} = |c_j|^2$. Secondly, one can directly apply the ignorance interpretation to the state $|\Psi^{\alpha\mu}\rangle$ of $\alpha\mu$. This state is pure and thus describes a homogeneous ensemble of N'' systems which all have the state $|\widetilde{\Psi}^{\alpha\mu}\rangle\langle\widetilde{\Psi}^{\alpha\mu}|$. These two routes thus lead to descriptions of ensembles which are different from one another. Hence, taking the state $\widetilde{W}^\mu$ as describing an inhomogeneous ensemble leads to an inconsistency. $\square$

More generally, d'Espagnat (1971, Sect. 6) (see also d'Espagnat (1966)) has proved that not every non-pure state can be taken as a mixed state describing an inhomogeneous ensemble. One therefore has to distinguish so-called proper mixtures and improper mixtures. A proper mixture can be taken as describing an inhomogeneous ensemble and an improper mixture cannot. The mixed state $W_{\rm mix}^\alpha$ defined in the main text is an example of a proper mixture. And the device state $\widetilde{W}^\mu$ is, according to the proof given in this footnote, an example of an improper mixture.

resent ignorance about the actual properties *and* about the actual state of systems.

Modal interpretations can thus be defined by means of this characteristic of rejecting ignorance with regard to states. Consider any theory and let *TS* denote the *Theoretical State* which this theory assigns to a system (for quantum mechanics *TS* is thus a density operator) and let *SoA* denote a description of a *State of Affairs* for that system (in quantum mechanics *SoA* is a set of definite properties of the system or a set of values assigned to magnitudes of that system). The definition I then propose is:

Definition of a modal interpretation

An interpretation of a theory is a modal interpretation iff:

(A) the theoretical state *TS* of a system α is interpreted in terms of one or more possible states of affairs $\{SoA_j\}_j$ from which exactly one describes the actual state of affairs, and

(B) if the theoretical state *TS* of a system α does not uniquely determine the actual state of affairs SoA_k (that is, if *TS* is interpreted in terms of two or more different possible states of affair), then α is not assigned a more accurate theoretical state *TS'* which does uniquely determine the actual state of affairs SoA_k.

This definition is my answer to the second question raised in the beginning of this section: the term 'modal interpretation' presently covers all interpretations which, more informal then the above definition, (A) interpret a state of a system in terms of ignorance with regard to the properties of the system, but (B) do not take this ignorance as ignorance with regard to a more accurate state of that system.[21]

The name 'modal' is in my opinion suited because one may understand it as pointing to the fact that modal interpretations interpret quantum mechanics by slightly changing the standard understanding of the modalities 'actuality' and 'possibility.' To illustrate this non-standard treatment, consider again the fact that modal interpretations maintain that after a Von Neumann measurement a device μ that actually possesses the reading $|R_j^{\mu}\rangle\langle R_j^{\mu}|$ may still have the state $\widetilde{W}^{\mu} = \sum_j |c_j|^2 |R_j^{\mu}\rangle\langle R_j^{\mu}|$. This means that the terms $\{|c_k|^2 |R_k^{\mu}\rangle\langle R_k^{\mu}|\}_{k\neq j}$ that refer to the non-actualised outcomes are not removed from the state of the device. This procedure of removing the non-actualised

---

[21] Given this definition, one may argue that the interpretation of statistical mechanics counts as a modal interpretation: the statistical state of, say, a gas is interpreted in terms of a number of possible mechanical states of the gas molecules (part (A) of the definition) but one never replaces in statistical mechanics the statistical state by a state which uniquely determines this mechanical state of the gas (part (B) of the definition). Private communication with Jos Uffink, 1997.

possibilities is, however, quite standard in statistical theories. Take, for instance, the weather forecast. Say, yesterday's prediction warned us that we may have a blizzard today with a ten per cent probability. If today the actual weather is quite sunny, the meteorologists, when preparing the weather forecast for tomorrow, start calculating with the actual sunny state of today's atmosphere. They thus ignore the non-actualised possibility of today's blizzard but use an updated state of the atmosphere which no longer contains references to that blizzard. Non-actualised possibilities are thus standardly removed from states. In modal interpretations the state is now not updated if a certain state of affairs becomes actual. The non-actualised possibilities are not removed from the description of a system and this state therefore codifies not only what is presently actual but also what was presently possible. These non-actualised possibilities can, as a consequence, in principle still affect the course of later events. This implies, if one translates it to the weather forecast, that blizzards that did not actually occur can still affect tomorrow's weather. Modal interpretations are thus called modal because they treat modalities non-standardly.

3.2 Starting points

Before presenting the different modal interpretations in the next chapter, I briefly list the general starting points which I adopt in developing how these interpretations give descriptions of reality (the first two points have already been mentioned in Section 2.2).

Firstly, physical magnitudes pertaining to a system α are represented by self-adjoint operators A^α defined on the Hilbert space $\mathscr{H}^\alpha$ associated with α. Secondly, a magnitude that is represented by an operator with a discrete spectral resolution $A^\alpha = \sum_j a_j \sum_k |a_{jk}^\alpha\rangle\langle a_{jk}^\alpha|$, can assume as a definite value only one of the eigenvalues $\{a_j\}_j$ of A^α. The property that this magnitude has value a_j is represented by the corresponding eigenprojection $\sum_k |a_{jk}^\alpha\rangle\langle a_{jk}^\alpha|$. The notation $[A^\alpha] = a_j$ captures that magnitude A^α has the definite value a_j and $[\sum_k |a_{jk}^\alpha\rangle\langle a_{jk}^\alpha|] = 1$ denotes that α possesses the property $\sum_k |a_{jk}^\alpha\rangle\langle a_{jk}^\alpha|$. It then follows that $[A^\alpha] = a_j$ if and only if $[\sum_k |a_{jk}^\alpha\rangle\langle a_{jk}^\alpha|] = 1$.

Thirdly, it can be the case that a magnitude A^α does not assume the eigenvalue a_j as a definite value, for instance, if A^α has the value $a_{j'} \neq a_j$. In that case α also does not possess the property $\sum_k |a_{jk}^\alpha\rangle\langle a_{jk}^\alpha|$. Let now $[A^\alpha] \neq a_j$ denote that A^α does not have value a_j and let $[\sum_k |a_{jk}^\alpha\rangle\langle a_{jk}^\alpha|] = 0$ denote that α does not possess the property $\sum_k |a_{jk}^\alpha\rangle\langle a_{jk}^\alpha|$. It follows that $[A^\alpha] \neq a_j$ if and only if $[\sum_k |a_{jk}^\alpha\rangle\langle a_{jk}^\alpha|] = 0$.

Fourthly, I do not commit myself to the position that it is always the case that a magnitude A^α either has definite value a_j or does not have that value. Instead I allow the possibility that it is indefinite whether or not A^α has value a_j. As a consequence, I also do not commit myself to the position that a property $\sum_k |a^\alpha_{jk}\rangle\langle a^\alpha_{jk}|$ is either possessed or not possessed; a property can also have the status of being indefinite (in the Sections 4.1, 5.6 and 14.1 I return to this possibility that properties and values of magnitudes are indefinite).

To capture these starting points more briefly, I adopt the following link between the values assignment and the property ascription to a system α:[22]

$$
\left.
\begin{array}{lll}
[A^\alpha] = a_j & \text{\emph{if and only if}} & \left[\sum_k |a^\alpha_{jk}\rangle\langle a^\alpha_{jk}|\right] = 1, \\[2ex]
[A^\alpha] \neq a_j & \text{\emph{if and only if}} & \left[\sum_k |a^\alpha_{jk}\rangle\langle a^\alpha_{jk}|\right] = 0, \\[2ex]
\begin{array}{l}\text{it is indefinite whether} \\ \text{or not } A^\alpha \text{ has value } a_j\end{array} & \text{\emph{if and only if}} & \sum_k |a^\alpha_{jk}\rangle\langle a^\alpha_{jk}| \text{ is indefinite.}
\end{array}
\right\} \quad (3.2)
$$

(I emphasise that this link is a starting point; in Section 5.6 I introduce generalisations of (3.2) in order to also accommodate magnitudes with continuous spectral resolutions.)

According to the general definition formulated in the previous section, a modal interpretation interprets the state of a system in terms of a number of states of affairs $\{SoA_j\}_j$. Since the physical state of affairs of a system α can be described by means of the properties of α or by means of the magnitudes pertaining to α, there is some freedom in capturing these states of affairs. I

---

[22] The link (3.2) is not as innocent as it may seem. Consider, for instance, two magnitudes A^α and $\tilde{A}^\alpha$ represented by operators which do not commute but which share the eigenprojection $|a^\alpha_1\rangle\langle a^\alpha_1|$. Then, if $[A^\alpha] = a_1$, it follows immediately from this link that $[\tilde{A}^\alpha] = \tilde{a}_1$. Now, according to quantum mechanical orthodoxy one should be cautious about simultaneously assigning definite values to magnitudes represented by non-commuting operators (although the orthodox eigenvalue–eigenstate link (2.13) allows such value assignments: if the state of α is $W^\alpha = |a^\alpha_1\rangle\langle a^\alpha_1|$, this link yields that both $[A^\alpha] = a_1$ and $[\tilde{A}^\alpha] = \tilde{a}_1$).

If one wants to avoid that $[A^\alpha] = a_1$ always implies $[\tilde{A}^\alpha] = \tilde{a}_1$, one should reject the link (3.2) and adopt instead, for instance, the following more restrictive link:

$$
\left.
\begin{array}{l}
[A^\alpha] = a_j \quad \text{iff} \quad \left[\sum_k |a^\alpha_{jk}\rangle\langle a^\alpha_{jk}|\right] = 1 \quad \text{and} \quad \left[\sum_k |a^\alpha_{j'k}\rangle\langle a^\alpha_{j'k}|\right] = 0 \ \text{ for all } \ j' \neq j, \\[2ex]
[A^\alpha] \neq a_j \quad \text{iff} \quad [A^\alpha] = a_{j'} \ \text{ with } \ a_{j'} \neq a_j.
\end{array}
\right\} \quad (3.2^*)
$$

This alternative link (which is more restrictive than the link (3.2) because now $[\sum_k |a^\alpha_{jk}\rangle\langle a^\alpha_{jk}|] = 1$ does not automatically imply that $[A^\alpha] = a_j$) is accepted by, for instance, Dieks (1988, 1989) and Clifton (1995a). I prefer my more liberal link (3.2*) because, given a set of projections with definite values, it assigns definite values to more magnitudes than the link (22). It is furthermore my position that it should not *a priori* be excluded that magnitudes represented by non-commuting operators can have simultaneously definite values; quantum mechanics only says that one cannot simultaneously measure such magnitudes and is silent about whether or not they can have simultaneously definite values. For a further discussion, see Reeder (1998), for instance.

now choose to describe them in terms of the ascribed properties. The value assignment to the magnitudes of α can then be derived by means of these ascribed properties and the link (3.2).

More precisely, I describe the state of affairs SoA_j as follows: A determination of the properties of a system α is according to my starting points equivalent to an assignment of the value 0 or 1 to a number of the eigenprojections of the magnitudes of α. These eigenprojections are projections on the Hilbert space $\mathscr{H}^\alpha$ associated with α. Hence, a determination of the properties of a system α can be given by an assignment of values to projections $\{Q^\alpha\}$ defined on $\mathscr{H}^\alpha$. Now let $\mathscr{DP}_j$ be the set of projections with definite values and let $[.]_j$ be the map from $\mathscr{DP}_j$ to the set $\{0,1\}$ which gives the definite value of the projections in $\mathscr{DP}_j$. The state of affairs SoA_j is thus captured by the ordered pair $\langle \mathscr{DP}_j, [.]_j \rangle$. The modal interpretations discussed in this book also give the probability p_j with which a state of affairs $\langle \mathscr{DP}_j, [.]_j \rangle$ is actually the case. So, an even more informative description is given by the ordered triple $\langle p_j, \mathscr{DP}_j, [.]_j \rangle$.

With all these notational means one can now characterise the rules of a modal interpretation by the map

$$W^\alpha \longmapsto \{\langle p_j, \mathscr{DP}_j, [.]_j \rangle\}_j. \tag{3.3}$$

This map should be read as follows: If α has a state W^α, then it is with probability p_j actually the case that, firstly, α possesses the properties represented by the projections $Q^\alpha \in \mathscr{DP}_j$ with $[Q^\alpha]_j = 1$, secondly, α does not possesses the properties represented by the projections $Q^\alpha \in \mathscr{DP}_j$ with $[Q^\alpha]_j = 0$, and, thirdly, the properties of α represented by projections $Q^\alpha \notin \mathscr{DP}_j$ are indefinite.

3.3 Demands, criteria and assumptions

In Section 2.2 I have given a preliminary discussion of what interpretations of quantum mechanics should provide. The points I mentioned were that an interpretation should describe reality by assigning values to magnitudes of systems and that this description of reality should meet the demands of being well developed, empirically adequate and metaphysically tenable. Now, since modal interpretations ascribe properties to systems by means of the map $W^\alpha \mapsto \{\langle p_j, \mathscr{DP}_j, [.]_j \rangle\}_j$, they automatically assign values to magnitudes via the link (3.2). So what is left is the question of when do modal interpretations meet the three given demands.

Unfortunately, the first demand that a modal interpretation should give a well-developed description of reality is quite open-ended. It is clear, of course, that such a description must satisfy some conditions. An obvious candidate is the condition that the property ascription $\{\langle p_j, \mathscr{DP}_j, [.]_j \rangle\}_j$ is well defined. It is also clear that a fully-blown description should include the correlations between the properties ascribed to different systems, as well as the dynamics of these properties. However, there also exist conditions for which it is not that obvious whether well-developed descriptions should meet them. Consider, for instance, the condition that a description of reality always assigns definite values to such key magnitudes as position, momentum or energy. Or take the condition that a description of reality is invariant under Lorentz transformations. Such conditions may seem obvious candidates to some authors but not to others. And even if there is consensus about a condition, say, that a well-developed description should be well defined, there need not be consensus about what such a condition implies: Most authors agree that this means that the sets of definite-valued properties $\{\mathscr{DP}_j\}_j$ must be closed under negation, conjunction and disjunction. But they again disagree about the question of whether this condition implies that the property ascription should satisfy the rule that $[Q^\alpha] = 1$ implies that $[\tilde{Q}^\alpha] = 1$ for every $\tilde{Q}^\alpha Q^\alpha = Q^\alpha$ (see Chapter 5 for a full discussion of the conditions for property ascriptions).

Due to this open-endedness, I take the position that any choice for a criterion for whether a description of reality is well developed or not, is arbitrary. I therefore do not propose one myself. Instead I develop modal interpretations as much as possible by trying to properly define the property ascription and by trying to determine the correlations between the ascribed properties as well as their dynamics. Afterwards, one can then review the results and reconsider the question of whether or not they give rise to a physically acceptable description of reality.

I use a number of assumptions to develop modal interpretations. I now present these assumptions and discuss their reasonableness.

The first assumption delimits the information necessary to give the property ascription to a system α. This property ascription is a map $W^\alpha \mapsto \{\langle p_j, \mathscr{DP}_j, [.]_j \rangle\}_j$ and because in physical theories the state of a system is meant to fully describe the physics of that system, it is natural to assume that W^α codifies all information about the property ascription $\{\langle p_j, \mathscr{DP}_j, [.]_j \rangle\}_j$. The standard formulation, for instance, fulfills such an assumption with regard to its predictions about measurements: the probabilities for the outcomes of a perfect measurement performed on α at time t depend solely on the state of α at t. I therefore assume that the property ascription $\{\langle p_j, \mathscr{DP}_j, [.]_j \rangle\}_j$ to α at time t also depends solely on the state of α on t. I call this assumption *Instantaneous Autonomy*: the state of α at time t codifies all information about the property ascription to α at t or, more briefly, the state of α determines autonomously the instantaneous property ascription to α. Slightly reformulated, this reads:

Instantaneous Autonomy

If two systems have equal states, then the instantaneous property ascriptions $\{\langle p_j, \mathcal{D}\mathcal{P}_j, [.]_j \rangle\}_j$ to those systems are also equal.

A necessary condition for this assumption is that the property ascription to a system α is a function of the state of α only. For if this property ascription were, for instance, a function of the state $W^{\alpha\beta}$ of α and a system β disjoint from α, one can violate Instantaneous Autonomy by changing the state $W^{\alpha\beta}$ while keeping W^{α} fixed.

In principle Instantaneous Autonomy does not need to hold. It could, for instance, be the case that the statistical property ascription to α is a function of the state of a composite of α and a second system β. Or the statistical property ascription could simply not be a function of a state of any system at all. In both cases the properties ascribed by modal interpretations would become rather difficult to determine. Consider, for example, an experimenter on earth who is preparing a system α in a specific state in order to examine the properties of this system. In the first case a change in the state of a second, possibly distant, system β, for instance, *le petit prince* who is rearranging his asteroid,[23] can make the properties of α change. In the second case there is no unique relation which fixes the properties of α. Hence, however precisely the experimenter prepares the state of α, he or she can never fix the property ascription to α. One can thus take Instantaneous Autonomy as the assumption that the state of a system uniquely fixes the property ascription to the system.

For composite systems which evolve freely (that is, composite systems which do not interact with an environment), the standard formulation also satisfies a dynamical version of autonomy. Consider, for instance, two or more measurements performed on a system α at different instants and assume that the composite ω of α and the measurement devices evolves freely. Then the statistical predictions about the correlations between the respective outcomes (and by correlations I mean not only correlations between outcomes which are simultaneously displayed by the devices but also *sequential correlations* between outcomes displayed at different times) depend solely on the state of ω at a given instant and the Hamiltonian H^{ω} of ω. The

---

[23] See de Saint-Exupéry (1943).

analogue of this dynamical autonomy for modal interpretations would be the following assumption:

Dynamical Autonomy for composite systems

If two composite systems evolve freely and have, during an interval, equal states and equal Hamiltonians, then the (simultaneous and sequential) correlations between the properties ascribed to these composites and their subsystems at different times in that interval are also equal.

One of the results reached in this book is that if one accepts Instantaneous Autonomy and the criterion of Empirical Adequacy (see below), then modal interpretations become contradictory if one also accepts this assumption of Dynamical Autonomy for composite systems. It is thus not possible to also assume Dynamical Autonomy. Instead, I formulate three weakened versions of the assumption of Dynamical Autonomy which can be accepted with varying success. The first version concerns only the correlations between properties ascribed to freely evolving composite systems as a whole (and thus not between properties ascribed to the subsystems of this composite). The second concerns the correlations between properties ascribed to freely evolving atomic systems. The third version concerns only the correlations between the initially possessed properties of an object system and the finally possessed outcomes of a measurement device if one measures the initially possessed properties of the object system:

Dynamical Autonomy for whole systems

If two, possibly composite, systems evolve freely and have, during an interval, equal states and equal Hamiltonians, then the correlations between the properties ascribed to the systems as a whole at different times in that interval are also equal.

Dynamical Autonomy for atomic systems

If two atomic systems evolve freely and have, during an interval, equal states and equal Hamiltonians, then the correlations between the properties ascribed to the atoms at different times in that interval are also equal.

Dynamical Autonomy for measurements

If two composite systems evolve freely and have, during an interval, equal states and equal Hamiltonians and if these composite systems consist of an object system and a measurement device and one is dealing with a measurement of the initially possessed properties of the object system, then the correlations between the initial properties of the object system and the finally possessed outcomes of the measurement device are also equal.

Necessary conditions for these weakened versions of Dynamical Autonomy are that the correlations between the ascribed properties are functions of only the state and the Hamiltonian of the respective freely evolving system.

The second demand that modal interpretations have to be empirically adequate is clear-cut: modal interpretations should yield a description of reality that confirms our observations that measurements usually have definite outcomes and that the probabilities and correlations with which these outcomes occur are correctly predicted by the standard formulation of quantum mechanics. Hence, modal interpretations are empirically adequate if they satisfy the following criterion:

Empirical Adequacy

The predictions about outcomes ascribed to measurement devices are equal to the predictions generated by the standard formulation of quantum mechanics.

In order to develop modal interpretations to fully-blown descriptions, I now also employ this criterion of Empirical Adequacy. More precisely, I require on two occasions that the predictions generated by modal interpretations about measurement outcomes satisfy Empirical Adequacy.

The reader might want to object that this criterion is not meant as a tool for *developing* modal interpretations, but is meant for *testing* them once they have been developed. However, sometimes modal interpretations lack enough structure to uniquely develop them to fully-blown theories. That is, it can happen that one cannot uniquely deduce one solution to a problem because the problem is underdetermined and thus allows many solutions. In such cases I now reverse the order of deduction and derive on the basis of the assumption that modal interpretations *are* fully-blown and empirically adequate theories, necessary conditions for modal interpretations.

To put it more pointedly: if I can't solve a problem because there exist too many solutions, I require that the correct solution is empirically adequate. On the basis of this requirement I then deduce necessary conditions for modal interpretations. One can take this reversed deduction as a secular version of the transcendental deduction developed by Kant (1787).

A further objection could be that if one indeed feeds in this criterion, modal interpretations are *by construction* empirically adequate. This is, however, not the case. I use Empirical Adequacy only to fix the correlations between measurement outcomes in two specific series of measurements. It is thus still very possible that in other series of measurements, modal interpretations yield predictions which differ from the ones given by the standard formulation.

So, by a limited use of this criterion, the task of being empirically adequate is still a genuine burden for modal interpretations.

From the second demand of empirical adequacy, one can argue that modal interpretations should in some sense re-establish the standard formulation of quantum mechanics. From a fundamental point of view, modal interpretations reject and overthrow the standard formulation, especially by denying the projection postulate. But at the same time they have to reproduce the predictions of this formulation, because these predictions are correct. In that sense they need to recover the standard formulation. More generally, modal interpretations should try to account for what physicists do in their laboratories and institutes. Physicists constantly use the standard formulation in their descriptions of quantum mechanical phenomena and it seems a bit pedantic to discard this as wrong. I therefore take the position that modal interpretations should try to explain or maybe even be the foundation of the standard formulation of quantum mechanics as a correct theory about measurement outcomes.

The third demand that a modal interpretation should yield a metaphysically tenable description of reality surpasses the first two demands because a fully developed and empirically adequate description of reality can still give a totally weird and unacceptable description of the properties of non-observed quantum systems. In Section 12.1 I argue that precisely because modal interpretations describe states of affairs which are in principle unobservable, one should be careful about discarding modal descriptions of reality as metaphysically untenable. Our criteria about what is tenable and what is not may be guided by our intuitions about the states of affairs that we can observe. And it seems to me that it is incorrect to impose intuitions about descriptions of what is observable on descriptions of what is, in principle, unobservable. The criteria I propose for metaphysical tenability are thus very sparse:

Consistency

The description of reality should be free of contradiction.

Internal Completeness

The description of reality by an interpretation should be complete with regard to the standards set by that interpretation: that is, an interpretation should deliver the description of reality that it promises to deliver.

This book is divided into three parts called *Formalism*, *Physics* and *Philosophy*. Each part is devoted to meeting one of the discussed demands.

In Part I the different modal interpretations are introduced and further developed. Part II is about their empirical adequacy. And in Part III I consider the metaphysical tenability of modal interpretations and the way in which they resurrect the standard formulation.

Part one

Formalism

In Chapter 4 the different modal interpretations are introduced. Their property ascription to a system α is characterised by a core property ascription $\{\langle p_j, C_j^{\alpha} \rangle\}_j$.

Chapter 5 treats the question of how the core property ascription to a system determines the full property ascription $\{\langle p, \mathcal{DP}_j, [.]_j \rangle\}_j$ to the system, and of how that full property ascription induces an assignment of values to the magnitudes of the system.

In Chapter 6 it is determined whether the properties that modal interpretations simultaneously ascribe to different systems can be correlated. A no-go theorem is derived which restricts the possibility of giving such correlations.

In Chapter 7 it is shown that the set of properties, which a system *possibly* possesses, evolves in a number of undesirable ways. This evolution is, for instance, discontinuous and unstable.

Chapter 8 is concerned with the evolution of the *actually* possessed properties of a system. It is proved for the case of freely evolving systems that this evolution is deterministic and for the case of interacting systems it is argued that this evolution cannot be uniquely fixed.

In Chapter 9 it is proved that the evolution of the actually possessed properties of systems violates a number of the Dynamical Autonomy assumptions presented in Section 3.3. It is shown that this allows the descriptions of reality by modal interpretations to be non-local in a quite explicit way.

4

The different versions

In this chapter I introduce the different versions of the modal interpretation, including the three on which I focus in this book. The property ascription of these versions is characterised by the map $W^\alpha \mapsto \{\langle p_j, C_j^\alpha \rangle\}_j$ which I call the core property ascription. In the next chapter I discuss how this core property ascription determines the full property ascription $W^\alpha \mapsto \{\langle p_j, \mathcal{DP}_j, [.]_j \rangle\}_j$.

4.1 The best modal interpretation

The best imaginable modal interpretation is, I guess, an interpretation which (A) ascribes at all times all the properties to a system which pertain to that system, and (B) ascribes these properties such that the classical logical relations between the negation, conjunction and disjunction of properties are satisfied.

The content of the first requirement (A) is clear: assuming that every projection onto a subspace of a Hilbert space $\mathcal{H}^\alpha$ represents one and only one property of α, it follows that all sets $\{\mathcal{DP}_j\}_j$ of definite-valued projections should contain all the projections in $\mathcal{H}^\alpha$ and that all the maps $\{[.]_j\}_j$ should be maps from all the projections in $\mathcal{H}^\alpha$ to the values $\{0, 1\}$.

The content of requirement (B) is, however, less clear because there is consensus neither about how to define the negation, conjunction and disjunction of properties in quantum mechanics nor about how to impose the logical relations. In Section 5.1 I present my choice for the definitions of the negation, conjunction and disjunction. But assume here, for the sake of argument, that requirement (B) implies that the negation $\neg$ of a property represented by the projection Q^α is represented by the projection

$$\neg Q^\alpha = \mathbb{I}^\alpha - Q^\alpha, \tag{4.1}$$

and that the conjunction $\wedge$ and disjunction $\vee$ of two properties represented by two pair-wise *commuting* projections Q_1^α and Q_2^α are given, respectively, by

$$Q_1^\alpha \wedge Q_2^\alpha = Q_1^\alpha Q_2^\alpha, \quad Q_1^\alpha \vee Q_2^\alpha = Q_1^\alpha + Q_2^\alpha - Q_1^\alpha Q_2^\alpha. \tag{4.2}$$

(In my notation I do not distinguish between the property Q^α and the projection Q^α which represents this property.) Requirement (B) then implies that the maps $\{[.]_j\}_j$ should satisfy the classical logical relations

$$
\begin{aligned}
[\neg Q^\alpha]_j &= 1 - [Q^\alpha]_j, \\
[Q_1^\alpha \wedge Q_2^\alpha]_j &= [Q_1^\alpha]_j [Q_2^\alpha]_j, \\
[Q_1^\alpha \vee Q_2^\alpha]_j &= [Q_1^\alpha]_j + [Q_2^\alpha]_j - [Q_1^\alpha]_j [Q_2^\alpha]_j,
\end{aligned}
\qquad
\left.
\begin{aligned}
[\mathbb{O}^\alpha]_j &= 0, \\
[\mathbb{I}^\alpha]_j &= 1,
\end{aligned}
\right\}
\tag{4.3}
$$

where Q^α is any projection in $\mathcal{DP}_j$, where Q_1^α and Q_2^α are any pair of commuting projections in $\mathcal{DP}_j$ and where $\mathbb{O}^\alpha$ and $\mathbb{I}^\alpha$ are, respectively, the null and unit operator. (The conditions (4.3) express the classical logical relations because they imply that if Q^α is possessed, then $\neg Q^\alpha$ is not possessed and *vice versa*, that if Q_1^α and Q_2^α are possessed, then $Q_1^\alpha \wedge Q_2^\alpha$ is also possessed, etc.)

Unfortunately, Kochen and Specker (1967) proved the theorem that for Hilbert spaces $\mathcal{H}^\alpha$ with a dimension strictly larger than 2, there do not exist maps $[.]_j$ from the set of all projections in $\mathcal{H}^\alpha$ to the values $\{0, 1\}$ (needed to satisfy requirement (A)) which obey the conditions (4.1), (4.2) and (4.3) (needed to satisfy (B)). Hence, Kochen and Specker proved that the best imaginable modal interpretation does not exist.[24] Interpretations of quantum mechanics thus all have to give up on at least one of the two requirements (A) and (B). The modal interpretations considered here now all drop the first requirement. A property pertaining to a system then need not be either possessed or not possessed but can have a third ontological status of being *indefinite*. To acquire some sort of feeling for this third status, one can consider the properties 'heads' and 'tails' during the toss of a coin. When the coin has fallen, either heads or tails is a possessed property and tails respectively heads is a property not possessed. However, when the coin is still flipping, one usually does not ascribe heads or tails to the coin since these properties apply to coins at rest on a surface. The properties 'heads' and 'tails' can thus be taken as not applicable or indefinite for flipping coins. Modal interpretations which take this route of dropping requirement (A) specify which properties are definite (possessed or not possessed) and which are indefinite.[25]

One can now raise the question of which of the existing modal interpretations can be regarded as the second best modal interpretation. And in

---

[24] See Bub (1997, Chap. 3) for an extensive discussion of the Kochen and Specker theorems.
[25] An example of a modal interpretation that gives up on requirement (B) is given by the one by Healey (1989).

attempts to answer this question, a number of authors have given motivations for specific versions.[26] In general, such a motivation consists of a proof that the considered version is the only one that satisfies a set of natural criteria. However, as I argued in Section 3.3, it is not that clear which criteria one should impose on interpretations. It therefore seems that these motivations do not really answer the question of which version is the second best modal interpretation, but that they only rephrase it to the question of which set of 'natural' criteria is the most natural. I sympathise with Feyerabend's (1975) slogan that anything goes, meaning that I think that any choice of criteria as the 'natural' ones is of limited value. Instead, my position is that the determination of the second best modal interpretation should be based on the success with which the different versions meet the more general demands discussed in Section 3.3. Hence, I firstly develop modal interpretations into fully-fledged descriptions of reality and investigate afterwards how they fare in meeting these demands.

In this book I limit myself to only three versions of the modal interpretation, namely the version by Kochen (1985) and Dieks (1988), the version presented in Vermaas and Dieks (1995) and the version proposed by Bacciagaluppi and Dickson (1997) and Dieks (1998b). I call these versions the *bi modal interpretation*, the *spectral modal interpretation* and the *atomic modal interpretation*, respectively. Also I briefly sketch the modal interpretation by Van Fraassen (1973) and the interpretation by Bub (1992). (I do not discuss the modal interpretation by Healey (1989). This interpretation is in many respects similar to the bi modal interpretation of Kochen and Dieks. However, because there are also substantial differences between these interpretations, it is better not to identify them.[27])

The reason why I limit myself to the bi, spectral and atomic modal interpretations, is that they can be seen as being part of one and the same programme towards an interpretation of quantum mechanics. Historically this programme started with the bi modal interpretation. The spectral modal interpretation followed in order to answer questions left open by the bi modal interpretation. And then the atomic modal interpretation was formulated to by-pass problems encountered in the spectral modal interpretation. There is also a methodological connection between these three interpretations: namely, they all aim at defining the properties possessed by a system from only the state of that system. That is, the map which gives the sets $\{\mathcal{DP}_j\}_j$

---

[26] Papers which motivate specific modal interpretations are Clifton (1995a,b), Dickson (1995a,b) and Dieks (1995). More general arguments are given in Bub and Clifton (1996), Bub (1997) and Zimba and Clifton (1998).

[27] Further papers on Healey's (1989) interpretation are Healey (1993a,b, 1994, 1995, 1998), Reeder (1995, 1998), Reeder and Clifton (1995) and Clifton (1996).

of definite properties of a system, is a function solely of the state of that system and not a function of factors other than the state. To illustrate this common feature, consider the orthodox interpretation (see Section 2.2) and consider Bohmian mechanics (Bohm 1952; Bohm and Hiley 1993). The orthodox interpretation ascribes by means of the eigenvalue–eigenstate link the properties $\{Q^\alpha \,|\, \mathrm{Tr}^\alpha(W^\alpha Q^\alpha) = 1\}$ to a system α. This ascription shares the common feature of the bi, spectral and atomic modal interpretations because these properties are defined solely from the state of the system α. On the other hand, Bohmian mechanics does not share this common property: Bohmian mechanics ascribes the positions $\{|\vec{r}^{\,\alpha}\rangle\langle\vec{r}^{\,\alpha}|\}$ at all times to an elementary particle and these positions cannot be defined solely from the state of that particle but must have their origin elsewhere.

To make this feature of 'being defined solely from the state' precise, one can use the notion of *implicit definability* as developed by Malament (1977). The idea is that if one object Y is defined solely from another object X, then any transformation of Y that is induced by a symmetry of X should be a symmetry of Y as well. For if there exists a symmetry of X which induces a transformation of Y that is not a symmetry of Y, then the definition of Y must contain, in addition to X, ingredients which account for the loss of symmetry.

In our case, X is a state W^α and Y is a set of properties $\{Q^\alpha\}$. The properties in this set are all represented by operators defined on the Hilbert space $\mathscr{H}^\alpha$. This means that any transformation of the state W^α is immediately also a transformation of a property Q^α. Now, transformations of W^α, which may be symmetries, are given by $W^\alpha \mapsto U^\alpha W^\alpha [U^\alpha]^\dagger$ with U^α a unitary operator on $\mathscr{H}^\alpha$. Thus, a property Q^α is definable solely from the state W^α if and only if Q^α is preserved under all the transformations $Q^\alpha \mapsto U^\alpha Q^\alpha [U^\alpha]^\dagger$ that preserve W^α. And a set of properties $\{Q^\alpha\}$ is definable from the state W^α if the set $\{U^\alpha Q^\alpha [U^\alpha]^\dagger\}$ of transformed properties is equal to the original set $\{Q^\alpha\}$, for all the unitary transformations U^α that preserve W^α. This criterion of definability is called 'implicit definability' by Malament:

Implicit Definability

A property Q^α is implicitly definable from the state W^α if and only if $U^\alpha Q^\alpha [U^\alpha]^\dagger = Q^\alpha$ for all U^α with $U^\alpha W^\alpha [U^\alpha]^\dagger = W^\alpha$.

A set of properties $\{Q^\alpha\}$ is implicitly definable from the state W^α if and only if $\{U^\alpha Q^\alpha [U^\alpha]^\dagger\} = \{Q^\alpha\}$ for all U^α with $U^\alpha W^\alpha [U^\alpha]^\dagger = W^\alpha$.

The set of properties $\{Q^\alpha \,|\, \mathrm{Tr}^\alpha(W^\alpha Q^\alpha) = 1\}$ ascribed by the orthodox interpretation is indeed implicitly definable from W^α. And the set of positions $\{|\vec{r}^{\,\alpha}\rangle\langle\vec{r}^{\,\alpha}|\}$ ascribed by Bohmian mechanics is generally not implicitly definable from W^α.

Core property ascription

The core property ascription $W^\alpha \mapsto \{\langle p_j, C_j^\alpha \rangle\}_j$ assigns with probability p_k the value 1 to the core projection C_k^α and generates the full property ascription to α.

The property ascriptions by the bi, spectral and atomic modal interpretations are now all implicitly definable from the states of systems, and this distinguishes them from the modal interpretations by Van Fraassen and by Bub. This common feature of the bi, spectral and atomic modal interpretations means that they form a programme towards an interpretation of quantum mechanics which stays close to quantum mechanics. That is, these interpretations ascribe properties to systems which are fixed by the states of the systems themselves. The ascribed properties are thus not fixed by something which is not part of the quantum formalism — they are not put in 'by hand,' for instance.

In the next sections I introduce the different interpretations. However, before doing so, it should be noted that the property ascriptions of modal interpretations are usually given in two steps. Firstly, authors give what I call a **core property ascription**. This core property ascription to a system α consists of a map $W^\alpha \mapsto \{\langle p_j, C_j^\alpha \rangle\}_j$ from the state of α to a set of ordered pairs $\langle p_j, C_j^\alpha \rangle$ containing a probability p_j and a corresponding core projection C_j^α. And this core property ascription implies that with probability p_k the core projection C_k^α has the value 1, that is, it implies that α at least possesses the property represented by C_k^α. Secondly, authors give rules for how this core property ascription $[C_k^\alpha] = 1$ generates the full property ascription $\langle \mathcal{DP}_k, [.]_k \rangle$. I present here the core property ascription of the different versions and in the next chapter I discuss how this core property ascription fixes the full property ascription.

4.2 Van Fraassen's Copenhagen modal interpretation

As I have said before, Van Fraassen formulated the first modal interpretation in 1972 which he called the *Copenhagen modal interpretation*.[28] Its core

---

[28] The name 'modal interpretation' indeed first appeared in Van Fraassen (1972). However, if authors refer to the modal interpretation by Van Fraassen, they are refering to the one presented in Van Fraassen (1973). This latter interpretation is also the one discussed in Van Fraassen (1991, Sect. 9.2 and 9.3 and the *Proofs and illustrations* of Sect. 9.3) (see footnote 1 of Chapter 9 of Van Fraassen (1991) for a short chronological overview). Other papers on Van Fraassen's modal interpretation are Van Fraassen (1976, 1981, 1990, 1997) and Leeds and Healey (1996).

property ascription[29] as formulated in Van Fraassen (1973), complies with the following rules. Firstly, if a system α has a state W^α, the core property ascription assigns the value 1 to some one-dimensional projection C^α in the support of W^α.[30] So, if the state of α is pure, say, $W^\alpha = |\psi^\alpha\rangle\langle\psi^\alpha|$, then the core property ascription is uniquely $[|\psi^\alpha\rangle\langle\psi^\alpha|] = 1$ since $|\psi^\alpha\rangle\langle\psi^\alpha|$ is the only one-dimensional projection in the support of W^α. However, if the state of α is not pure, there are many different core property ascriptions possible. If the state is, for instance, $W^\alpha = \frac{2}{3}|w_1^\alpha\rangle\langle w_1^\alpha| + \frac{1}{3}|w_2^\alpha\rangle\langle w_2^\alpha|$, then any projection $|\psi^\alpha\rangle\langle\psi^\alpha|$, with $|\psi^\alpha\rangle$ equal to a superposition $c_1|w_1^\alpha\rangle + c_2|w_2^\alpha\rangle$, may be the core projection.

Secondly, if the core property ascription to a composite $\alpha\beta$ is $[C^{\alpha\beta}] = 1$, the core property ascription to the subsystem α assigns the value 1 to a one-dimensional projection C^α not in the support of W^α but in the (usually smaller) support of the density operator $\mathrm{Tr}^\beta(C^{\alpha\beta})$. Hence, if, say, $W^{\alpha\beta} = \sum_j w_j |e_j^\alpha\rangle\langle e_j^\alpha| \otimes |f_j^\beta\rangle\langle f_j^\beta|$ (with $\langle e_j^\alpha|e_k^\alpha\rangle = \delta_{jk}$ and $\langle f_j^\beta|f_k^\beta\rangle = \delta_{jk}$) and if the core projection of $\alpha\beta$ is given by $|e_5^\alpha\rangle\langle e_5^\alpha| \otimes |f_5^\beta\rangle\langle f_5^\beta|$, then the core projection of α is a one-dimensional projection in the support of $\mathrm{Tr}^\beta(|e_5^\alpha\rangle\langle e_5^\alpha| \otimes |f_5^\beta\rangle\langle f_5^\beta|) = |e_5^\alpha\rangle\langle e_5^\alpha|$ (the core projection C^α is thus uniquely $|e_5^\alpha\rangle\langle e_5^\alpha|$) and not an arbitrary one-dimensional projection in the support of $W^\alpha = \sum_j w_j |e_j^\alpha\rangle\langle e_j^\alpha|$.

Thirdly, at the conclusion of a measurement,[31] the core projection of the measurement device is with probability $\mathrm{Tr}^\mu(\widetilde{W}^\mu |R_k^\mu\rangle\langle R_k^\mu|)$ equal to $|R_k^\mu\rangle\langle R_k^\mu|$, where $\{|R_j^\mu\rangle\langle R_j^\mu|\}_j$ represent the pointer readings of the device. Fourthly, if after a Von Neumann measurement of a magnitude A^α, the state of the composite $\alpha\mu$ is equal to $|\widetilde{\Psi}^{\alpha\mu}\rangle = \sum_j c_j |a_j^\alpha\rangle \otimes |R_j^\mu\rangle$, then the core property ascriptions to α and to the device μ are with probability $|c_k|^2$ simultaneously $[|a_k^\alpha\rangle\langle a_k^\alpha|] = 1$ and $[|R_k^\mu\rangle\langle R_k^\mu|] = 1$, respectively.

The Copenhagen modal interpretation manages to solve the measurement problem by ascribing, more or less by construction, pointer readings to devices after measurements. However, a drawback of this interpretation is that its property ascription is usually not that informative because the set of one-dimensional projections in the support of a non-pure state contains a non-denumerable number of elements. Consider, to illustrate this, a spin $\frac{1}{2}$-particle σ with a state $W^\sigma = w_1 |u_{\hat{v}}^\sigma\rangle\langle u_{\hat{v}}^\sigma| + w_2 |d_{\hat{v}}^\sigma\rangle\langle d_{\hat{v}}^\sigma|$ (the projection $|u_{\hat{v}}^\sigma\rangle\langle u_{\hat{v}}^\sigma|$

---

[29] In Van Fraassen's terminology, the core property C^α is the *value state* of α.

[30] The support of a state with spectral resolution $W = \sum_j w_j P_j$ is the subspace of $\mathcal{H}$ corresponding to the projection $\sum_{\{j|w_j \neq 0\}} P_j$ (the sum contains only the eigenprojections with non-zero eigenvalues).

[31] Van Fraassen (1991, Sect. 7.4 and page 284) defines measurements as a special class of interactions between systems, and this definition does not refer to (human) observers. Hence, a measurement is neither a primitive notion in the Copenhagen modal interpretation, nor an anthropomorphic one.

represents spin up in the $\vec{v}$ direction, $|d_{\vec{v}}^{\sigma}\rangle\langle d_{\vec{v}}^{\sigma}|$ represents spin down in the $\vec{v}$ direction and $\langle u_{\vec{v}}^{\sigma}|d_{\vec{v}}^{\sigma}\rangle = 0$). The property ascription to this particle is then equivalent to the idle statement that σ possesses with an unspecified probability spin up or spin down in an unspecified direction. In the modal interpretations to be introduced next, the set of core projections is much more limited and thus much more informative.

The core property ascription to a system by the Copenhagen modal interpretation is not always implicitly definable from the state of the system. The core property ascription to a system, which is not a measurement device at the end of a measurement, is implicitly definable from its state: the set of one-dimensional projections in the support of a state is clearly preserved by any transformation that preserves that state. However, the ascription of the core properties $\{|\mathbf{R}_j^{\mu}\rangle\langle\mathbf{R}_j^{\mu}|\}_j$ to a measurement device μ at the end of a measurement need not be implicitly definable from W^{μ}.[32]

The biorthogonal decomposition

A biorthogonal decomposition of a vector $|\Psi^{\alpha\beta}\rangle$ with respect to the bisection of $\alpha\beta$ in α and β, is given by

$$|\Psi^{\alpha\beta}\rangle = \sum_j c_j \, |c_j^{\alpha}\rangle \otimes |c_j^{\beta}\rangle \qquad (4.4)$$

with $\{|c_j^{\alpha}\rangle\}_j$ and $\{|c_j^{\beta}\rangle\}_j$ sets of pair-wise orthogonal vectors in $\mathscr{H}^{\alpha}$ and $\mathscr{H}^{\beta}$, respectively. A biorthogonal decomposition has a spectrum $\{\lambda_d\}_d$ which is the set of all the different values in the set $\{|c_j|^2\}_j$. Every element λ_d of this spectrum generates a projection on $\mathscr{H}^{\alpha}$ and a projection on $\mathscr{H}^{\beta}$ defined as $P^{\alpha}(\lambda_d) = \sum_{j\in I_d} |c_j^{\alpha}\rangle\langle c_j^{\alpha}|$ and $P^{\beta}(\lambda_d) = \sum_{j\in I_d} |c_j^{\beta}\rangle\langle c_j^{\beta}|$, respectively (the index-set I_d contains all the indices j with $|c_j|^2 = \lambda_d$). It holds that $|\Psi^{\alpha\beta}\rangle$ uniquely fixes this spectrum and the corresponding projections. That is, all possible biorthogonal decompositions of $|\Psi^{\alpha\beta}\rangle$ with respect to the bisection of $\alpha\beta$ in α and β give the same values $\{\lambda_d\}_d$ and the same corresponding projections $\{P^{\alpha}(\lambda_d)\}_d$ and $\{P^{\beta}(\lambda_d)\}_d$.

---

[32] Without going into full detail: If one considers a measurement of the second kind (the measurement given by interaction (10.1) on page 174, for instance), then the final state of the measurement device allows unitary transformations which do not preserve the set $\{|\mathbf{R}_j^{\mu}\rangle\langle\mathbf{R}_j^{\mu}|\}_j$.

4.3 The bi modal interpretation

The first of the modal interpretations considered in this book is the one proposed by Kochen (1985) and Dieks (1988).[33] This interpretation ascribes properties only to the subsystems of a composite ω with a pure state. Let α and β be two disjoint systems which together form such a composite ω. That is, ω is equal to $\alpha\beta$ and the Hilbert space $\mathcal{H}^{\omega}$ associated with ω is equal to the tensor product $\mathcal{H}^{\alpha\beta} = \mathcal{H}^{\alpha} \otimes \mathcal{H}^{\beta}$ of the Hilbert spaces associated with α and β. And the pure state of ω is given by a vector $|\Psi^{\alpha\beta}\rangle$ in $\mathcal{H}^{\alpha} \otimes \mathcal{H}^{\beta}$. Kochen and Dieks now ascribe properties to α and β by means of the so-called polar or **biorthogonal decomposition** of this state vector $|\Psi^{\alpha\beta}\rangle$.

Let me start by introducing this biorthogonal decomposition. Take a vector $|\Psi^{\alpha\beta}\rangle$ in $\mathcal{H}^{\alpha} \otimes \mathcal{H}^{\beta}$ and let the sets $\{|e_j^{\alpha}\rangle\}_j$ and $\{|e_k^{\beta}\rangle\}_k$ be orthonormal bases for $\mathcal{H}^{\alpha}$ and $\mathcal{H}^{\beta}$, respectively. A well-known fact is that the set $\{|e_j^{\alpha}\rangle \otimes |e_k^{\beta}\rangle\}_{j,k}$ is then an orthonormal basis for $\mathcal{H}^{\alpha\beta}$ and that the vector $|\Psi^{\alpha\beta}\rangle$ can be decomposed as

$$|\Psi^{\alpha\beta}\rangle = \sum_{j,k} c_{jk} |e_j^{\alpha}\rangle \otimes |e_k^{\beta}\rangle. \tag{4.5}$$

This decomposition contains two independent summations and the coefficients c_{jk} are equal to $(\langle e_j^{\alpha}| \otimes \langle e_k^{\beta}|)|\Psi^{\alpha\beta}\rangle$. A lesser-known fact is that for each vector $|\Psi^{\alpha\beta}\rangle$ there exist orthonormal bases for $\mathcal{H}^{\alpha}$ and for $\mathcal{H}^{\beta}$, let's call them $\{|c_j^{\alpha}\rangle\}_j$ and $\{|c_k^{\beta}\rangle\}_k$, respectively, such that the values $(\langle c_j^{\alpha}| \otimes \langle c_k^{\beta}|)|\Psi^{\alpha\beta}\rangle$ are zero for all $j \neq k$. The above decomposition can thus be simplified to one which contains only one summation:

$$|\Psi^{\alpha\beta}\rangle = \sum_{j} c_j |c_j^{\alpha}\rangle \otimes |c_j^{\beta}\rangle. \tag{4.6}$$

This second decomposition is called a biorthogonal decomposition of $|\Psi^{\alpha\beta}\rangle$ with respect to the bisection of $\alpha\beta$ in α and β. Because the proof of its existence stems via Von Neumann (1955, Chap. VI.2) from Schmidt (1907),[34] it is also called the Schmidt decomposition.

In order to formulate the core property ascription by Kochen and Dieks, I need further mathematical machinery. Firstly, define the spectrum $\{\lambda_d\}_d$ of a biorthogonal decomposition of $|\Psi^{\alpha\beta}\rangle$ as the set of all the *different* values $|c_j|^2$, where the c_js are the coefficients in (4.6). Secondly, construct for each spectrum value λ_d the index-set I_d of all the indices j for which it holds that $|c_j|^2 = \lambda_d$. Then let $n[I_d]$ be the number of indices in I_d. A biorthogonal decomposition is now called non-degenerate if each index-set I_d contains

---

[33] See also, for instance, Dieks (1989, 1993, 1994a).
[34] See Schrödinger (1935) for another proof of the existence of the biorthogonal decomposition.

only one index (such that $|c_j|^2 \neq |c_k|^2$ if and only if $j \neq k$) and is called degenerate if some index-sets contain two or more indices. Each value λ_d of the spectrum selects with its index-set I_d a number of vectors $|c_j^\alpha\rangle$ in $\mathcal{H}^\alpha$ and $|c_j^\beta\rangle$ in $\mathcal{H}^\beta$, namely those with an index $j \in I_d$. Finally, define the projections $P^\alpha(\lambda_d)$ and $P^\beta(\lambda_d)$ as the projections on the subspaces spanned by the thus selected vectors. That is,

$$P^\alpha(\lambda_d) := \sum_{j \in I_d} |c_j^\alpha\rangle\langle c_j^\alpha|, \qquad P^\beta(\lambda_d) := \sum_{j \in I_d} |c_j^\beta\rangle\langle c_j^\beta|. \qquad (4.7)$$

The projections $\{P^\alpha(\lambda_d)\}_d$ are pair-wise orthogonal. So, if $\mathcal{H}^\alpha$ is a n-dimensional Hilbert space ($n = 1, 2, \ldots \infty$), the set $\{P^\alpha(\lambda_d)\}_d$ contains at most n elements. Further, all projections $P^\alpha(\lambda_d)$ are one-dimensional if the biorthogonal decomposition is non-degenerate and some projections are multidimensional if the decomposition is degenerate. The same remarks apply to the projections $\{P^\beta(\lambda_d)\}_d$.

Kochen and Dieks then propose the following:[35] if $\alpha\beta$ is in the state $|\Psi^{\alpha\beta}\rangle$, the core projections of α and β are given by, respectively, the projections $\{P^\alpha(\lambda_d)\}_d$ and $\{P^\beta(\lambda_d)\}_d$ generated by the *non-zero* members of the spectrum of a biorthogonal decomposition of $|\Psi^{\alpha\beta}\rangle$, and the projections $P^\alpha(\lambda_a)$ and $P^\beta(\lambda_b)$ have with probability

$$p(P^\alpha(\lambda_a), P^\beta(\lambda_b)) = n[I_a]\lambda_a \delta_{ab} \qquad (4.8)$$

simultaneously the value 1.

Thus, to sum up, in order to ascribe properties one has to take a composite ω in a pure state $|\Psi^\omega\rangle$, bisect ω into two disjoint subsystems α and β and then determine the core properties of α and β by means of a biorthogonal decomposition of $|\Psi^\omega\rangle$. The property ascription is thus defined by means of a bisection of a composite system and by means of a biorthogonal decomposition of the state of that composite. I therefore call the modal interpretation by Kochen and Dieks the *bi modal interpretation*.

The bi modal interpretation yields the same positive results for Von Neumann measurements as the Copenhagen modal interpretation. Consider the final state $|\tilde{\Psi}^{\alpha\mu}\rangle = \sum_j c_j |a_j^\alpha\rangle \otimes |R_j^\mu\rangle$ of the object system α and device μ after the measurement. The decomposition of this state is exactly a biorthogonal decomposition so if it is not degenerate, the projections $\{P^\alpha(\lambda_d)\}_d$ and $\{P^\mu(\lambda_d)\}_d$ are equal to $\{|a_j^\alpha\rangle\langle a_j^\alpha|\}_j$ and $\{|R_j^\mu\rangle\langle R_j^\mu|\}_j$, respectively. The core pro-

---

[35] In fact Kochen (1985) considered the property ascription to α and β only in the case that the state vector $|\Psi^{\alpha\beta}\rangle$ has a non-degenerate biorthogonal decomposition. The general formulation of the bi modal interpretation, which includes the degenerate case, is given by Dieks (1993).

perty ascription is therefore $[|a_a^\alpha\rangle\langle a_a^\alpha|] = 1$ and $[|R_b^\mu\rangle\langle R_b^\mu|] = 1$ with probability $|c_a|^2\delta_{ab}$. The device thus possesses its readings with the right probabilities.

The bi modal interpretation gives, on the other hand, a much more detailed description of the properties of systems than the Copenhagen modal interpretation. Firstly, the bi modal interpretation ascribes at most n possible core properties $\{P^\alpha(\lambda_d)\}_d$ to a system α, where n is finite if $\mathcal{H}^\alpha$ is finite-dimensional and where n is denumerable infinite if $\mathcal{H}^\alpha$ is infinite-dimensional. In contrast, the Copenhagen modal interpretation ascribes, in general, a non-denumerable number of possible core properties to a system. The bi modal interpretation gives, moreover, at all times the probabilities that a system α possesses one of its core properties $\{P^\alpha(\lambda_d)\}_d$, whereas the Copenhagen modal interpretation does not. So, to illustrate this, consider again the spin $\frac{1}{2}$-particle σ and assume that it is part of a composite $\sigma\tau$ with a pure state. If this state is given by $|\Psi^{\sigma\tau}\rangle = c_1\,|u_{\vec{v}}^\sigma\rangle\otimes|e_1^\tau\rangle + c_2\,|d_{\vec{v}}^\sigma\rangle\otimes|e_2^\tau\rangle$ (which is a non-degenerate biorthogonal decomposition by taking $\langle e_1^\tau|e_2^\tau\rangle = 0$ and $|c_1|^2 \neq |c_2|^2$), the bi modal interpretation yields that in the $\vec{v}$ direction σ possesses with probability $|c_1|^2$ spin up and with probability $|c_2|^2$ spin down. The reduced state of σ is $W^\sigma = |c_1|^2\,|u_{\vec{v}}^\sigma\rangle\langle u_{\vec{v}}^\sigma| + |c_2|^2\,|d_{\vec{v}}^\sigma\rangle\langle d_{\vec{v}}^\sigma|$, so the Copenhagen modal interpretation still yields that σ possesses some spin is some direction with an unspecified probability.

Another advantage is that the bi modal interpretation gives correlations between the properties of two disjoint systems α and β with a pure composite state. With the joint probabilities (4.8) and the standard definition of conditional probabilities, one has

$$p(P^\alpha(\lambda_a)/P^\beta(\lambda_b)) = p(P^\beta(\lambda_b)/P^\alpha(\lambda_a)) = \delta_{ab}. \tag{4.9}$$

So, if $[P^\alpha(\lambda_j)] = 1$, then $[P^\beta(\lambda_j)] = 1$ with probability 1, and *vice versa*.

There are, however, also questions which remain unanswered in the bi modal interpretation. Firstly, if a system α is not part of a composite in a pure state, does α then still possess properties? Secondly, if one has a number of systems which are ascribed properties, say, the systems α, β, γ, $\alpha\beta$, $\alpha\gamma$, and $\beta\gamma$ which are part of a composite $\omega = \alpha\beta\gamma$ with a pure state, can one then correlate these properties? The first question is answered in neither Kochen (1985) nor the early writings of Dieks. With respect to the second question one can go two ways: either one denies that correlations between the properties of sets of systems always need to exist or one admits that the bi modal interpretation should be supplemented by such correlations. In Kochen (1985) one can find evidence that he takes the first way, Dieks takes the second.

In Kochen's account of the bi modal interpretation one can deny the need for correlations between the properties of all possible subsystems of a composite because for Kochen properties have a truly relational character. In order to see how Kochen arrives at this relational character, I briefly discuss how he constructs the bi modal interpretation. Firstly, Kochen considers the state $|\tilde{\Psi}^{\alpha\mu}\rangle = \sum_j c_j |a_j^\alpha\rangle \otimes |R_j^\mu\rangle$ that one obtains after a Von Neumann measurement.[36] He notes that in the usual analysis of the measurement process, this state is interpreted as that α possibly possesses one of the properties $\{|a_j^\alpha\rangle\langle a_j^\alpha|\}_j$ and that the actual possessed property $|a_k^\alpha\rangle\langle a_k^\alpha|$ is determined by the observation that the device μ possesses the reading $|R_k^\mu\rangle\langle R_k^\mu|$. Then he notes that the biorthogonal decomposition of this state yields these possible possessed properties of $\{|a_j^\alpha\rangle\langle a_j^\alpha|\}_j$ of α as well as the readings $\{|R_j^\mu\rangle\langle R_j^\mu|\}_j$ of the device μ. Secondly, Kochen constructs the bi modal interpretation by proposing that any pure state $|\Psi^{\alpha\beta}\rangle$ of a composite of two disjoint systems α and β should receive this interpretation.[37] That is, if the (non-degenerate) biorthogonal decomposition of this state is $|\Psi^{\alpha\beta}\rangle = \sum_j c_j |c_j^\alpha\rangle \otimes |c_j^\beta\rangle$, then α possibly possesses one of the properties $\{|c_j^\alpha\rangle\langle c_j^\alpha|\}_j$ and the actual possessed property $|c_a^\alpha\rangle\langle c_a^\alpha|$ is determined by the observation that β possesses the property $|c_a^\beta\rangle\langle c_a^\beta|$. However, because we (usually) cannot observe atomic events directly, Kochen replaces the criterion that α possesses $|c_a^\alpha\rangle\langle c_a^\alpha|$ iff an 'official human observer' observes that β possesses $|c_a^\beta\rangle\langle c_a^\beta|$, by the criterion that α possesses $|c_a^\alpha\rangle\langle c_a^\alpha|$ iff β *witnesses* that α possesses $|c_a^\alpha\rangle\langle c_a^\alpha|$. And β witnesses that α possesses $|c_a^\alpha\rangle\langle c_a^\alpha|$ if β possesses itself the property $|c_a^\beta\rangle\langle c_a^\beta|$. Conversely, $|\Psi^{\alpha\beta}\rangle = \sum_j c_j |c_j^\alpha\rangle \otimes |c_j^\beta\rangle$ is also interpreted as that β possibly possesses one of the properties $\{|c_j^\beta\rangle\langle c_j^\beta|\}_j$ and that the actually possessed property $|c_k^\beta\rangle\langle c_k^\beta|$ of β is witnessed by the actually possessed property $|c_k^\alpha\rangle\langle c_k^\alpha|$ of α.

It is now this witnessing relation which may be taken as introducing perspectives into the bi modal interpretation. For, in order to be able to determine the actual properties of a system α, one has to find a composite system $\omega = \alpha\beta$ with a pure state and to bisect that composite to fix the system β which witnesses the actual property of α. The property of α thus comes about in relation to the bisection of ω into α and β. In Kochen's words: 'The world from this view becomes one of perspectives from different systems, with no privileged role for any one, and of properties which acquire a relational character by being realized only upon being witnessed by other systems.'[38]

If one accepts such perspectivalism, one indeed can deny that the bi modal

[36] See Kochen (1985, Sect. 2). I have reproduced his reasoning (specifically the second half of the line which runs from page 154 to page 155) in terms of my language of possessed properties.
[37] See Kochen (1985, pages 160-1).
[38] Quotation from Kochen (1985, page 164).

interpretation always needs to supply correlations between the properties of systems. Consider again the composite $\omega = \alpha\beta\gamma$ with a pure state and take the two subsystems $\alpha\beta$ and $\alpha\gamma$. These two subsystem have properties which are realised with respect to different perspectives: $\alpha\beta$ has properties relative to the perspective 'ω bisected in $\alpha\beta$ and γ' and $\alpha\gamma$ has properties relative to the perspective 'ω bisected in $\alpha\gamma$ and β.' Hence, if one argues that one cannot adopt simultaneously two different perspectives (one could, for instance, say that there doesn't exist a system which simultaneously witnesses the properties of $\alpha\beta$ and of $\alpha\gamma$), one can deny that the bi modal interpretation ever needs to say something sensible about the joint occurrence of the properties of $\alpha\beta$ and $\alpha\gamma$.

On the other hand, if one accepts perspectivalism, an unpleasant consequence is that one may lose the ability to correlate the outcomes of measurements. Consider, for instance, the philosophically popular correlations between the measurement outcomes obtained in the two wings of the Einstein, Podolsky and Rosen (1935). Let α and β be the two particles and let μ and v be the two space-like separated devices with which one performs measurements on α and β, respectively. One can now with the bi modal interpretation ascribe outcomes to the two devices at the end of the measurements. Just bisect, $\alpha\beta\mu v$ firstly into μ and $\alpha\beta v$ and secondly into v and $\alpha\beta\mu$, respectively. But since these bisections correspond to two different perspectives, one cannot determine correlations between the outcomes.

In Dieks' account of the bi modal interpretation the properties ascribed to systems do not have a relational character. Dieks (1994a, Sect. IV) therefore started to look for more general correlations like, for instance, the correlations between the properties of two systems α and $\alpha\beta$, part of a composite $\omega = \alpha\beta\gamma$ with a pure state. This eventually led to the formulation of modal interpretations which generalise the bi modal interpretation, starting with the spectral modal interpretation.

I end here by noting that the spectral modal interpretation proved in retrospect that the property ascription of the bi modal interpretation can also be formulated without invoking the somewhat enigmatic biorthogonal decomposition. In the MATHEMATICS it is proved that the core property ascription of the bi modal interpretation is equivalent to the following: if a composite $\alpha\beta$ of two disjoint systems α and β has a pure state $|\Psi^{\alpha\beta}\rangle$, such that the states of α and β are given by the partial traces $W^{\alpha} = \mathrm{Tr}^{\beta}(|\Psi^{\alpha\beta}\rangle\langle\Psi^{\alpha\beta}|)$ and $W^{\beta} = \mathrm{Tr}^{\alpha}(|\Psi^{\alpha\beta}\rangle\langle\Psi^{\alpha\beta}|)$, respectively, then the core projections of α are the eigenprojections $\{P_j^{\alpha}\}_j$ of W^{α} corresponding to the non-zero eigenvalues $\{w_j^{\alpha}\}_j$ and the core projections of β are the eigenprojections $\{P_k^{\beta}\}_k$ of W^{β}

corresponding to the non-zero eigenvalues $\{w_k^\beta\}_k$. Also the projections P_a^α and P_b^β have with probability

$$p(P_a^\alpha, P_b^\beta) = \text{Tr}^{\alpha\beta}(|\Psi^{\alpha\beta}\rangle\langle\Psi^{\alpha\beta}| [P_a^\alpha \otimes P_b^\beta]) \tag{4.10}$$

simultaneously the value 1. In this book I now use this more straightforward formulation of the bi modal interpretation instead of the one by means of the biorthogonal decomposition.

MATHEMATICS

The core property ascription of the bi modal interpretation is well defined. Firstly, for every vector $|\Psi^{\alpha\beta}\rangle$ there exists a biorthogonal decomposition. Secondly, the core property ascription to α and β constructed by the biorthogonal decomposition of $|\Psi^{\alpha\beta}\rangle$ can be captured by the map

$$|\Psi^{\alpha\beta}\rangle \longmapsto \{\langle n[I_a]\lambda_a\, \delta_{ab}, P^\alpha(\lambda_a), P^\beta(\lambda_b)\rangle\}_{a,b} \tag{4.11}$$

(a triple $\langle n[I_a]\lambda_a\, \delta_{ab}, P^\alpha(\lambda_a), P^\beta(\lambda_b)\rangle$ represents that with probability $n[I_a]\lambda_a\, \delta_{ab}$ the core projections $P^\alpha(\lambda_a)$ and $P^\beta(\lambda_b)$ simultaneously have the value 1) and it can be proved that this map is uniquely defined by $|\Psi^{\alpha\beta}\rangle$. Thirdly, the probabilities $\{n[I_a]\lambda_a\, \delta_{ab}\}_{a,b}$ define a classical probability measure on the joint core property ascription to α and β.

I start by proving that for every vector $|\Psi^{\alpha\beta}\rangle$ there exists a biorthogonal decomposition.[39] The proof is essentially the one given by Bacciagaluppi (1996b, Sect. 2.3.1).

Proof: Take any vector $|\Psi^{\alpha\beta}\rangle$. This vector defines, by means of a partial trace, the density operator W^α and, according to the spectral theorem (see the box on page 10), W^α has a set of eigenvectors $\{|w_j^\alpha\rangle\}_j$ which forms an orthonormal basis for $\mathcal{H}^\alpha$. Take now any orthonormal basis $\{|e_j^\beta\rangle\}_j$ for $\mathcal{H}^\beta$ and decompose $|\Psi^{\alpha\beta}\rangle$ as

$$|\Psi^{\alpha\beta}\rangle = \sum_j \sum_k c_{jk}\, |w_j^\alpha\rangle \otimes |e_k^\beta\rangle. \tag{4.12}$$

Consider the sum over the index j and delete all terms for which it holds that $c_{jk} = 0$ for all k. Define the non-normalised vectors $|\tilde{e}_j^\beta\rangle = \sum_k c_{jk}\, |e_k^\beta\rangle$ for all terms j which are left. One can then rewrite the decomposition as

$$|\Psi^{\alpha\beta}\rangle = \sum_j^{(|\tilde{e}_j^\beta\rangle\neq0)} |w_j^\alpha\rangle \otimes |\tilde{e}_j^\beta\rangle. \tag{4.13}$$

---

[39] The proof is valid only for the case where all sums are finite. The additions needed to include the case of infinite sums are left to those who are mathematically more skilled than I.

The partial trace $W^\alpha = \mathrm{Tr}^\beta(|\Psi^{\alpha\beta}\rangle\langle\Psi^{\alpha\beta}|)$ is thus equal to

$$W^\alpha = \sum_{j,j'}^{(\langle\tilde{e}_j^\beta|\neq 0,\,\langle\tilde{e}_{j'}^\beta|\neq 0)} \langle\tilde{e}_{j'}^\beta|\tilde{e}_j^\beta\rangle|w_j^\alpha\rangle\langle w_{j'}^\alpha| \tag{4.14}$$

and because the vectors $\{|w_j^\alpha\rangle\}_j$ are eigenvectors of W^α, the cross terms in this sum vanish. So $\langle\tilde{e}_{j'}^\beta|\tilde{e}_j^\beta\rangle = 0$ for all $j \neq j'$ and it follows that all vectors $\{|\tilde{e}_j^\beta\rangle\}_j$ are pair-wise orthogonal. Hence, the decomposition

$$|\Psi^{\alpha\beta}\rangle = \sum_j^{(\langle\tilde{e}_j^\beta|\neq 0)} (\langle\tilde{e}_j^\beta|\tilde{e}_j^\beta\rangle)^{\frac{1}{2}}|w_j^\alpha\rangle \otimes |\hat{e}_j^\beta\rangle, \tag{4.15}$$

with $\{|\hat{e}_j^\beta\rangle\}_j$ normalised vectors given by $|\hat{e}_j^\beta\rangle = (\langle\tilde{e}_j^\beta|\tilde{e}_j^\beta\rangle)^{-\frac{1}{2}}|\tilde{e}_j^\beta\rangle$, is a biorthogonal decomposition of $|\Psi^{\alpha\beta}\rangle$. □

Secondly, I prove the uniqueness of the map (4.11) by proving that it is equivalent to the map

$$|\Psi^{\alpha\beta}\rangle \longmapsto \{\langle\mathrm{Tr}^{\alpha\beta}(|\Psi^{\alpha\beta}\rangle\langle\Psi^{\alpha\beta}|\,[P_a^\alpha \otimes P_b^\beta]), P_a^\alpha, P_b^\beta\rangle\}_{a,b}, \tag{4.16}$$

with $\{P_a^\alpha\}_a$ and $\{P_b^\beta\}_b$ the eigenprojections of W^α and W^β, respectively. This last map is obviously unique because the state $|\Psi^{\alpha\beta}\rangle$ uniquely determines its partial traces W^α and W^β and these partial traces uniquely determine their respective eigenprojections $\{P_a^\alpha\}_a$ and $\{P_b^\beta\}_b$. Hence, if the maps (4.11) and (4.16) are equivalent, the first map (4.11) is unique as well.

Proof: The probability $n[I_a]\lambda_a\,\delta_{ab}$ in the map (4.11) can be expressed as a function of $|\Psi^{\alpha\beta}\rangle$ and the projections $P^\alpha(\lambda_a)$ and $P^\beta(\lambda_b)$: namely,

$$n[I_a]\lambda_a\,\delta_{ab} = \mathrm{Tr}^{\alpha\beta}(|\Psi^{\alpha\beta}\rangle\langle\Psi^{\alpha\beta}|\,[P^\alpha(\lambda_a) \otimes P^\beta(\lambda_b)]). \tag{4.17}$$

(To check this equality, substitute a biorthogonal decomposition of $|\Psi^{\alpha\beta}\rangle$ as well as the definitions of the projections $P^\alpha(\lambda_a)$ and $P^\beta(\lambda_b)$.) So, the map (4.11) is equivalent to the map

$$|\Psi^{\alpha\beta}\rangle \longmapsto \{\langle\mathrm{Tr}^{\alpha\beta}(|\Psi^{\alpha\beta}\rangle\langle\Psi^{\alpha\beta}|\,[P^\alpha(\lambda_a) \otimes P^\beta(\lambda_b)]), P^\alpha(\lambda_a), P^\beta(\lambda_b)\rangle\}_{a,b}. \tag{4.18}$$

I now show that the projections $\{P^\alpha(\lambda_a)\}_a$ are the eigenprojections of W^α that correspond to the non-zero eigenvalues of W^α. Take a biorthogonal decomposition $|\Psi^{\alpha\beta}\rangle = \sum_j c_j|c_j^\alpha\rangle \otimes |c_j^\beta\rangle$. The partial trace W^α is then

$$W^\alpha = \sum_{j,j'} c_j\bar{c}_{j'}\,\langle c_{j'}^\beta|c_j^\beta\rangle\,|c_j^\alpha\rangle\langle c_{j'}^\alpha| = \sum_j |c_j|^2\,|c_j^\alpha\rangle\langle c_j^\alpha| \tag{4.19}$$

because the vectors $\{|c_j^\beta\rangle\}_j$ are pair-wise orthogonal. Using firstly the definition of the spectrum $\{\lambda_d\}_d$ of the biorthogonal decomposition and secondly the definition of the projections $\{P^\alpha(\lambda_d)\}_d$, one can rewrite W^α as

$$W^\alpha = \sum_d \lambda_d \sum_{j \in I_d} |c_j^\alpha\rangle\langle c_j^\alpha| = \sum_d \lambda_d\, P^\alpha(\lambda_d). \tag{4.20}$$

The right-hand expression is a spectral resolution of W^α because the projections $\{P^\alpha(\lambda_d)\}_d$ are pair-wise orthogonal. So, given the uniqueness of the spectral resolution, the projections $\{P^\alpha(\lambda_d)\}_d$ are the eigenprojections of W^α and the spectral values $\{\lambda_d\}_d$ are the eigenvalues of W^α. Hence, the set of projections $\{P^\alpha(\lambda_d)\}_d$ corresponding to the non-zero values $\{\lambda_d\}_d$ is equal to the set of all the eigenprojections $\{P_a^\alpha\}_a$ of W^α which correspond to the non-zero eigenvalues of W^α.

Analogously one can show that the set of projections $\{P^\beta(\lambda_d)\}_d$ corresponding to the non-zero values $\{\lambda_d\}_d$ is equal to the set of all the eigenprojections $\{P_b^\beta\}_b$ of the partial trace W^β which correspond to the non-zero eigenvalues of W^β.

With this identification of the projections $\{P^\alpha(\lambda_a)\}_a$ with the eigenprojections $\{P_a^\alpha\}_a$ and of the projections $\{P^\beta(\lambda_b)\}_b$ with the eigenprojections $\{P_b^\beta\}_b$, it follows that the map (4.18) can be rewritten as the map (4.16).   □

From this proof of the equivalence of the maps (4.11) and (4.16), it immediately follows that the core property ascription of the bi modal interpretation can be rephrased in terms of the eigenprojections of the partial traces W^α and W^β and in terms of the joint probabilities (4.10), as was done at the end of the main text of Section 4.3.

Finally, I prove that the probabilities $\{n[I_a]\lambda_a\,\delta_{ab}\}_{a,b}$ define a classical probability measure.

Proof: In the bi modal interpretation the elementary events are given by the joint core property ascription $[P^\alpha(\lambda_a)] = [P^\beta(\lambda_b)] = 1$ with $\lambda_a \neq 0$ and $\lambda_b \neq 0$. Label these elementary events with $\langle a, b\rangle$. The probability measure over the set of elementary events is then $p_{\langle a,b\rangle} = n[I_a]\lambda_a\,\delta_{ab}$. This measure is indeed positive or equal to 0. Furthermore, it yields 1 when summed over all possible elementary events:

$$\overset{(\lambda_a \neq 0, \lambda_b \neq 0)}{\sum_{a,b}} p_{\langle a,b\rangle} = \overset{(\lambda_a \neq 0, \lambda_b \neq 0)}{\sum_{a,b}} n[I_a]\lambda_a\,\delta_{ab} = \overset{(\lambda_a \neq 0)}{\sum_a} n[I_a]\lambda_a = \overset{(|c_j|^2 \neq 0)}{\sum_j} |c_j|^2 = 1.$$

$$\tag{4.21}$$

The last equality holds because $|\Psi^{\alpha\beta}\rangle$ is normalised.   □

4.4 The spectral modal interpretation

The second modal interpretation I consider in this book is the one developed by Vermaas and Dieks (1995). This interpretation ascribes properties to any system α by taking the eigenprojections of the state W^α as the core projections[40] and correlates the properties ascribed to any set of disjoint systems.

Consider, firstly, the spectral resolution of the state W^α of a system α. States are always trace class operators,[41] so the spectral resolution of W^α is discrete:

$$W^\alpha = \sum_j w_j P_j^\alpha. \tag{4.22}$$

In this resolution the eigenvalues $\{w_j\}_j$ are pair-wise distinct (that is, $w_j \neq w_k$ for $j \neq k$) and the eigenprojections $\{P_j^\alpha\}_j$ are pair-wise orthogonal ($P_j^\alpha P_k^\alpha = 0$ for $j \neq k$). If the state W^α is degenerate with regard to the eigenvalue w_j, the corresponding eigenprojection P_j^α is a multi-dimensional projection. (Throughout this book I adopt the convention that P_j^α denotes the possibly multi-dimensional eigenprojection of W^α that corresponds to the eigenvalue w_j.)

According to Vermaas and Dieks, the core projections of α are now given by the eigenprojections $\{P_j^\alpha\}_j$ that correspond to the non-zero eigenvalues of W^α. And the probability that one of these eigenprojections, say, P_a^α, has the value 1 is equal to

$$p(P_a^\alpha) = \mathrm{Tr}^\alpha(W^\alpha P_a^\alpha). \tag{4.23}$$

Vermaas and Dieks also give correlations between the properties ascribed to sets of disjoint systems. Take N disjoint systems α, β, γ, etc. The joint probability that the core property ascriptions to these systems are simultaneously $[P_a^\alpha] = 1$, $[P_b^\beta] = 1$, $[P_c^\gamma] = 1$, ..., is then equal to

$$p(P_a^\alpha, P_b^\beta, P_c^\gamma, \ldots) = \mathrm{Tr}^\omega(W^\omega\, [P_a^\alpha \otimes P_b^\beta \otimes P_c^\gamma \otimes \cdots]), \tag{4.24}$$

where W^ω is the state of the composite $\omega = \alpha\beta\gamma\cdots$. From this joint probability one can determine the correlations between the properties of α, β, γ, etc.

Because the core property ascription of the modal interpretation by Vermaas and Dieks is formulated by means of the spectral resolution of states, I call it the *spectral modal interpretation*.

---

[40] This first feature of the modal interpretation by Vermaas and Dieks (1995) is shared with the modal interpretations by Krips (1987) and by Clifton (1995a).

[41] A state W is trace class because its trace norm $\|W\|_1$ (see footnote 7) is finite: $\|W\|_1 = 1$ because $\sum_j w_j \mathrm{Tr}(P_j) = 1$. The spectral resolution of W is therefore discrete (see the box on page 10).

In the case of a Von Neumann measurement with the final state $|\tilde{\Psi}^{\alpha\mu}\rangle = \sum_j c_j |a_j^\alpha\rangle \otimes |R_j^\mu\rangle$ the spectral modal interpretation solves the measurement problem as well. Given this final state, the device μ has a final state $\widetilde{W}^\mu = \sum_j |c_j|^2 |R_j^\mu\rangle\langle R_j^\mu|$ and if this decomposition is not degenerate, it is a spectral resolution. Hence, the core property ascription yields that the reading $|R_a^\mu\rangle\langle R_a^\mu|$ is possessed with probability $|c_a|^2$.

The spectral modal interpretation is a generalisation of the bi modal interpretation. That is, if one accepts the two constraints to (A) only ascribe properties to systems α which are part of a composite ω with a pure state, and (B) only correlate the properties of two systems α and β if their composite $\alpha\beta$ has a pure state $|\Psi^{\alpha\beta}\rangle$, then the spectral modal interpretation becomes equivalent to the bi modal interpretation. This fact has already been demonstrated by the reformulation of the bi modal interpretation given at the end of Section 4.3 (but see Lemma 6.2 on page 96 for an explicit proof). On the other hand, it can be proved that the spectral modal interpretation follows naturally from the bi modal interpretation. That is, if one accepts (A) the assumption of Instantaneous Autonomy, (B) the assumption of Dynamical Autonomy for measurements and (C) the criterion of Empirical Adequacy (see Section 3.3), one can uniquely derive the spectral modal interpretation from the bi modal interpretation (see Appendix A for the proof).

As a generalisation, the spectral modal interpretation manages to (partly) answer the questions raised for the bi modal interpretation. Firstly, if a system α is not part of a composite that is in a pure state, this system still possesses properties. Secondly, the spectral modal interpretation gives correlations between the properties of arbitrary sets of disjoint systems.

The project started with the bi modal interpretation and continued with the spectral modal interpretation would now be successfully concluded if one could also determine correlations for *non-disjoint* systems. Unfortunately, it has been proved (Vermaas 1997) that the spectral modal interpretation cannot be completed with such correlations. This no-go result is given in Section 6.3.

The impossibility of correlating the properties of non-disjoint systems seems to reintroduce perspectivalism. In the bi modal interpretation perspectivalism means that one can only correlate the properties of two systems α and β if they are disjoint and divide a composite ω with a pure state $|\Psi^\omega\rangle$ into two parts. Here, in the spectral modal interpretation, one can apparently only correlate the properties of N systems if they are disjoint and divide a composite ω into N parts. Hence, if one defines a (generalised) perspective as a subdivision of a composite in N disjoint systems, it seems to be the case that there only exist correlations between the properties of systems which can be viewed from one and the same perspective.

The third modal interpretation to be considered in this book can be seen as an interpretation in which only one generalised perspective is accepted, namely the one corresponding to the subdivision of systems into atoms.

The core property ascription to a system by the spectral modal interpretation is implicitly definable from the state of that system. Every unitary transformation U^α which preserves the state W^α also preserves the eigenprojections of that state (see the MATHEMATICS). Every individual eigenprojection P_j^α is thus implicitly definable from W^α. And since the bi modal interpretation is a special case of the spectral modal interpretation, this proves that the core property ascription of the bi modal interpretation is also implicitly definable from the states of systems.

MATHEMATICS

Consider any unitary operator U^α which satisfies $U^\alpha W^\alpha [U^\alpha]^\dagger = W^\alpha$. If one substitutes the spectral resolution into this equality, one obtains

$$\sum_j w_j U^\alpha P_j^\alpha [U^\alpha]^\dagger = \sum_j w_j P_j^\alpha. \tag{4.25}$$

The decompositions on each side of the equation are both spectral resolutions of W^α and, since the eigenprojection of W^α corresponding to, say, the eigenvalue w_k is uniquely fixed by W^α, one obtains that $U^\alpha P_k^\alpha [U^\alpha]^\dagger = P_k^\alpha$ for every k. This proves that any unitary transformation U^α which preserves W^α also preserves the core properties ascribed by the bi and the spectral modal interpretations.

4.5 The atomic modal interpretation

In the modal interpretation introduced by Bacciagaluppi and Dickson (1997) and Dieks (1998b), it is assumed that there exists in nature a special set of disjoint systems $\{\alpha_q\}_q$ which are the building blocks of all physical systems. These building blocks are for obvious reasons called atomic systems. And because these atomic systems play a privileged rôle in the property ascription to systems, I call the interpretation by Bacciagaluppi, Dickson and Dieks the *atomic modal interpretation*.

The core property ascription to an atom α_q in the atomic modal interpretation is equal to the core property ascription in the spectral modal interpretation. Thus, the core projections are the eigenprojections $\{P_j^{\alpha_q}\}_j$ of the state of α_q and $P_a^{\alpha_q}$ has the value 1 with probability

$$p(P_a^{\alpha_q}) = \mathrm{Tr}^{\alpha_q}(W^{\alpha_q}P_a^{\alpha_q}). \tag{4.26}$$

Moreover, the joint probability that the core property ascriptions to a set of different atoms α_q, α_r, α_s, ... are simultaneously $[P_a^{\alpha_q}] = 1$, $[P_b^{\alpha_r}] = 1$, etc. is equal to

$$p(P_a^{\alpha_q}, P_b^{\alpha_r}, P_c^{\alpha_s}, \dots) = \mathrm{Tr}^{\alpha_q\alpha_r\alpha_s\cdots}(W^{\alpha_q\alpha_r\alpha_s\cdots}[P_a^{\alpha_q} \otimes P_b^{\alpha_r} \otimes P_c^{\alpha_s} \otimes \cdots]). \tag{4.27}$$

The core property ascription to a system which is a composite of atoms (called a molecular system, obviously) in the atomic modal interpretation is, however, different from the one in the spectral modal interpretation. Consider such a system $\beta = \alpha_q\alpha_r\alpha_s\cdots$ consisting of the atoms α_q, α_r, etc. The core projections of β are now the products $\{P_j^{\alpha_q} \otimes P_k^{\alpha_r} \otimes P_l^{\alpha_s} \otimes \cdots\}_{j,k,l,\dots}$ of the eigenprojections of the states of the atoms in β.[42] The core property ascription to β assigns the value 1 to the projection $P_a^{\alpha_q} \otimes P_b^{\alpha_r} \otimes P_c^{\alpha_s} \otimes \cdots$ *if and only if* the core property ascriptions to the atoms in β assign simultaneously the value 1 to $P_a^{\alpha_q}$, $P_b^{\alpha_r}$, etc. Let $P_{jkl\dots}^{\beta}$ be shorthand notation for $P_j^{\alpha_q} \otimes P_k^{\alpha_r} \otimes P_l^{\alpha_s} \otimes \cdots$. The joint probability for the simultaneous core property ascription to β and its atoms is thus (use the joint probability (4.27))

$$p(P_a^{\alpha_q}, P_b^{\alpha_r}, \dots, P_{a'b'c'\dots}^{\beta}) = \delta_{aa'}\,\delta_{bb'} \cdots \mathrm{Tr}^{\beta}(W^{\beta}P_{abc\dots}^{\beta}). \tag{4.28}$$

It follows that the core property ascription to only β is $[P_{abc\dots}^{\beta}] = 1$ with probability

$$p(P_{abc\dots}^{\beta}) = \mathrm{Tr}^{\beta}(W^{\beta}P_{abc\dots}^{\beta}). \tag{4.29}$$

The atomic modal interpretation thus gives correlations between the properties of collections of atoms and between the properties of collections of one molecule and its atoms. In Section 6.4 I derive correlations for collections containing N molecules.

The core property ascription to atoms by the atomic modal interpretation is, as I said above, equal to the core property ascription by the spectral modal interpretation. It thus follows that this property ascription to atoms is implicitly definable from the states of those atoms.

---

[42] This rule that the core projections of molecular systems are products of the core projections of the constituting atoms has been used before in Healey's (1989) modal interpretation.

4.6 Bub's fixed modal interpretation

The last modal interpretation I briefly discuss is the one by Bub (1992).[43]
In this interpretation it is assumed that for every system α there exists a
preferred magnitude which has always a definite value. This magnitude is
fixed in the sense that it is independent of time or of the state of α. Let F^α
denote the preferred magnitude and let's call Bub's interpretation the *fixed
modal interpretation* because this magnitude is fixed.

Consider now the special case that the preferred magnitude of a system
α has a discrete spectral resolution given by $F^\alpha = \sum_j f_j \sum_k |f^\alpha_{jk}\rangle\langle f^\alpha_{jk}|$. In
order that this magnitude has a definite value via the link (3.2) (or via the
alternative link (22) given in footnote 22) the core properties in the fixed
modal interpretation always equal the eigenprojections $\{\sum_k |f^\alpha_{jk}\rangle\langle f^\alpha_{jk}|\}_j$ of
F^α. And the probability that, say, the eigenprojection $\sum_k |f^\alpha_{ak}\rangle\langle f^\alpha_{ak}|$ has the
value 1 is equal to

$$p(\sum_k |f^\alpha_{ak}\rangle\langle f^\alpha_{ak}|) = \mathrm{Tr}^\alpha(W^\alpha \sum_k |f^\alpha_{ak}\rangle\langle f^\alpha_{ak}|). \tag{4.30}$$

If one then applies the link (3.2) (or the link (22)) to this core property
ascription, one obtains that with probability $\mathrm{Tr}^\alpha(W^\alpha \sum_k |f^\alpha_{ak}\rangle\langle f^\alpha_{ak}|)$ the value
of the preferred magnitude is given by $[F^\alpha] = f_a$.

The fixed modal interpretation solves the measurement problem if one
assumes that at the conclusion of a measurement the preferred magnitude
F^μ of the measurement device μ has the pointer readings $\{|\mathrm{R}^\mu_j\rangle\langle \mathrm{R}^\mu_j|\}_j$ as
its eigenprojections. For if the preferred magnitude has a spectral resolution
$F^\mu = \sum_j f_j |\mathrm{R}^\mu_j\rangle\langle \mathrm{R}^\mu_j|$, then the core property ascription (4.30) yields that at the
end of the measurement the device has the reading $|\mathrm{R}^\mu_a\rangle\langle \mathrm{R}^\mu_a|$ with probability
$\mathrm{Tr}^\mu(\widetilde{W}^\mu |\mathrm{R}^\mu_a\rangle\langle \mathrm{R}^\mu_a|)$, where $\widetilde{W}^\mu$ is the final state of the device.

Note that there is quite a difference between the way in which the fixed
modal interpretation (and the Copenhagen modal interpretation) solve the
measurement problem and the way in which the bi, spectral and atomic
modal interpretations hope to do this. These latter interpretations propose
precise rules which lay down the core property ascription to every system
exactly. Hence, it is then a non-trivial matter whether these rules, when
they are applied to measurement devices at the conclusion of measurements,
yield that measurement devices assume their readings. The Copenhagen and
fixed modal interpretations, on the other hand, propose rules which leave
substantial room for the precise core property ascription to systems. In the

---

[43] See also Bub (1997). Other papers on Bub's interpretation are Bub (1993, 1995, 1998a,b), Bub and
Clifton (1996), Bacciagaluppi and Dickson (1997), Bacciagaluppi (1998) and Dickson and Clifton
(1998).

Copenhagen modal interpretation the core properties of a system can be given by any projection in the support of the state of the system. While in the fixed modal interpretation the core property ascription only depends on the choice of the preferred magnitude. Hence, in these two interpretations one can simply declare that at the conclusion of measurements the core properties of the devices are their readings. This makes the question of whether they solve the measurement problem rather trivial.

If for an elementary particle one takes position as the preferred magnitude, then the core property ascription of the fixed modal interpretation becomes equivalent to the ascription of positions by Bohmian mechanics (Bohm 1952; Bohm and Hiley 1993). Bohmian mechanics can thus be seen as a special case of the fixed modal interpretation. In fact, if one drops the requirement that the preferred magnitude is fixed in time, many modal interpretations are special cases of the fixed modal interpretation. If, for instance, the preferred magnitude of a system α has at all times the same eigenprojections as the state of that system, so $F^\alpha = \sum_j f_j P_j^\alpha$, then the fixed modal interpretation yields the spectral modal interpretation.[44] (Note that $F^\alpha = \sum_j f_j P_j^\alpha$ is indeed time-dependent because the eigenprojections $\{P_j^\alpha\}_j$ of an evolving state are usually time-dependent.)

The core property ascription to a system α by the fixed modal interpretation is, in general, not implicitly definable from its state. If the state of α is given by $W^\alpha = |\psi^\alpha\rangle\langle\psi^\alpha|$ with $|\psi^\alpha\rangle$ some non-trivial superposition $\sum_{j,k} c_{jk} |f_{jk}^\alpha\rangle$, then a unitary transformation which preserves this state generally changes the set of core properties $\{\sum_k |f_{jk}^\alpha\rangle\langle f_{jk}^\alpha|\}_j$. The core property ascription thus cannot be defined from solely the state of that system, as is the case in the bi, spectral and atomic modal interpretations. So, this preferred magnitude should have (part of) its origin elsewhere. One could, for instance, just choose it — F^α is then simply put in 'by hand' — or one may define it from other ingredients of the quantum formalism, say, from the Hamiltonian.

4.7 Some measurement schemes

In the next four chapters I develop the bi, spectral and atomic modal interpretations further. In order to do so, I occasionally consider schemes of measurements for which it holds that these interpretations ascribe outcomes after the measurements. In this section I present these schemes. (A thorough

---

[44] The proofs in Bub and Clifton (1996) and Bub (1997) that modal interpretations follow uniquely from a set of natural criteria (see footnote 26) make use of this general characterisation of modal interpretations.

discussion of the ability of modal interpretations to solve the measurement problem is given in Chapter 10.)

I have shown that after a Von Neumann measurement (2.9), the bi and spectral modal interpretations ascribe the properties $\{|R_j^\mu\rangle\langle R_j^\mu|\}_j$ to the measurement device μ (I ignore for a moment complications due to degeneracies). If one assumes that the pointer reading magnitude is represented by $M^\mu = \sum_j m_j |R_j^\mu\rangle\langle R_j^\mu|$, one indeed obtains that the measurement yields pointer readings as outcomes. But how plausible is this assumption? The magnitude represented by M^μ is a magnitude pertaining to the measurement device as a whole but it seems that in typical measurements the outcome is not a property possessed by the device as a whole. Consider, for instance, a Geiger counter consisting of a mechanism and a pointer on top of it which is supposed to indicate the number of detected particles per second. We observe that such a measurement has an outcome by observing that the pointer assumes a specific reading, but we don't observe that the Geiger counter as a whole assumes a reading. Hence, the pointer reading magnitude need not be a magnitude pertaining to the device as a whole, but is rather a magnitude pertaining to a subsystem (the pointer) of the device.

From this remark, it seems that the bi and spectral modal interpretations do not solve the measurement problem for the Von Neumann measurement (2.9); these two interpretations ascribe specific properties to the measurement device as a whole, but that does not prove that they ascribe readings to the pointer. Let's therefore start again and assume that in a Von Neumann measurement the interaction between an object system and the device is given by

$$|\Psi^{\alpha\mu}\rangle = \left[\sum_j c_j |a_j^\alpha\rangle\right] \otimes |D_0^\mu\rangle \longmapsto |\tilde{\Psi}^{\alpha\mu}\rangle = \sum_j c_j |a_j^\alpha\rangle \otimes |D_j^\mu\rangle, \qquad (4.31)$$

where the vectors $\{|D_j^\mu\rangle\}_j$ are pair-wise orthogonal device vectors in $\mathscr{H}^\mu$. The pointer of the device is a subsystem of the device so let's divide μ into this pointer π and a remainder $\bar{\mu}$ equal to μ/π. Assume that the vectors $\{|D_j^\mu\rangle\}_j$ are given by $|D_j^\mu\rangle = |\bar{D}_j^{\bar{\mu}}\rangle \otimes |R_j^\pi\rangle$, where the vectors $\{|\bar{D}_j^{\bar{\mu}}\rangle\}_j$ are again pair-wise orthogonal and the vectors $\{|R_j^\pi\rangle\}_j$ are the eigenvectors of the pointer reading magnitude M^π. This pointer reading is then indeed a magnitude of the pointer itself. The final state of the composite $\alpha\bar{\mu}\pi$ is $\sum_j c_j |a_j^\alpha\rangle \otimes |\bar{D}_j^{\bar{\mu}}\rangle \otimes |R_j^\pi\rangle$ and the final state of the composite $\alpha\pi$ thus becomes

$$\tilde{W}^{\alpha\pi} = \sum_j |c_j|^2 |a_j^\alpha\rangle\langle a_j^\alpha| \otimes |R_j^\pi\rangle\langle R_j^\pi|. \qquad (4.32)$$

If one applies the bi modal interpretation to this physically more realistic model of a Von Neumann measurement, one still obtains that π possesses the reading $|\text{R}_a^\pi\rangle\langle\text{R}_a^\pi|$ with probability $|c_a|^2$ and that α possesses the eigenstate $|a_b^\alpha\rangle\langle a_b^\alpha|$ of the measured magnitude with probability $|c_b|^2$. However, the nice correlation that π possesses $|\text{R}_a^\pi\rangle\langle\text{R}_a^\pi|$ with probability 1 if and only if α possesses $|a_a^\alpha\rangle\langle a_a^\alpha|$ can no longer be derived since the composite of α and π is not in a pure state.

So, if one splits the measurement device into a pointer π and some additional mechanism $\bar{\mu}$, the bi modal interpretation can still ascribe readings after a Von Neumann measurement but no longer correlates these readings with the properties possessed by the object system after the measurement. In the proof that the bi modal interpretation generalises to the spectral modal interpretation (Appendix A) and in the derivation of the evolution of the core properties for the bi modal interpretation (Chapter 8), I need, however, to use these correlations. I therefore consider measurements where the device consists of only a pointer (so $\mu = \pi$ and $|\text{D}_j^\mu\rangle = |\text{R}_j^\pi\rangle$ for all j). It is then again possible to derive the correlations between the finally possessed pointer readings and the finally possessed properties of the object system. From a physical point of view, I think one should reject the possibility that such a device exists and this rejection proves its value when in Chapter 10 the measurement problem is considered in general. However, there is nothing in quantum mechanics which rules out the theoretical possibility that such measurement devices exist. Hence, although physically implausible, it is theoretically allowed to consider devices for which the bi modal interpretation correlates the final pointer readings with the final properties of the object system.

The spectral modal interpretation fares better in solving the measurement problem for the more realistic Von Neumann measurement (4.31). Apply the spectral modal interpretation to the final state (4.32). One obtains that the pointer π possesses its readings and that α possesses the eigenstates $\{|a_j^\alpha\rangle\langle a_j^\alpha|\}_j$ and the joint probability that α and π simultaneously possess their properties is $p(|a_a^\alpha\rangle\langle a_a^\alpha|, |\text{R}_b^\pi\rangle\langle\text{R}_b^\pi|) = \delta_{ab}|c_a|^2$. Hence, for the spectral modal interpretation one need not introduce physically implausible devices if one wants to consider measurements which correlate the readings with the final properties of the object system.

There also exist measurement devices containing a pointer π and a mechanism $\bar{\mu}$ for which the atomic modal interpretation fully solves the measurement problem. Let the pointer π consist of the atoms $\{\alpha_q\}_{q=1}^n$ and assume that all vectors $|\text{D}_j^\mu\rangle$ are given by $|\bar{\text{D}}_j^{\bar{\mu}}\rangle \otimes |\text{D}_j^{\alpha_1}\rangle \otimes \cdots \otimes |\text{D}_j^{\alpha_n}\rangle$, where $\{|\bar{\text{D}}_j^{\bar{\mu}}\rangle\}_j$ and $\{|\text{D}_j^{\alpha_1}\rangle\}_j, \ldots, \{|\text{D}_j^{\alpha_n}\rangle\}_j$ are sets of pair-wise orthogonal vectors. Let the pointer reading projections $\{R_j^\pi\}_j$ be given by

$$R_j^\pi = |D_j^{\alpha_1}\rangle\langle D_j^{\alpha_1}| \otimes \cdots \otimes |D_j^{\alpha_n}\rangle\langle D_j^{\alpha_n}|. \qquad (4.33)$$

Then, if one considers a Von Neumann measurement (4.31) on an atomic system α and applies the atomic modal interpretation to the final state $|\tilde{\Psi}^{\alpha\mu}\rangle$, one obtains that the pointer π possesses the readings $\{R_j^\pi\}_j$, that α possesses the properties $\{|a_j^\alpha\rangle\langle a_j^\alpha|\}_j$ and that the correlations are again $p(|a_a^\alpha\rangle\langle a_a^\alpha|, R_b^\pi) = \delta_{ab}|c_a|^2$.

So, to conclude, for the bi, spectral and atomic modal interpretations there exist (sometimes unrealistic) measurement schemes for which these interpretations ascribe readings to the pointer and correlate these readings to the final properties of the object system.

5

The full property ascription

The different modal interpretations all advance a core property ascription $\{\langle p_j, C_j^\alpha \rangle\}_j$. This chapter is about how this core property ascription fixes the full property ascription $\{\langle p_j, \mathscr{DP}_j, [.]_j \rangle\}_j$. I start with some logic and algebra. I then present two existing proposals for determining the full property ascriptions as well as four conditions one can impose on them. Lastly I give my own proposal and end by discussing how the full property ascription leads to a value assignment to magnitudes.

5.1 Some logic and algebra

To prepare the ground for discussing the full property ascription, I firstly define the logical connectives (negation, conjunction and disjunction) for properties. Then because, as discussed in Section 4.1, all the modal interpretations which I consider give up on the idea that the full property ascription assigns definite values to all properties, I secondly define two types of subsets of properties: Boolean algebras and faux-Boolean algebras.

I have already assumed that every property Q^α pertaining to a system α is represented by one and only one projection Q^α defined on the Hilbert space $\mathscr{H}^\alpha$ associated with α. Each projection Q^α in its turn corresponds one-to-one to the subspace of $\mathscr{H}^\alpha$, denoted by $\mathscr{Q}^\alpha$, onto which it projects. One thus has a bijective mapping from a property Q^α to a projection Q^α to a subspace $\mathscr{Q}^\alpha$. For the set of subspaces of a Hilbert space one can in a natural way define an orthocomplement, a meet and a join. I now choose[45] to define the logical connectives for properties pertaining to a

---

[45] I define the logical connectives for properties by means of the lattice of the subspaces of a Hilbert space but I am aware that this is open to debate. An alternative way to define the logical connectives is by taking the set of properties of a system as a partial Boolean algebra, as, for instance, Bacciagaluppi (1996b, 2000) does. A difference between these two approaches is that the conjunction and disjunction of two incompatible properties (that is, two properties represented by projections

---

The logical connectives

The negation $\neg Q^\alpha$ of a property Q^α is represented by the projection $\mathbb{I}^\alpha - Q^\alpha$ onto the subspace $\mathscr{Q}^\alpha_\perp$. The conjunction $Q^\alpha_1 \wedge Q^\alpha_2$ of two properties is the property represented by the projection onto the subspace $\mathscr{Q}^\alpha_1 \cap \mathscr{Q}^\alpha_2$. The disjunction $Q^\alpha_1 \vee Q^\alpha_2$ is the property represented by the projection onto the subspace $\mathscr{Q}^\alpha_1 \oplus \mathscr{Q}^\alpha_2$.

---

system by means of this orthocomplement, meet and join for subspaces and the bijective mapping between properties and subspaces (by means of an isomorphism, more shortly).

The set of subspaces $\{\mathscr{Q}^\alpha\}$ of a Hilbert space $\mathscr{H}^\alpha$ has the algebraic structure[46] of an orthocomplemented lattice[47] if one defines the orthocomplement $\neg$, the meet $\wedge$ and the join $\vee$ in the following (standard) way. The orthocomplement $\neg\mathscr{Q}^\alpha$ of $\mathscr{Q}^\alpha$ is the subspace $\mathscr{Q}^\alpha_\perp$ containing all the vectors orthogonal to all the vectors in $\mathscr{Q}^\alpha$. The meet $\mathscr{Q}^\alpha_1 \wedge \mathscr{Q}^\alpha_2$ of two subspaces $\mathscr{Q}^\alpha_1$ and $\mathscr{Q}^\alpha_2$ is the subspace $\mathscr{Q}^\alpha_1 \cap \mathscr{Q}^\alpha_2$ of all the vectors in the intersection of $\mathscr{Q}^\alpha_1$ and $\mathscr{Q}^\alpha_2$. And the join $\mathscr{Q}^\alpha_1 \vee \mathscr{Q}^\alpha_2$ of two subspaces is the subspace $\mathscr{Q}^\alpha_1 \oplus \mathscr{Q}^\alpha_2$ of all the vectors spanned by all the vectors in $\mathscr{Q}^\alpha_1$ and $\mathscr{Q}^\alpha_2$.

Using the above mentioned bijection, I define the **logical connectives** as follows. The negation $\neg Q^\alpha$ of a property Q^α is the property represented by the projection onto the subspace $\mathscr{Q}^\alpha_\perp$. This projection $\neg Q^\alpha$ is equal to $\mathbb{I}^\alpha - Q^\alpha$.

---

which do not commute) are defined in the lattice approach but not in the partial Boolean algebra approach.

I prefer the lattice approach because I see no reason why modal interpretations should not ascribe simultaneously incompatible properties to a system. Incompatible properties are incompatible in the sense that they cannot be simultaneously measured. However, they are not incompatible in the sense that they cannot be simultaneously possessed by systems. Hence, if in principle they can be simultaneously possessed, it should also be possible in principle to consider their conjunctions and disjunctions. The differences between the lattice and partial Boolean algebra approach are for modal interpretations, however, rather academic since Bacciagaluppi and I reach essentially the same conclusions about the full property ascription.

[46] All the algebraic notions used in this chapter are introduced and defined in the main text for sets of projections on Hilbert space. For those interested in the general definitions of these notions, I add footnotes. The discussion in this section is based on Redhead (1987, Sects. 1.4, 7.4 and App. III) from which I also have taken the general definitions.

[47] An orthocomplemented lattice is defined in four steps. Firstly, a *partially ordered set* $\mathscr{S}$ is a set with a relation $\leq$ which is reflexive ($a \leq a$ for all $a \in \mathscr{S}$), antisymmetric (if $a \leq b$ and $b \leq a$, then $a = b$ for all $a, b \in \mathscr{S}$) and transitive (if $a \leq b$ and $b \leq c$, then $a \leq c$ for all $a, b, c \in \mathscr{S}$). Secondly, a *lattice* $\mathscr{L}$ is a partially ordered set with a meet $a \wedge b$ and a join $a \vee b$ in $\mathscr{L}$ for every pair $a, b \in \mathscr{L}$. The meet $a \wedge b$ is the highest element with respect to the ordering $\leq$ of $\mathscr{L}$ with $[a \wedge b] \leq a$ and $[a \wedge b] \leq b$. The join $a \vee b$ is the lowest element with $a \leq [a \vee b]$ and $b \leq [a \vee b]$). A lattice contains a zero element 0 and a unit element 1 which satisfy $0 \leq a$ and $a \leq 1$ for all $a \in \mathscr{L}$.

Thirdly, a *complemented lattice* $\mathscr{L}$ is a lattice which contains a complement $\neg a$ for every element $a \in \mathscr{L}$ which satisfies $a \wedge \neg a = 0$ and $a \vee \neg a = 1$. Finally, an *orthocomplemented lattice* $\mathscr{L}$ is a complemented lattice where the complement $\neg a$ is unique for every element $a \in \mathscr{L}$ and satisfies $\neg[\neg a] = a$ as well as $a \leq b$ iff $\neg b \leq \neg a$ for all $a, b \in \mathscr{L}$.

The conjunction $Q_1^\alpha \wedge Q_2^\alpha$ of two properties is the property represented by the projection onto the subspace $\mathcal{Q}_1^\alpha \cap \mathcal{Q}_2^\alpha$. Their disjunction $Q_1^\alpha \vee Q_2^\alpha$ is the property represented by the projection onto the subspace $\mathcal{Q}_1^\alpha \oplus \mathcal{Q}_2^\alpha$.

The set of properties of a system also obtains with these definitions the algebraic structure of an orthocomplemented lattice. As a consequence, the set of all properties acquires the non-classical (and unwelcome) structure of a non-distributive lattice, that is, the conjunction and the disjunction do not satisfy the relations

$$\left.\begin{aligned} Q_1^\alpha \wedge (Q_2^\alpha \vee Q_3^\alpha) &= (Q_1^\alpha \wedge Q_2^\alpha) \vee (Q_1^\alpha \wedge Q_3^\alpha), \\ Q_1^\alpha \vee (Q_2^\alpha \wedge Q_3^\alpha) &= (Q_1^\alpha \vee Q_2^\alpha) \wedge (Q_1^\alpha \vee Q_3^\alpha). \end{aligned}\right\} \tag{5.1}$$

One can regain this distributiveness by considering subsets of the set of all properties. Let $\{S_j^\alpha\}_j$ be a set of properties represented by pair-wise orthogonal projections, so $S_j^\alpha S_k^\alpha = \delta_{jk} S_j^\alpha$. Then the Boolean algebra[48] obtained by closing this set $\{S_j^\alpha\}_j$ under negation, conjunction and disjunction, and denoted by $\mathcal{B}(\{S_j^\alpha\}_j)$, satisfies the relations (5.1). Furthermore, restricted to such a Boolean algebra, the projections representing the properties and their negations, conjunctions and disjunctions, are related by the following simple relations:

$$\neg Q^\alpha = \mathbb{I}^\alpha - Q^\alpha, \quad Q_1^\alpha \wedge Q_2^\alpha = Q_1^\alpha Q_2^\alpha, \quad Q_1^\alpha \vee Q_2^\alpha = Q_1^\alpha + Q_2^\alpha - Q_1^\alpha Q_2^\alpha. \tag{5.2}$$

Finally, value assignments [.] to the properties in Boolean algebras can be given by homomorphisms[49] to the set $\{0, 1\}$ satisfying the classical rules

$$\left.\begin{aligned} [\neg Q^\alpha] &= 1 - [Q^\alpha], \\ [Q_1^\alpha \wedge Q_2^\alpha] &= [Q_1^\alpha][Q_2^\alpha], \qquad\qquad [0^\alpha] = 0, \\ [Q_1^\alpha \vee Q_2^\alpha] &= [Q_1^\alpha] + [Q_2^\alpha] - [Q_1^\alpha][Q_2^\alpha], \qquad [\mathbb{I}^\alpha] = 1. \end{aligned}\right\} \tag{5.3}$$

A slight extension of the notion of a Boolean algebra is given by a **faux-Boolean algebra**,[50] denoted by $\mathcal{F}(\{S_j^\alpha\}_j)$. A faux-Boolean algebra, generated by a set of properties $\{S_j^\alpha\}_j$ represented by pair-wise orthogonal projections, is defined as the closure under negation, conjunction and disjunction of the union of the set $\{S_j^\alpha\}_j$ and the set $\{\widehat{S}^\alpha\}$ containing all the projections which are orthogonal to all projections $\{S_j^\alpha\}_j$. In general, a faux-Boolean algebra

---

[48] A *Boolean algebra* $\mathcal{B}$ is a *distributive* complemented lattice, that is, a complemented lattice $\mathcal{L}$ which satisfies $a \wedge [b \vee c] = [a \wedge b] \vee [a \wedge c]$ and $a \vee [b \wedge c] = [a \vee b] \wedge [a \vee c]$ for all $a, b, c \in \mathcal{L}$.

[49] A *homomorphism* h between two Boolean algebras $\mathcal{B}$ and $\mathcal{B}'$ is a map $h : \mathcal{B} \to \mathcal{B}'$ which satisfies $h(\neg a) = \neg h(a)$ and $h(a \wedge b) = h(a) \wedge h(b)$ for all $a, b \in \mathcal{B}$.

[50] Faux-Boolean algebras were introduced by Dickson (1995a,b) and also appear as the X-form sets in Zimba and Clifton (1998) and as the lattice $\mathcal{L}_{e_{r_1} e_{r_2} \cdots e_{r_k}}$ in Bub (1995). That is, the X-form set generated by the set of mutually orthogonal projections $\{X\}$ is equal to $\mathcal{F}(\{X\})$ and $\mathcal{L}_{e_{r_1} e_{r_2} \cdots e_{r_k}}$ generated by the mutually orthogonal one-dimensional projections $\{e_{r_j}\}_{j=1}^k$ is equal to $\mathcal{F}(\{e_{r_j}\}_{j=1}^k)$.

Faux-Boolean algebra

Let $\{S_j^\alpha\}_j$ be a set of pair-wise orthogonal projections and let $\{\widehat{S}^\alpha\}$ be the set of all the projections orthogonal to all the projections $\{S_j^\alpha\}_j$. The faux-Boolean algebra $\mathscr{F}(\{S_j^\alpha\}_j)$ generated by $\{S_j^\alpha\}_j$ is then the closure under negation, conjunction and disjunction of the union $\{S_j^\alpha\}_j \cup \{\widehat{S}^\alpha\}$.

---

does not satisfy relations (5.1) nor are the logical connectives of its members given by the simple rules (5.2). However, value assignments can still be given to the properties in a faux-Boolean algebra by homomorphisms that satisfy the classical rules (5.3).

In this chapter I present three proposals for the full property ascription by modal interpretations. According to these proposals the sets $\{\mathscr{DP}_j\}_j$ are either Boolean or faux-Boolean algebras because then the value assignments can satisfy the rules (5.3).

In this chapter I also prove various claims. In order to facilitate these proofs, I show in the MATHEMATICS of this section that any member Q^α of $\mathscr{B}(\{S_j^\alpha\}_j)$ or of $\mathscr{F}(\{S_j^\alpha\}_j)$ can be written as

$$Q^\alpha = \sum_{j \in I_{Q^\alpha}} S_j^\alpha + Q_\perp^\alpha. \tag{5.4}$$

Here, $Q_\perp^\alpha$ is a projection orthogonal to all the projections $\{S_j^\alpha\}_j$ and the index-set I_{Q^α} is defined by the relation $k \in I_{Q^\alpha}$ if and only if $Q^\alpha S_k^\alpha = S_k^\alpha$. For $\mathscr{B}(\{S_j^\alpha\}_j)$ the projection $Q_\perp^\alpha$ is either $\mathbb{O}^\alpha$ or $\mathbb{I}^\alpha - \sum_j S_j^\alpha$. For $\mathscr{F}(\{S_j^\alpha\}_j)$ the projection $Q_\perp^\alpha$ can be $\mathbb{O}^\alpha$ or any projection orthogonal to all projections $\{S_j^\alpha\}_j$. With this decomposition the negation, the conjunction and the disjunction become

$$
\left.
\begin{aligned}
\neg Q^\alpha &= \sum_{j \notin I_{Q^\alpha}} S_j^\alpha + [\mathbb{I}^\alpha - \sum_j S_j^\alpha - Q_\perp^\alpha], \\
Q^\alpha \wedge \tilde{Q}^\alpha &= \sum_{j \in I_{Q^\alpha} \cap I_{\tilde{Q}^\alpha}} S_j^\alpha + Q_\perp^\alpha \wedge \tilde{Q}_\perp^\alpha, \\
Q^\alpha \vee \tilde{Q}^\alpha &= \sum_{j \in I_{Q^\alpha} \cup I_{\tilde{Q}^\alpha}} S_j^\alpha + Q_\perp^\alpha \vee \tilde{Q}_\perp^\alpha.
\end{aligned}
\right\} \tag{5.5}
$$

MATHEMATICS

Firstly, I prove that any member of a faux-Boolean algebra $\mathscr{F}(\{S_j\}_j)$ can be decomposed as $Q = \sum_{j \in I_Q} S_j + Q_\perp$ (I here suppress the superscript α). Let $\{S_j\}_j$ be a set of pair-wise orthogonal projections and let $\{\widehat{S}\}$ be the set of projections orthogonal to all the projections $\{S_j\}_j$. Let Ω be the set $\{Q|Q = \sum_{j\in I} S_j + Q_\perp\}$ with $Q_\perp$ a projection orthogonal to all $\{S_j\}_j$ and with I an index-set such that $\sum_{j\in I} S_j$ is a sum over a number of projections in $\{S_j\}_j$. I prove that Ω is equal to $\mathscr{F}(\{S_j\}_j)$.

Proof: Firstly, $\Omega \subset \mathscr{F}(\{S_j\}_j)$: any member of Ω can be written as $[\vee_{j\in I} S_j] \vee Q_\perp$. And since $Q_\perp \in \{\widehat{S}\}$, it follows that any member of Ω is in the closure of $\{S_j\}_j \cup \{\widehat{S}\}$ under disjunction. Any member of Ω is thus in $\mathscr{F}(\{S_j\}_j)$.

Secondly, to prove that $\mathscr{F}(\{S_j\}_j) \subset \Omega$, note that $\{S_j\}_j \cup \{\widehat{S}\}$ is a subset of Ω. I prove that Ω is closed under negation, conjunction and disjunction. It then follows that the closure of $\{S_j\}_j \cup \{\widehat{S}\}$, that is, $\mathscr{F}(\{S_j\}_j)$, is also a subset of Ω.

The negation $\neg Q$ of $Q = \sum_{j\in I_Q} S_j + Q_\perp$ is given by $\mathbb{I} - Q$ and can be written as $\neg Q = \sum_{j\notin I_Q} S_j + [\mathbb{I} - \sum_j S_j - Q_\perp]$. The projection $\mathbb{I} - \sum_j S_j - Q_\perp$ is orthogonal to all $\{S_j\}_j$ so $\neg Q$ is a member of Ω.

The conjunction $Q \wedge \widetilde{Q}$ of $Q = \sum_{j\in I_Q} S_j + Q_\perp$ and $\widetilde{Q} = \sum_{j\in I_{\widetilde{Q}}} S_j + \widetilde{Q}_\perp$ is a projection onto the subspace $\mathscr{Q} \cap \widetilde{\mathscr{Q}}$. Q projects onto the subspace $\sum_{j\in I_Q} \mathscr{S}_j + \mathscr{Q}_\perp$, $\widetilde{Q}$ projects on the subspace $\sum_{j\in I_{\widetilde{Q}}} \mathscr{S}_j + \widetilde{\mathscr{Q}}_\perp$. Inspection yields that the meet of these subspaces is given by $\sum_{j\in I_Q \cap I_{\widetilde{Q}}} \mathscr{S}_j + \mathscr{Q}_\perp \cap \widetilde{\mathscr{Q}}_\perp$ such that $Q \wedge \widetilde{Q}$ is equal to $\sum_{j\in I_Q \cap I_{\widetilde{Q}}} S_j + Q_\perp \wedge \widetilde{Q}_\perp$. The projection $Q_\perp \wedge \widetilde{Q}_\perp$ is orthogonal to all $\{S_j\}_j$ so $Q \wedge \widetilde{Q}$ is a member of Ω.

Analogously the disjunction $Q \vee \widetilde{Q}$ is the projection onto the subspace $\sum_{j\in I_Q \cup I_{\widetilde{Q}}} \mathscr{S}_j + \mathscr{Q}_\perp \oplus \widetilde{\mathscr{Q}}_\perp$. This projection is given by $\sum_{j\in I_Q \cup I_{\widetilde{Q}}} S_j + Q_\perp \vee \widetilde{Q}_\perp$ and because $Q_\perp \vee \widetilde{Q}_\perp$ is orthogonal to all $\{S_j\}_j$, $Q \vee \widetilde{Q}$ is a member of Ω.

Hence, Ω is closed under negation, conjunction and disjunction. $\qquad \square$

With the above proof expressions (5.5) for the negation, conjunction and disjunction of members of a faux-Boolean algebra are also proved.

The proof that the members of a Boolean algebra $\mathscr{B}(\{S_j\}_j)$ also satisfy expressions (5.4) and (5.5) proceeds analogously and is therefore omitted.

5.2 The full property ascriptions by Kochen and by Clifton

We are now in the position to review two proposals for how the core property ascription $\{\langle p_j, C_j^\alpha \rangle\}_j$ fixes the full property ascription $\{\langle p_j, \mathscr{D}\mathscr{P}_j, [.]_j \rangle\}_j$.

Firstly, Kochen (1985, page 161) considered the bi modal interpretation and proposed that the sets $\{\mathscr{D}\mathscr{P}_j\}_j$ of a system α are all equal to the Boolean algebra $\mathscr{B}(\{P_j^\alpha\}_j)$ generated by the core projections $\{P_j^\alpha\}_j$ of α (these core projections are eigenprojections of the state W^α, and because eigenprojections are pair-wise orthogonal, they generate a proper Boolean algebra $\mathscr{B}(\{P_j^\alpha\}_j))$. The value assignments $\{[.]_j\}_j$ are homomorphisms from this set $\mathscr{B}(\{P_j^\alpha\}_j)$ to the values $\{0, 1\}$. For $\langle p_k, \mathscr{D}\mathscr{P}_k, [.]_k \rangle$, corresponding to the core property P_k^α, the homomorphism $[.]_k$ obviously assigns the value 1 to the core projection P_k^α, the value 1 to any projection Q^α in $\mathscr{D}\mathscr{P}_k$ with $P_k^\alpha Q^\alpha = P_k^\alpha$, and the value 0 to all other projections in $\mathscr{D}\mathscr{P}_k$. This value assignment can be captured by the function $[Q^\alpha]_k = \mathrm{Tr}^\alpha(P_k^\alpha Q^\alpha)/\mathrm{Tr}^\alpha(P_k^\alpha)$. So, the proposal is that, for all k, the core property ascription $\langle p_k = \mathrm{Tr}^\alpha(W^\alpha P_k^\alpha), C_k^\alpha = P_k^\alpha \rangle$ induces the full property ascription:

$$\textbf{Kochen:} \quad \left\langle\, p_k = \mathrm{Tr}^\alpha(W^\alpha P_k^\alpha),\, \mathscr{D}\mathscr{P}_k = \mathscr{B}(\{P_j^\alpha\}_j),\, [Q^\alpha]_k = \frac{\mathrm{Tr}^\alpha(P_k^\alpha Q^\alpha)}{\mathrm{Tr}^\alpha(P_k^\alpha)} \,\right\rangle.$$

$$(5.6)$$

Or in words: the core property ascription is $[P_k^\alpha] = 1$ with probability $p_k = \mathrm{Tr}^\alpha(W^\alpha P_k^\alpha)$ and this core ascription induces the projections in the set $\mathscr{D}\mathscr{P}_k = \mathscr{B}(\{P_j^\alpha\}_j)$ to have the value $[Q^\alpha]_k = \mathrm{Tr}^\alpha(P_k^\alpha Q^\alpha)/\mathrm{Tr}^\alpha(P_k^\alpha)$.

Kochen motivates his proposal by pointing out that on this proposal the sets of definite-valued projections $\{\mathscr{D}\mathscr{P}_j\}_j$ are closed under the logical connectives and that the value assignments $\{[.]_j\}_j$ satisfy the rules (5.3) of classical logic. So, if Q^α is possessed, then $\neg Q^\alpha$ is not possessed, and if Q^α and $\tilde{Q}^\alpha$ are possessed, then $Q^\alpha \wedge \tilde{Q}^\alpha$ is also possessed, etc. The advantage of Kochen's proposal is thus that the full property ascription satisfies the following condition:

Closure

If Q^α and $\tilde{Q}^\alpha$ are members of $\mathscr{D}\mathscr{P}_k$, then so are $\neg Q^\alpha$, $Q^\alpha \wedge \tilde{Q}^\alpha$ and $Q^\alpha \vee \tilde{Q}^\alpha$. The value assignment to $\mathscr{D}\mathscr{P}_k$ satisfies the classical rules $[\neg Q^\alpha]_k = 1 - [Q^\alpha]_k$, $[Q^\alpha \wedge \tilde{Q}^\alpha]_k = [Q^\alpha]_k [\tilde{Q}^\alpha]_k$ and $[Q^\alpha \vee \tilde{Q}^\alpha]_k = [Q^\alpha]_k + [\tilde{Q}^\alpha]_k - [Q^\alpha]_k [\tilde{Q}^\alpha]_k$.

Secondly, Clifton (1995a, Sect. 3) gave a full property ascription for the spectral modal interpretation. He proposed that the sets $\{\mathscr{D}\mathscr{P}_j\}_j$ are all equal to the set

$$\{Q^\alpha | Q^\alpha = Q_1^\alpha + Q_2^\alpha, \text{ with } Q_1^\alpha \in \mathscr{B}(\{P_j^\alpha\}_j) \text{ and with } Q_2^\alpha \in \mathscr{N}(W^\alpha)\}. \quad (5.7)$$

Here, $\{P_j^\alpha\}_j$ is the set of the eigenprojections of W^α corresponding to the non-zero eigenvalues, and $\mathscr{N}(W^\alpha)$ is the set of projections onto subspaces of the null space of W^α. Clifton's value assignments $\{[.]_j\}_j$ are equal to the ones by Kochen, so are again captured by the homomorphisms $[Q^\alpha]_k = \mathrm{Tr}^\alpha(P_k^\alpha Q^\alpha)/\mathrm{Tr}^\alpha(P_k^\alpha)$. It can now be proved (see the MATHEMATICS) that the set (5.7) is equal to the faux-Boolean algebra $\mathscr{F}(\{P_j^\alpha\}_j)$. Hence, this second proposal is, for all k, equal to:

Clifton: $\left\langle p_k = \mathrm{Tr}^\alpha(W^\alpha P_k^\alpha),\ \mathscr{DP}_k = \mathscr{F}(\{P_j^\alpha\}_j),\ [Q^\alpha]_k = \dfrac{\mathrm{Tr}^\alpha(P_k^\alpha Q^\alpha)}{\mathrm{Tr}^\alpha(P_k^\alpha)} \right\rangle.$

$$(5.8)$$

Since faux-Boolean algebras are closed under the logical connectives and since the value assignments are again homomorphisms, Clifton's proposal shares with Kochen's the advantage of satisfying the Closure condition. But Clifton's proposal adds a second advantage. Consider a property Q^α for which quantum mechanics predicts that a perfect measurement of this property has either with certainty a positive outcome or with certainty a negative outcome. That is, consider a property Q^α for which the Born probability $\mathrm{Tr}^\alpha(W^\alpha Q^\alpha)$ is either 1 or 0. Clifton's proposal is now such that this property is always definite-valued. Clifton stresses this advantage by pointing out that 'the existence of the value $[Q^\alpha]$ provides the best causal explanation of why we are bound to find a 0 or 1 if Q^α is measured in state W^α.'[51] Clifton's proposal thus satisfies Closure and the following condition:

Certainty

If a projection Q^α has a Born probability $\mathrm{Tr}^\alpha(W^\alpha Q^\alpha)$ equal to 0 or 1, then $Q^\alpha \in \mathscr{DP}_k$ for all k with value $[Q^\alpha]_k = 0$ or $[Q^\alpha]_k = 1$, respectively.

Kochen's full property ascription does not satisfy this Certainty condition.

MATHEMATICS

Kochen's full property ascription satisfies Closure.

Proof: Firstly, $\mathscr{B}(\{P_j^\alpha\}_j)$ is by definition closed under $\neg$, $\wedge$ and $\vee$. Secondly, the value assignment functions $\{[.]_j\}_j$ satisfy the rules of the Closure condition:

Any member Q^α of $\mathscr{B}(\{P_j^\alpha\}_j)$ can, according to (5.4), be decomposed as $Q^\alpha = \sum_{j \in I_Q} P_j^\alpha + Q_\perp^\alpha$. Substitution of this decomposition in the function $[.]_k$ yields that $[Q^\alpha]_k = 1$ if $k \in I_Q$ and $[Q^\alpha]_k = 0$ if $k \notin I_Q$. The value $1 - [Q^\alpha]_k$ is thus equal to 0 if $k \in I_Q$ and equal to 1 if $k \notin I_Q$. Substitution of the decomposition (5.5) of $\neg Q^\alpha$ in $[.]_k$ yields that $[\neg Q^\alpha]_k = 0$ if $k \in I_Q$ and $[\neg Q^\alpha]_k = 1$ if $k \notin I_Q$. Hence, the rule $[\neg Q^\alpha]_k = 1 - [Q^\alpha]_k$ holds.

The product $[Q^\alpha]_k [\widetilde{Q}^\alpha]_k$ is equal to 1 if $k \in I_Q \cap I_{\widetilde{Q}}$ and equal to 0 if $k \notin I_Q \cap I_{\widetilde{Q}}$. Substitution of the decomposition (5.5) of $Q^\alpha \wedge \widetilde{Q}^\alpha$ in $[.]_k$ yields

---

[51] Quotation from Clifton (1995a, page 43) with the notation adjusted to the present notation.

that $[Q^\alpha \wedge \tilde{Q}^\alpha]_k$ is also 1 if $k \in I_Q \cap I_{\tilde{Q}}$ and 0 if $k \notin I_Q \cap I_{\tilde{Q}}$. Hence, the rule $[Q^\alpha \wedge \tilde{Q}^\alpha]_k = [Q^\alpha]_k [\tilde{Q}^\alpha]_k$ holds as well.

Analogously one can derive that $[Q^\alpha \vee \tilde{Q}^\alpha]_k = [Q^\alpha]_k + [\tilde{Q}^\alpha]_k - [Q^\alpha]_k [\tilde{Q}^\alpha]_k$ holds. $\qquad\square$

Kochen's property ascription does not satisfy Certainty.

Proof: Consider the bi modal interpretation and take the state $|\Psi^{\alpha\beta}\rangle = |\psi^\alpha\rangle \otimes |\psi^\beta\rangle$. Let $\mathscr{H}^\alpha$ be a three-dimensional Hilbert space. The reduced state of α is $W^\alpha = |\psi^\alpha\rangle\langle\psi^\alpha|$ and the definite properties of α are according to Kochen with probability 1 equal to $\{\mathbb{O}^\alpha, |\psi^\alpha\rangle\langle\psi^\alpha|, \mathbb{I}^\alpha - |\psi^\alpha\rangle\langle\psi^\alpha|, \mathbb{I}^\alpha\}$. The projection $|\psi^\alpha\rangle\langle\psi^\alpha| + |\phi^\alpha\rangle\langle\phi^\alpha|$ with $\langle\psi^\alpha|\phi^\alpha\rangle = 0$ has now Born probability 1 but is not definite-valued. $\qquad\square$

According to Clifton's proposal the sets $\{\mathscr{DP}_j\}_j$ are equal to the set (5.7). It can be proved that this set is the faux-Boolean algebra $\mathscr{F}(\{P_j^\alpha\}_j)$. Firstly, I show that every member of the set (5.7) is a member of $\mathscr{F}(\{P_j^\alpha\}_j)$. Secondly, I show the converse.

Proof: Any member Q^α of the set (5.7) is a sum $Q_1^\alpha + Q_2^\alpha$ with Q_1^α in $\mathscr{B}(\{P_j^\alpha\}_j)$ and Q_2^α in $\mathscr{N}(W^\alpha)$. According to decomposition (5.4) of a member of $\mathscr{B}(\{P_j^\alpha\}_j)$, Q^α can be further expanded as $Q^\alpha = \sum_{j \in I_Q} P_j^\alpha + Q_\perp^\alpha + Q_2^\alpha$ with $Q_\perp^\alpha$ equal to either $\mathbb{O}^\alpha$ or $\mathbb{I}^\alpha - \sum_j P_j^\alpha$. Let $\tilde{Q}^\alpha$ be the sum $Q_\perp^\alpha + Q_2^\alpha$. Then $\tilde{Q}^\alpha$ is either Q_2^α or $\mathbb{I}^\alpha - \sum_j P_j^\alpha + Q_2^\alpha$. In both cases $\tilde{Q}^\alpha$ is orthogonal to all $\{P_j^\alpha\}_j$ and thus, according to decomposition (5.4), a member of $\mathscr{F}(\{P_j^\alpha\}_j)$. Also all $\{P_j^\alpha\}_j$ are members of $\mathscr{F}(\{P_j^\alpha\}_j)$. Q^α is thus a sum of pair-wise orthogonal projections which are individually all members of $\mathscr{F}(\{P_j^\alpha\}_j)$ and since $\mathscr{F}(\{P_j^\alpha\}_j)$ is closed under $\vee$, it follows that $Q^\alpha = [\vee_{j \in I_Q} P_j^\alpha] \vee \tilde{Q}^\alpha$ is also a member of $\mathscr{F}(\{P_j^\alpha\}_j)$.

Any member Q^α of $\mathscr{F}(\{P_j^\alpha\}_j)$ can according to decomposition (5.4), be written as $Q^\alpha = \sum_{j \in I_Q} P_j^\alpha + Q_\perp^\alpha$. The projection $\sum_{j \in I_Q} P_j^\alpha$ is a member of $\mathscr{B}(\{P_j^\alpha\}_j)$. $Q_\perp^\alpha$ is orthogonal to all $\{P_j^\alpha\}_j$ and is thus a member of $\mathscr{N}(W^\alpha)$. Hence, Q^α is a sum of two projections $Q_1^\alpha \in \mathscr{B}(\{P_j^\alpha\}_j)$ and $Q_2^\alpha \in \mathscr{N}(W^\alpha)$ and thus a member of the set (5.7). $\qquad\square$

To prove that Clifton's full property ascription satisfies Closure, consider the above proof that Kochen's property ascription satisfies Closure. By changing 'Boolean' into 'faux-Boolean' this proof also delivers the goods for Clifton's proposal.

To prove that Clifton's property ascription satisfies Certainty, note that a projection Q^α with Born probability 0 is a projection in the null space of W^α. So Q^α is a member of $\mathcal{N}(W^\alpha)$ and thus a member of the set (5.7). A projection Q^α with Born probability 1 is a projection onto a subspace which includes the support of W^α. Such a projection can be written as $Q^\alpha = \sum_j P_j^\alpha + Q_2^\alpha$ with Q_2^α a member of $\mathcal{N}(W^\alpha)$. The projection $\sum_j P_j^\alpha$ is a member of $\mathcal{B}(\{P_j^\alpha\}_j)$ so Q^α is a member of the set (5.7).

5.3 Conditions on full property ascriptions

The presentation of the proposals by Kochen and by Clifton introduced two conditions on the full property ascription: the Closure and the Certainty conditions. One can now impose at least two more conditions. And, given my link (3.2) between the property ascription to systems and the value assignment to magnitudes, it can be proved that these further conditions must be necessarily satisfied by the full property ascription if one assumes that (A) the full property ascription satisfies Closure with respect to the negation $\neg$ (that is, if $Q^\alpha \in \mathcal{DP}_k$, then $\neg Q^\alpha \in \mathcal{DP}_k$ and $[\neg Q^\alpha]_k = 1 - [Q^\alpha]_k$) and (B) any value assignment to a magnitude A^α with two possible values a_1 and a_2, satisfies the rule[52]

$$[A^\alpha] = a_1 \quad implies\ that \quad [A^\alpha] \neq a_2. \tag{5.9}$$

The first of the two further conditions hangs together with the assumption that if a magnitude A^α has a value a_j, then A^α does not have any other value $a_{j'} \neq a_j$ (this is a generalisation of the rule (5.9) to magnitudes A^α with n eigenvalues). This assumption holds if the full property ascription satisfies the condition that if the property $\sum_k |a_{jk}^\alpha\rangle\langle a_{jk}^\alpha|$ is possessed, then the properties $\sum_k |a_{j'k}^\alpha\rangle\langle a_{j'k}^\alpha|$ for all $j' \neq j$ are not possessed. More generally, this assumption holds if possession of the property Q^α means that any property $\widetilde{Q}^\alpha$ with $\widetilde{Q}^\alpha Q^\alpha = 0$ is not possessed. Possession of Q^α thus excludes that such properties $\widetilde{Q}^\alpha$ are possessed or are indefinite. I therefore call this condition

Exclusion

If Q^α is a member of $\mathcal{DP}_k$ and $[Q^\alpha]_k = 1$, then all the projections in the set $\{\widetilde{Q}^\alpha | \widetilde{Q}^\alpha Q^\alpha = 0\}$ are also members of $\mathcal{DP}_k$ with $[\widetilde{Q}^\alpha]_k = 0$.

If one accepts rule (5.9), the necessity of this condition is proved as follows. Let a projection Q^α have the value 1, and consider any projection $\widetilde{Q}^\alpha$ in the

---

[52] The rule (5.9) may seem trivial: if A^α can have only two values, what else does $[A^\alpha] = a_1$ mean than $[A^\alpha] \neq a_2$? However, since the link (3.2) between the property ascription and the value assignment to magnitudes allows that it is sometimes indefinite whether a magnitude has a specific value, there exists the logical possibility that it is simultaneously the case that A^α has value a_1 and that it is indefinite whether or not A^α has value a_2. So, although natural, the rule (5.9) is non-trivial.

set $\{\tilde{Q}^\alpha | \tilde{Q}^\alpha Q^\alpha = 0\}$. Since $\tilde{Q}^\alpha$ is orthogonal to Q^α, one can construct a two-valued magnitude A^α which has both Q^α and $\tilde{Q}^\alpha$ as its eigenprojections, namely $A^\alpha = a_1 Q^\alpha + a_2 \tilde{Q}^\alpha$. By the link (3.2) it then follows that the property ascription $[Q^\alpha] = 1$ is equivalent to the value assignment $[A^\alpha] = a_1$. By rule (5.9) this value assignment implies that $[A^\alpha] \neq a_2$ and by the link (3.2) this is equivalent to $[\tilde{Q}^\alpha]_k = 0$.

Healey (1989, Sect. 2.2) makes a case for a second condition which he called[53]

Weakening

If Q^α is a member of $\mathscr{D}\mathscr{P}_k$ and $[Q^\alpha]_k = 1$, then all the projections in the set $\{\tilde{Q}^\alpha | \tilde{Q}^\alpha Q^\alpha = Q^\alpha\}$ are also members of $\mathscr{D}\mathscr{P}_k$ with $[\tilde{Q}^\alpha]_k = 1$.

The necessity of this condition is proved as follows.[54] Firstly, it has already been shown from rule (5.9) that the full property ascription must necessarily satisfy Exclusion. Secondly, one can prove that if the full property ascription is closed with regard to the negation, then the necessity of Exclusion implies the necessity of Weakening (and *vice versa*). This proof is given in the MATHEMATICS.

Kochen's and Clifton's full property ascriptions fail to satisfy these two further conditions (see also the MATHEMATICS). As a result of this failure one is faced with the following strange consequences. Assume that somebody asks you if it is already five o'clock.[55] You can now have a look at your watch and answer 'yes' if it is exactly displaying that it is five ($[M^\pi] = 17.00$ if the reading magnitude of your watch is given by $M^\pi = \sum_{j=00.00}^{23.59} j \, |R_j^\pi\rangle\langle R_j^\pi|$). However, if your watch indicates that it is four o'clock, you cannot safely answer 'no:' given what you know about the properties of your watch ($[M^\pi] = 16.00$, or $[|R_{16.00}^\pi\rangle\langle R_{16.00}^\pi|] = 1$ by the link (3.2)), it is only possible to answer that it is four or that the present time is not different to four ($[\mathbb{I}^\pi - |R_{16.00}^\pi\rangle\langle R_{16.00}^\pi|] = 0$, which is warranted because Kochen's and Clifton's proposals satisfy Closure). But you cannot give the usual answer that it is not yet five (represented by $[M^\pi] \neq 17.00$ or $[|R_{17.00}^\pi\rangle\langle R_{17.00}^\pi|] = 0$) because, if Exclusion does not hold, the property ascription $[|R_{16.00}^\pi\rangle\langle R_{16.00}^\pi|] = 1$ does not imply that $[|R_{17.00}^\pi\rangle\langle R_{17.00}^\pi|] = 0$.

---

[53] The name 'weakening' refers to the fact that the property ascription $[\tilde{Q}^\alpha] = 1$ is less precise or weaker than the property ascription $[Q^\alpha] = 1$ if $\tilde{Q}^\alpha Q^\alpha = Q^\alpha$.

[54] Healey motivates Weakening by a requirement which he calls the *property inclusion condition* and which says that if a magnitude A^α has a value restricted to a set Γ, then the value of A^α is also restricted to any larger set $\Delta \supset \Gamma$ (see Healey (1989, page 67)).

[55] This example is a variation of Problem 1 of Arntzenius (1990, Sect. 3)).

Or assume that somebody asks if it is tea-time yet (and let it be tea-time between, say, 15.30 and 16.30 hours). If your watch shows that it is four, you can certainly answer that it is four o'clock. But it is uncertain whether you can confirm that it is indeed time for tea. Let the property 'it is tea-time' be represented by the projection $Q^{\pi}_{\text{tea-time}} = \sum_{j=15.30}^{16.30} |R^{\pi}_j\rangle\langle R^{\pi}_j|$, that is, by the disjunction of all the projections that represent a time in the tea-time interval. You can then confirm that it is tea-time if this projection has value 1. However, if Weakening is not satisfied, your observation that $[|R^{\pi}_{16.00}\rangle\langle R^{\pi}_{16.00}|] = 1$ need not imply that $[Q^{\pi}_{\text{tea-time}}] = 1.$[56]

MATHEMATICS

I start by proving that if the sets $\{\mathscr{DP}_j\}_j$ are closed under negation $\neg$ and if the value assignment functions $\{[.]_j\}_j$ satisfy the rule $[\neg Q^{\alpha}]_j = 1 - [Q^{\alpha}]_j$, then Exclusion implies Weakening.

Proof: Let $Q^{\alpha} \in \mathscr{DP}_j$ and let $[Q^{\alpha}]_j = 1$. Consider a projection $\tilde{Q}^{\alpha}$ element of the set $\{\tilde{Q}^{\alpha}|\tilde{Q}^{\alpha}Q^{\alpha} = Q^{\alpha}\}$. The negation $\neg\tilde{Q}^{\alpha}$ satisfies $\neg\tilde{Q}^{\alpha}Q^{\alpha} = 0$. By Exclusion it thus follows that $[\neg\tilde{Q}^{\alpha}] = 0$ and by closure under $\neg$ it follows that $\neg\neg\tilde{Q}^{\alpha} = \tilde{Q}^{\alpha}$ is also definite-valued. The rule $[\neg Q^{\alpha}]_j = 1 - [Q^{\alpha}]_j$ now implies that $[\tilde{Q}^{\alpha}]_j = 1 - [\neg\tilde{Q}^{\alpha}]_j = 1$. Hence, if $Q^{\alpha} \in \mathscr{DP}_j$ and $[Q^{\alpha}]_j = 1$, then every projection in $\{\tilde{Q}^{\alpha}|\tilde{Q}^{\alpha}Q^{\alpha} = Q^{\alpha}\}$ is also a member of $\mathscr{DP}_j$ with $[\tilde{Q}^{\alpha}]_j = 1$. □

The proof that if the full property ascription is closed under negation, then Weakening implies Exclusion goes analogously and is therefore left to the reader.

To prove that neither Kochen's nor Clifton's full property ascription

---

[56] The proofs that the full property ascription should satisfy Exclusion and Weakening, hinge on the link (3.2) between the property ascription and the value assignment to magnitudes. Hence, if one rejects this link and, instead, adopts the more restrictive link (22) given in footnote 22, one need not be worried about the fact that the full property ascriptions by Kochen and by Clifton violate Exclusion and Weakening. Moreover, if one indeed accepts this restrictive link, one can prove that these two full property ascriptions escape the strange consequences illustrated by the example of the watch.

Firstly, it follows from the second line of the restricted link (22) that if $[A^{\alpha}] = a_j$, then $[A^{\alpha}] \neq a_k$ for all $a_k \neq a_j$. (Rule (5.9) is thus automatically satisfied.) So, if somebody asks if it is five o'clock and your watch indicates $[M^{\pi}] = 16.00$, then one may safely answer 'no' because $[M^{\pi}] = 16.00$ implies that $[M^{\pi}] \neq j$ for all $j \neq 16.00$. Secondly, it follows from the first line of the restricted link (22) that if $[M^{\pi}] = 16.00$, then $[|R^{\pi}_j\rangle\langle R^{\pi}_j|] = \delta_{j16.00}$. Because the full property ascriptions by Kochen and by Clifton satisfy Closure, it follows that $[Q^{\pi}_{\text{tea-time}}] = 1$. Hence, if your watch indicates that it is four o'clock, then you may also safely conclude that it is time for tea.

On the other hand, with the restrictive link (22), a new strange conclusion is that if you observe that your watch possesses the property $|R^{\pi}_{16.00}\rangle\langle R^{\pi}_{16.00}|$, you may not conclude that it indicates $[M^{\pi}] = 16.00$, that is, that it is four. This conclusion is only correct if it is also the case that the watch does not possess the properties $|R^{\pi}_j\rangle\langle R^{\pi}_j|$ for all values $j \neq 16.00$. So, if only $|R^{\pi}_{16.00}\rangle\langle R^{\pi}_{16.00}|$ is possessed, it is unclear what you observe; in that case, I guess, you will conclude that you need a new watch.

satisfies Exclusion and Weakening, take the state $W^\alpha = \sum_{j=1}^{\infty} 2^{-j} |e_j^\alpha\rangle\langle e_j^\alpha|$ with $\{|e_j^\alpha\rangle\}_j$ an orthonormal basis for an infinite-dimensional Hilbert space $\mathcal{H}^\alpha$. The full property ascription to α for both Kochen and Clifton is then such that $\mathscr{DP}_k = \mathscr{B}(\{|e_j^\alpha\rangle\langle e_j^\alpha|\}_j)$ for all k. Assume now that $[|e_1^\alpha\rangle\langle e_1^\alpha|] = 1$ (which is the case with probability $\frac{1}{2}$) and consider the full property ascription $\mathscr{DP}_1$ generated by this core property ascription. The projection $\frac{1}{2}(|e_2^\alpha\rangle + |e_3^\alpha\rangle)(\langle e_2^\alpha| + \langle e_3^\alpha|)$ is not a member of $\mathscr{DP}_1$, so Exclusion is violated. And the projection $|e_1^\alpha\rangle\langle e_1^\alpha| + \frac{1}{2}(|e_2^\alpha\rangle + |e_3^\alpha\rangle)(\langle e_2^\alpha| + \langle e_3^\alpha|)$ is also not a member of $\mathscr{DP}_1$, so Weakening is violated as well.

5.4 A new proposal

The proposals by Kochen and Clifton do not only fail to satisfy Exclusion and Weakening, they are also applicable only to modal interpretations which put forward pair-wise orthogonal core projections.[57] This latter fact makes these proposals unsuitable for application to the Copenhagen modal interpretation, for instance. Finally, Kochen and Clifton define sets $\{\mathscr{DP}_j\}_j$ which all contain the same projections (so, $\mathscr{DP}_k = \mathscr{DP}_l$ for all $k \neq l$). They, however, do not argue why this should be the case.

In an attempt to overcome these disadvantages[58] I develop a new proposal which is applicable to any modal interpretation and which always satisfies Closure, Exclusion and Weakening. And for those modal interpretations in which the core projections $\{C_j^\alpha\}_j$ project only in the support of the state W^α (the Copenhagen, bi and spectral modal interpretations), it also satisfies the Certainty condition. Thus only for the atomic and fixed modal interpretations, does the new proposal fail to satisfy Certainty. The newly determined sets $\{\mathscr{DP}_j\}_j$ do not, in general, contain the same projections.

I propose to determine the sets $\{\mathscr{DP}_j\}_j$ by assuming that the full property ascription is closed under negation and by assuming that the rule (5.9) for two-valued magnitudes holds. That is, I propose to construct every set $\mathscr{DP}_k$ by starting with the core property ascription $[C_k^\alpha] = 1$ and then adding all

---

[57] Kochen's proposal cannot be applied to modal interpretations with pair-wise non-orthogonal core projections since the closure of a set of such projections under orthocomplement, meet and join does not result in a Boolean algebra. Clifton's proposal cannot be applied to such modal interpretations since faux-Boolean algebras are not defined for pair-wise non-orthogonal projections.

[58] On the basis of the remarks made in footnote 56, one can dispute whether it is disadvantageous that the proposals by Kochen and by Clifton violate Exclusion and Weakening: if one adopts the restrictive link (22) to relate the property ascriptions and value assignments to magnitudes, there is no reason why full property ascriptions need to satisfy these two conditions. However, with the restrictive link (22), the full property ascriptions by Kochen and by Clifton are still not applicable to modal interpretations with pair-wise non-orthogonal core projections. So, even if one adopts this alternative link, there are still reasons to try to improve on Kochen's and Clifton's proposals.

In addition, I still prefer the link (3.2) because, given a set of definite-valued projections, this link assigns values to more magnitudes than the restrictive link (22).

those projections to $\mathscr{DP}_k$ whose membership is necessary by closure under
negation and by the rule (5.9).

Given my starting point that the relation between the property ascription
and the value assignment to magnitudes is fixed by the link (3.2), rule
(5.9) implies that the full property ascription satisfies Exclusion (see the
previous section). Conversely, given the link (3.2), Exclusion implies that the
full property ascription satisfies rule (5.9) (the proof is left to the reader).
Hence, my proposal is equivalent to adding all the projections to $\mathscr{DP}_k$ whose
membership is necessary by closure under negation and by Exclusion.

To execute this new construction of the full property ascription, consider
the set $\mathscr{DP}_k$ induced by the core property ascription $[C_k^\alpha] = 1$. Let's start by
adding all those projections such that Exclusion is satisfied. That is, add to
$\mathscr{DP}_k$ all the projections $\{Q^\alpha\}$ orthogonal to C_k^α and let their values be 0. So,

$$\mathscr{DP}_k \supseteq \{C_k^\alpha\} \cup \{Q^\alpha | Q^\alpha C_k^\alpha = 0\}$$
$$\text{with} \quad [C_k^\alpha]_k = 1 \quad \text{and} \quad [Q^\alpha]_k = 0 \text{ if } Q^\alpha C_k^\alpha = 0. \tag{5.10}$$

Then add all the projections such that $\mathscr{DP}_k$ is closed under negation and
that the value assignment function $[.]_k$ satisfies $[\neg Q^\alpha]_k = 1 - [Q^\alpha]_k$. The
negation $\neg C_k^\alpha$ of the core projection C_k^α is already added to $\mathscr{DP}_k$ because
$\neg C_k^\alpha$ is in the set $\{Q^\alpha | Q^\alpha C_k^\alpha = 0\}$. The value of $\neg C_k^\alpha$ satisfies by (5.10) the
relation $[\neg C^\alpha]_k = 1 - [C^\alpha]_k$. The negations $\neg Q^\alpha$ of the other projections in
the set $\{Q^\alpha | Q^\alpha C_k^\alpha = 0\}$ are, however, not yet in $\mathscr{DP}_k$. These negations $\{\neg Q^\alpha\}$
form the set $\{\tilde{Q}^\alpha | \tilde{Q}^\alpha C_k^\alpha = C_k^\alpha\}$ because $Q^\alpha C_k^\alpha = 0$, and their values should be
1 because the projections Q^α have the value 0. I therefore also add all the
projections in $\{\tilde{Q}^\alpha | \tilde{Q}^\alpha C_k^\alpha = C_k^\alpha\}$ to $\mathscr{DP}_k$ with their required values, so

$$\mathscr{DP}_k \supseteq \{Q^\alpha | Q^\alpha C_k^\alpha = C_k^\alpha\} \cup \{Q^\alpha | Q^\alpha C_k^\alpha = 0\}$$
$$\text{with} \quad [Q^\alpha]_k = 1 \text{ if } Q^\alpha C_k^\alpha = C_k^\alpha \quad \text{and} \quad [Q^\alpha]_k = 0 \text{ if } Q^\alpha C_k^\alpha = 0. \tag{5.11}$$

(Note that C_k^α is in the set $\{Q^\alpha | Q^\alpha C_k^\alpha = C_k^\alpha\}$; thus C_k^α is also, according to
(5.11), a member of $\mathscr{DP}_k$ with the value 1.)

The Exclusion condition does not force one to add more members to $\mathscr{DP}_k$.
Every projection in $\{Q^\alpha | Q^\alpha C_k^\alpha = C_k^\alpha\} \cup \{Q^\alpha | Q^\alpha C_k^\alpha = 0\}$ which has the value
1 is a member of the set $\{Q^\alpha | Q^\alpha C_k^\alpha = C_k^\alpha\}$. Also every projection orthogonal
to a member of $\{Q^\alpha | Q^\alpha C_k^\alpha = C_k^\alpha\}$ is in the set $\{Q^\alpha | Q^\alpha C_k^\alpha = 0\}$ and has, as
required by Exclusion, the value 0. Hence, if one identifies $\mathscr{DP}_k$ with the set
$\{Q^\alpha | Q^\alpha C_k^\alpha = C_k^\alpha\} \cup \{Q^\alpha | Q^\alpha C_k^\alpha = 0\}$, it satisfies Exclusion and is closed under
negation.

It can be proved that, with this identification, $\mathscr{DP}_k$ becomes equal to the
faux-Boolean algebra $\mathscr{F}(C_k^\alpha)$ and that the full property ascription is, for all

k, equal to:

New proposal: $\left\langle \, p_k = p(C_k^\alpha), \; \mathscr{D}\mathscr{P}_k = \mathscr{F}(C_k^\alpha), \; [Q^\alpha]_k = \dfrac{\mathrm{Tr}^\alpha(C_k^\alpha Q^\alpha)}{\mathrm{Tr}^\alpha(C_k^\alpha)} \, \right\rangle.$ (5.12)

Or in words: the core property ascription is $[C_k^\alpha] = 1$ with probability $p_k = p(C_k^\alpha)$ and this core ascription induces the projections in the set $\mathscr{D}\mathscr{P}_k = \mathscr{F}(C_k^\alpha)$ to have the value $[Q^\alpha]_k = \mathrm{Tr}^\alpha(C_k^\alpha Q^\alpha)/\mathrm{Tr}^\alpha(C_k^\alpha)$.

MATHEMATICS

I start by proving that the property ascription (5.11) with $\supseteq$ replaced by a '=' sign, is equal to the property ascription (5.12).

Proof: Firstly, any member $\tilde{Q}^\alpha$ of $\{Q^\alpha | Q^\alpha C_k^\alpha = C_k^\alpha \text{ or } 0\}$ is either orthogonal to C_k^α or can be written as a sum $\tilde{Q}^\alpha = C_k^\alpha + \tilde{Q}_\perp^\alpha$ with $\tilde{Q}_\perp^\alpha$ orthogonal to C_k^α. In both cases $\tilde{Q}^\alpha$ is a member of $\mathscr{F}(C_k^\alpha)$. Conversely, any member $\tilde{Q}^\alpha$ of $\mathscr{F}(C_k^\alpha)$ can be written (see decomposition (5.4)) as either $\tilde{Q}^\alpha = \tilde{Q}_\perp^\alpha$ or $\tilde{Q}^\alpha = C_k^\alpha + \tilde{Q}_\perp^\alpha$ with $\tilde{Q}_\perp^\alpha$ orthogonal to C_k^α. Hence, $\tilde{Q}^\alpha$ is a member of $\{Q^\alpha | Q^\alpha C_k^\alpha = C_k^\alpha \text{ or } 0\}$.   □

The new full property ascription satisfies Exclusion by construction. Because it is also by construction closed under negation $\neg$, the new full property ascription satisfies Weakening as well (Exclusion and closure under $\neg$ imply Weakening according to the MATHEMATICS of the previous section). The new full property ascription also satisfies Closure for the conjunction $\wedge$ and the disjunction $\vee$ (the proof is again a variation of the first proof given in the MATHEMATICS of Section 5.2). Finally, for all modal interpretations in which the core projections $\{C_j^\alpha\}_j$ project in the support of the state W^α, the new full property ascription satisfies Certainty.

Proof: Assume that all core projections C_k^α project in the support of W^α. This support is the space onto which the projection $\sum_{\{j | w_j \neq 0\}} P_j^\alpha$ (the sum of all the eigenprojections of W^α corresponding to non-zero eigenvalues) projects. Hence $\sum_{\{j | w_j \neq 0\}} P_j^\alpha C_k^\alpha = C_k^\alpha$ for all k.

Consider now a projection Q^α with $\mathrm{Tr}^\alpha(W^\alpha Q^\alpha) = 0$. Q^α is then a member of $\mathscr{N}(W^\alpha)$ and thus orthogonal to the projection $\sum_{\{j | w_j \neq 0\}} P_j^\alpha$ onto the support of W^α. It follows that Q^α is orthogonal to every core projection C_k^α since $Q^\alpha C_k^\alpha = Q^\alpha \sum_{\{j | w_j \neq 0\}} P_j^\alpha C_k^\alpha = 0$. Hence, Q^α is a member of $\mathscr{F}(C_k^\alpha)$ for all k. Application of the value assignment function gives $[Q^\alpha]_k = 0$ for all k.

Consider next a projection Q^α with $\mathrm{Tr}^\alpha(W^\alpha Q^\alpha) = 1$. Q^α then projects on a space which includes the support of W^α so $Q^\alpha \sum_{\{j | w_j \neq 0\}} P_j^\alpha = \sum_{\{j | w_j \neq 0\}} P_j^\alpha$. Multiplying each side of this relation with C_k^α at the right-hand side yields

$Q^{\alpha}C_k^{\alpha} = C_k^{\alpha}$ so Q^{α} can, for every k, be written as $Q^{\alpha} = C_k^{\alpha} + Q_{\perp}^{\alpha}$ with $Q_{\perp}^{\alpha}$ orthogonal to C_k^{α}. Hence, Q^{α} is a member of $\mathscr{F}(C_k^{\alpha})$ for all k and application of the value assignment function gives $[Q^{\alpha}]_k = 1$ for all k. $\quad\square$

Finally, I prove that the new full property ascription does not satisfy Certainty if it is applied to the atomic or the fixed modal interpretation.

Proof: Let $\alpha\beta$ be a composite of two atoms α and β and let the state of $\alpha\beta$ be $|\Psi^{\alpha\beta}\rangle = \frac{\sqrt{3}}{2}|\psi_1^{\alpha}\rangle \otimes |\phi_1^{\beta}\rangle + \frac{1}{2}|\psi_2^{\alpha}\rangle \otimes |\phi_2^{\beta}\rangle$, where $\langle\psi_1^{\alpha}|\psi_2^{\alpha}\rangle = 0$ and $\langle\phi_1^{\beta}|\phi_2^{\beta}\rangle = 0$. The core projection of $\alpha\beta$ is then, according to the atomic modal interpretation, either $C_1^{\alpha\beta} = |\psi_1^{\alpha}\rangle\langle\psi_1^{\alpha}| \otimes |\phi_1^{\beta}\rangle\langle\phi_1^{\beta}|$ or $C_2^{\alpha\beta} = |\psi_2^{\alpha}\rangle\langle\psi_2^{\alpha}| \otimes |\phi_2^{\beta}\rangle\langle\phi_2^{\beta}|$. The projection $|\Psi^{\alpha\beta}\rangle\langle\Psi^{\alpha\beta}|$ clearly has Born probability 1 but this projection is in neither $\mathscr{DP}_1$ nor $\mathscr{DP}_2$.

This conclusion follows also for the fixed modal interpretation if the preferred magnitude of $\alpha\beta$ is given by $F^{\alpha\beta} = \sum_{j=1}^{2} f_j |\psi_j^{\alpha}\rangle\langle\psi_j^{\alpha}| \otimes |\phi_j^{\beta}\rangle\langle\phi_j^{\beta}|$. $\quad\square$

5.5 Results

The new proposal (5.12) can be applied to any modal interpretation and the resulting full property ascription satisfies for a number of modal interpretations all the given conditions. Only for the atomic and fixed modal interpretations does the proposal slip by not satisfying Certainty.

More specifically, application of the new proposal to the Copenhagen modal interpretation yields that the full property ascription to a system α with state W^{α} is given by $\mathscr{F}(|\psi^{\alpha}\rangle\langle\psi^{\alpha}|)$ with $|\psi^{\alpha}\rangle\langle\psi^{\alpha}|$ a projection in the support of W^{α}. This property ascription reproduces the full property ascription one would obtain on the basis of the eigenvalue–eigenstate link if the state of α had been $|\psi^{\alpha}\rangle\langle\psi^{\alpha}|$. (Application of this link to the state $|\psi^{\alpha}\rangle\langle\psi^{\alpha}|$ yields that all projections with $Q^{\alpha}|\psi^{\alpha}\rangle\langle\psi^{\alpha}| = |\psi^{\alpha}\rangle\langle\psi^{\alpha}|$ have the value 1. Application of proposal (5.12) to the core property ascription $[|\psi^{\alpha}\rangle\langle\psi^{\alpha}|] = 1$ yields the same.) Hence, as already pointed out by Van Fraassen (1991, page 282), it seems *as if* the ignorance interpretation (see Section 3.1) is correct. That is, the core property ascription to a system is $[|\psi^{\alpha}\rangle\langle\psi^{\alpha}|] = 1$, modal interpretations (with the new full property ascription) do ascribe properties *as if* α were in a pure state $|\psi^{\alpha}\rangle\langle\psi^{\alpha}|$.

Application of the new proposal to the bi modal interpretation yields that the definite-valued projections of a system α, which is part of a composite $\alpha\beta$ with a pure state $|\Psi^{\alpha\beta}\rangle$, are with probability $p(P_k^{\alpha}) = \mathrm{Tr}^{\alpha}(W^{\alpha}P_k^{\alpha})$ the projections in the set $\mathscr{DP}_k = \mathscr{F}(P_k^{\alpha})$. In general, this set contains more projections than the set $\mathscr{DP}_k = \mathscr{B}(\{P_j^{\alpha}\}_j)$ proposed by Kochen. (By considering the relevant definitions, one can conclude that any member of $\mathscr{B}(\{P_j^{\alpha}\}_j)$ is also a member of $\mathscr{F}(P_k^\alpha)$. On the other hand, a projection Q^α orthogonal to P_k^α is a member of $\mathscr{F}(P_k^\alpha)$ but need not be a member of $\mathscr{B}(\{P_j^\alpha\}_j)$.)

A similar remark holds for the spectral modal interpretation and Clifton's proposal. The new proposal applied to the spectral modal interpretation yields that the definite-valued projections of a system α with a state W^α are with probability $p(P_k^\alpha) = \mathrm{Tr}^\alpha(W^\alpha P_k^\alpha)$ the projections in the set $\mathscr{DP}_k = \mathscr{F}(P_k^\alpha)$ and this set contains, in general, more projections than the set $\mathscr{DP}_k = \mathscr{F}(\{P_j^\alpha\}_j)$ proposed by Clifton. (Any member of $\mathscr{F}(\{P_j^\alpha\}_j)$ is a member of $\mathscr{F}(P_k^\alpha)$. Also a projection Q^α orthogonal to P_k^α is a member $\mathscr{F}(P_k^\alpha)$ but need not be a member of $\mathscr{F}(\{P_j^\alpha\}_j)$.)

A further result is that Clifton's proposal can be understood as equivalent to the new proposal (5.12) if one restricts attention to those properties which are with certainty (that is, with probability 1) assigned a definite value. Consider a system α with a state W^α and let $\mathscr{DP}_{p=1}$ be the set of projections that have with probability $p = 1$ a definite value. This set is clearly equal to the meet of the sets $\{\mathscr{DP}_j\}_j$ since if a projection is a member of all the sets $\{\mathscr{DP}_j\}_j$, it has a definite value independent of the specific core property ascription to α. So, if one restricts the full property ascription (5.12) to this set $\mathscr{DP}_{p=1}$, one obtains the property ascription

$$\left\langle\, p_k = p(C_k^\alpha),\ \mathscr{DP}_{p=1} = \cap_j \mathscr{F}(C_j^\alpha),\ [Q^\alpha]_k = \frac{\mathrm{Tr}^\alpha(C_k^\alpha Q^\alpha)}{\mathrm{Tr}^\alpha(C_k^\alpha)} \,\right\rangle. \tag{5.13}$$

It can be proved that if the core projections are pair-wise orthogonal, the meet of the faux-Boolean algebras $\{\mathscr{F}(P_j^\alpha)\}_j$ generated by the individual core projections is equal to the faux-Boolean algebra $\mathscr{F}(\{C_j^\alpha\}_j)$ generated by the set of all core projections. In the spectral modal interpretation the core projections are indeed pair-wise orthogonal so the full property ascription (5.12) restricted to the set $\mathscr{DP}_{p=1}$ is in the spectral modal interpretation given by (use that $p(P_k^\alpha) = \mathrm{Tr}^\alpha(W^\alpha P_k^\alpha)$):

$$\left\langle\, p_k = \mathrm{Tr}^\alpha(W^\alpha P_k^\alpha),\ \mathscr{DP}_{p=1} = \mathscr{F}(\{P_j^\alpha\}_j),\ [Q^\alpha]_k = \frac{\mathrm{Tr}^\alpha(P_k^\alpha Q^\alpha)}{\mathrm{Tr}^\alpha(P_k^\alpha)} \,\right\rangle. \tag{5.14}$$

This restricted property ascription is equivalent to Clifton's proposal.[59]

[59] Clifton (1995a) motivates his proposal for the spectral modal interpretation by proving that any full property ascription to a system α which assigns definite values to a set of projections different from $\mathscr{F}(\{P_j^\alpha\}_j)$ must violate at least one of six desirable requirements. In Vermaas (1998a, Sect. 9) it is argued that these six requirements should only be imposed on the set of projections which have definite values with probability 1. Clifton's proof thus shows only that the set $\mathscr{DP}_{p=1}$ should be equal to $\mathscr{F}(\{P_j^\alpha\}_j)$. As such, the new proposal (5.12) complies with this proof via the restricted full property ascription (5.14). The new proposal yields, however, that there also exist projections which have a definite value with a probability strictly smaller than 1 (all the projections in the set $\cup_j \mathscr{F}(P_j^\alpha) - \mathscr{F}(\{P_j^\alpha\}_j)$. In this way the new proposal (5.12) assigns definite values to more projections than there are in the set $\mathscr{F}(\{P_j^\alpha\}_j)$.

Application of the new proposal (5.12) to the atomic and fixed modal interpretations yields a full property ascription that, as I said, violates Certainty.
This implies that there can exist a projection Q^α which is indefinite-valued
while a measurement of Q^α has with certainty a positive or with certainty
a negative outcome. Hence, in the atomic or fixed modal interpretation, a
measurement of Q^α which has with certainty a positive or with certainty a
negavite outcome, need not reveal a pre-existing value of Q^α. Consequently,
one cannot explain the certain outcome of this measurement by means of
a pre-existing value $[Q^\alpha]$, as was proposed in the quotation of Clifton on
page 69). A question is now what does a measurement with a certain outcome reveal. This question is taken up for the atomic modal interpretation
in Section 13.4.

MATHEMATICS

For pair-wise orthogonal projections $\{C_j^\alpha\}_j$ the sets $\cap_j \mathcal{F}(C_j^\alpha)$ and $\mathcal{F}(\{C_j^\alpha\}_j)$
are identical.

Proof: Let $Q^\alpha \in \cap_j \mathcal{F}(C_j^\alpha)$. Then $Q^\alpha \in \mathcal{F}(C_j^\alpha)$ for all j, and one can write
Q^α with (5.4) as $Q_\perp^\alpha$ or $C_j^\alpha + Q_\perp^\alpha$ with $Q_\perp^\alpha$ orthogonal to C_j^α. It follows that
$Q^\alpha C_j^\alpha = 0$ or $Q^\alpha C_j^\alpha = C_j^\alpha$ for all j. Define now the index-set I_Q as $k \in I_Q$ iff
$Q^\alpha C_k^\alpha = C_k^\alpha$. One can then derive that

$$Q^\alpha = Q^\alpha(\sum_j C_j^\alpha + \mathbb{I}^\alpha - \sum_j C_j^\alpha) = \sum_{j \in I_Q} C_j^\alpha + Q^\alpha(\mathbb{I}^\alpha - \sum_j C_j^\alpha). \tag{5.15}$$

The projections Q^α and $\mathbb{I}^\alpha - \sum_j C_j^\alpha$ commute, so $Q^\alpha(\mathbb{I}^\alpha - \sum_j C_j^\alpha)$ is a projection.
Moreover, $Q^\alpha(\mathbb{I}^\alpha - \sum_j C_j^\alpha)$ is orthogonal to all projections in $\{C_j^\alpha\}_j$. Hence,
by decomposition (5.4), Q^α is a member of $\mathcal{F}(\{C_j^\alpha\}_j)$.

By inspection it follows that any member Q^α of $\mathcal{F}(\{C_j^\alpha\}_j)$ is a member of
$\mathcal{F}(C_k^\alpha)$ for all k. Hence, Q^α is also a member of the meet $\cap_j \mathcal{F}(C_j^\alpha)$. □

5.6 Definite-valued magnitudes

The new full property ascription (5.12) specifies how modal interpretations
ascribe properties to systems. In this final section I discuss how this property
ascription induces a value assignment to the magnitudes of those systems.

As I have already said on a number of occasions, I adopt the link (3.2) as
my starting point to relate the ascription of properties to the assignment of
values to magnitudes.[60] So, let A^α be a magnitude represented by an operator

---

[60] And as I have said in a number of footnotes, some authors choose to adopt the more restrictive link
(22) given in footnote 22.

with a discrete spectral resolution $A^\alpha = \sum_j a_j \sum_k |a_{jk}^\alpha\rangle\langle a_{jk}^\alpha|$. Then the value assignment to A^α is related to the full property ascription (5.12) by the link

$$
\left.
\begin{array}{lll}
[A^\alpha] = a_j & \text{if and only if} & [\sum_k |a_{jk}^\alpha\rangle\langle a_{jk}^\alpha|] = 1, \\[2ex]
[A^\alpha] \neq a_j & \text{if and only if} & [\sum_k |a_{jk}^\alpha\rangle\langle a_{jk}^\alpha|] = 0, \\[2ex]
\begin{array}{l}\text{it is indefinite whether}\\ \text{or not } A^\alpha \text{ has value } a_j\end{array} & \text{if and only if} & \sum_k |a_{jk}^\alpha\rangle\langle a_{jk}^\alpha| \text{ is indefinite.}
\end{array}
\right\} \quad (5.16)
$$

This link has now two drawbacks. Firstly, it applies only to magnitudes represented by operators with a discrete spectral resolution. Many important magnitudes in quantum mechanics are, however, represented by operators with a continuous spectral resolution (position and energy of a free particle, for instance) and one would also like to assign values to those magnitudes, but the link (5.16) is silent about such values. Secondly, even when applied only to magnitudes represented by operators with a discrete spectral resolution, the link seems not to capture all that can be said about the values of magnitudes. Consider, for instance, a magnitude $A^\alpha = \sum_j a_j |a_j^\alpha\rangle\langle a_j^\alpha|$ and assume that the state of α is given by $|\psi^\alpha\rangle = c_1 |a_1^\alpha\rangle + c_2 |a_2^\alpha\rangle$. The new full property ascription (5.12) to α in, say, the spectral modal interpretation is in this case with probability 1 equal to $\mathscr{DP}_1 = \mathscr{F}(|\psi^\alpha\rangle\langle\psi^\alpha|)$. The link yields that it is then indefinite whether or not A^α has the value a_1 or the value a_2, and that $[A^\alpha]$ is not equal to a_3, a_4, etc. From these value assignments it now seems acceptable to conclude that the value of A^α is in this case in some sense confined to the set $\{a_1, a_2\}$. However, the link (5.16) is again silent about such a confinement.

In an attempt to overcome these two drawbacks I develop in this section a generalisation of the link (5.16) which yields information about the values of *all* the magnitudes of a system. The two drawbacks then vanish because the generalised link applies equally to magnitudes represented by operators with continuous spectral resolutions and to magnitudes A^α for which the link (5.16) only yields that $[A^\alpha]$ is not equal to specific values. However, this generalised link turns out to suffer from a new problem, namely that value assignments become genuinely *inexact*. That is, the value assignment to a magnitude is in general not given by the assignment of one exact eigenvalue, as is the case if one uses the link (5.16), but is captured by the assignment of a whole set of values. And this assignment of, say, the set Γ to the magnitude A^α need not mean that A^α has a single exact value x, where x is an element of the set Γ, but may mean that A^α has an inexact value restricted to the set Γ. Hence, the assignment of values to magnitudes confronts one with a dilemma. Either one stays with the link (5.16) which assigns exact values to some magnitudes represented by operators with a discrete spectral resolution and one swallows the two drawbacks. Or one adopts the generalised link which applies to all possible magnitudes and one swallows that value assignments can be inexact. Here, I do not attempt to resolve this dilemma; I only explore the second option of defining the envisaged generalised link.[61]

Consider therefore a magnitude represented by a self-adjoint, hypermaximal operator A^α defined on $\mathcal{H}^\alpha$. Let $\{\Gamma\}$ represent the Borel sets on the real line $\mathbb{R}$ and let $\{E_A^\alpha(\Gamma)\}_\Gamma$ be the projections which form the spectral family of A^α. The spectral resolution of A^α is then given by[62]

$$A^\alpha = \int_{\lambda=-\infty}^{\infty} \lambda \, dE_A^\alpha((-\infty, \lambda]). \tag{5.17}$$

The spectral family $\{E_A^\alpha(\Gamma)\}_\Gamma$ has the properties:

$$\left. \begin{aligned} E_A^\alpha(\mathbb{R} - \Gamma) &= \mathbb{I}^\alpha - E_A^\alpha(\Gamma), \\ E_A^\alpha(\Gamma \cap \Delta) &= E_A^\alpha(\Gamma)E_A^\alpha(\Delta), \\ E_A^\alpha(\Gamma \cup \Delta) &= E_A^\alpha(\Gamma) + E_A^\alpha(\Delta) - E_A^\alpha(\Gamma)E_A^\alpha(\Delta), \end{aligned} \quad \begin{aligned} E_A^\alpha(\varnothing) &= \mathbb{O}^\alpha, \\ E_A^\alpha(\mathbb{R}) &= \mathbb{I}^\alpha. \end{aligned} \right\} \tag{5.18}$$

Being projections, the members of the spectral family can have definite values. The question now is how can a value of a member $E_A^\alpha(\Gamma)$ induce a value assignment to A^α. In order to answer this question I return for a moment to operators with a discrete spectral resolution. For such operators integral (5.17) simplifies to the sum

$$A^\alpha = \sum_j a_j E_A^\alpha(a_j). \tag{5.19}$$

A projection $E_A^\alpha(a_j)$ is thus the eigenprojection $\sum_k |a_{jk}^\alpha\rangle\langle a_{jk}^\alpha|$ of A^α corresponding to the eigenvalue a_j and a general spectral projection $E_A^\alpha(\Gamma)$ of A^α is equal to

$$E_A^\alpha(\Gamma) = \sum_{a_j \in \Gamma} E_A^\alpha(a_j). \tag{5.20}$$

Consider the special case that one eigenprojection of such an operator

---

[61] A way out of the dilemma might be to aim at a generalised link which does not apply to all magnitudes of a system but which only assigns exact values to some magnitudes represented by operators with a discrete spectral resolution and to some magnitudes represented by operators with a continuous spectral resolution. Such a less ambitious generalisation overcomes the first drawback of the link (5.16) (but not the second) and saves one from assigning inexact values. Private communication with Rob Clifton, 1998.

[62] See footnote 8 for the spectral resolutions of hypermaximal self-adjoint operators.

with a discrete spectral resolution has the value 1, say $[E^\alpha_A(a_k)] = 1$. In such a case one can straightforwardly derive relations between the value assignment to the spectral projections $\{E^\alpha_A(\Gamma)\}_\Gamma$ and the value assignment to A^α itself. Firstly, because the full property ascription (5.12) satisfies the Exclusion condition, it follows that all the other eigenprojections of A^α have a value 0, so $[E^\alpha_A(a_j)] = \delta_{jk}$ for all j. Secondly, because the full property ascription (5.12) satisfies Closure, it follows that all the spectral projections $\{E^\alpha_A(\Gamma)\}_\Gamma$ have values. Moreover, these values are $[E^\alpha_A(\Gamma)] = 1$ if $a_k \in \Gamma$ and $[E^\alpha_A(\Gamma)] = 0$ if $a_k \notin \Gamma$. Thirdly, it follows from the (ungeneralised) link (5.16) that $[A^\alpha] = a_k$. This in turn implies that $[A^\alpha] \in \Gamma$ if $a_k \in \Gamma$ and $[A^\alpha] \notin \Gamma$ if $a_k \notin \Gamma$. Hence, the value assignments to the projections $\{E^\alpha_A(\Gamma)\}_\Gamma$ and to A^α itself satisfy the relations

$$\left.\begin{array}{llll} [A^\alpha] \in \Gamma & \text{\textit{if and only if}} & [E^\alpha_A(\Gamma)] = 1, \\ [A^\alpha] \notin \Gamma & \text{\textit{if and only if}} & [E^\alpha_A(\Gamma)] = 0. \end{array}\right\} \quad (5.21)$$

These relations are still valid only for magnitudes represented by operators with a discrete spectral resolution for which one eigenprojection has the value 1, but on first sight they look like a proper basis for the envisaged generalised link. However, one immediately runs into trouble if one applies these relations to a magnitude represented by an operator with a discrete spectral resolution for which there does not exist an eigenprojection with a value 1. To see this, consider a system with a state

$$W^\alpha = \frac{E^\alpha_A(a_1) + E^\alpha_A(a_2)}{\text{Tr}^\alpha(E^\alpha_A(a_1) + E^\alpha_A(a_2))}. \quad (5.22)$$

The full property ascription (5.12) to α in, say, the spectral modal interpretation is then with probability 1 given by $\mathscr{DP}_1 = \mathscr{F}(E^\alpha_A(a_1) + E^\alpha_A(a_2))$, such that $[E^\alpha_A(a_1) + E^\alpha_A(a_2)] = 1$ and the individual eigenprojections $E^\alpha_A(a_1)$ and $E^\alpha_A(a_2)$ are not definite-valued. The projection $E^\alpha_A(a_1) + E^\alpha_A(a_2)$ is equal to the spectral projection $E^\alpha_A(\{a_1, a_2\})$ and if one applies relations (5.21), it follows that $[A^\alpha] \in \{a_1, a_2\}$. This value assignment in turn implies that either $[A^\alpha] = a_1$ or $[A^\alpha] = a_2$ and by using again relations (5.21), one can derive that either $[E^\alpha_A(a_1)] = 1$ or $[E^\alpha_A(a_2)] = 1$. But this last conclusion is in contradiction with the full property ascription (5.12) which yields that $E^\alpha_A(a_1)$ and $E^\alpha_A(a_2)$ have no definite values. Hence, the relations (5.21) cannot be taken as the generalised link.

The assumptions used in the derivation of this contradiction were: (A) the new full property ascription (5.12), (B) the sum rule (5.20) to identify the spectral projection $E^\alpha_A(\{a_1, a_2\})$, (C) the equivalence of $[E^\alpha_A(\{a_1, a_2\})] = 1$ and $[A^\alpha] \in \{a_1, a_2\}$, and (D) the equivalence of $[A^\alpha] = a_j$ and $[E^\alpha_A(a_j)] = 1$.

I have already adopted (A) and (D), and assumption (B) is a mathematical fact. So, the only way left for me to escape the contradiction is by denying (C). And one can deny (C) by assuming that $[E_A^\alpha(\Gamma)] = 1$ need not mean that A^α has an exact definite value x, where x is an element of the set Γ, but that $[E_A^\alpha(\Gamma)] = 1$ in general means that A^α has an *inexact* value restricted to the set Γ. It then follows that $[E_A^\alpha(a_1) + E_A^\alpha(a_2)] = 1$ does not imply that either $[E_A^\alpha(a_1)] = 1$ or $[E_A^\alpha(a_2)] = 1$, such that the derivation of the contradiction is blocked.

The trouble now is that it is difficult to make sense of this assumption that the value assignment $[E_A^\alpha(\Gamma)] = 1$ may mean that a magnitude A^α has an inexact value restricted to a set Γ. At the end of this section I say more about the meaning of inexact value assignments. But note here that these assignments already show up if one tries the generalise the link (5.16) to all magnitudes represented by operators with a discrete spectral resolution; thus it is not specifically the value assignment to magnitudes represented by operators with a continuous spectral resolution which forces one to accept inexact value assignments.

If one does accept the assumption that $[E_A^\alpha(\Gamma)] = 1$ may mean that A^α has an inexact value restricted to the set Γ, one can continue with formulating a generalised link. Let's write this inexact value assignment as $[A^\alpha] \in^* \Gamma$ (the * is added to indicate that $[A^\alpha]$ need not be an exact value x element of Γ). The value assignment $[E_A^\alpha(\Gamma)] = 0$ should then analogously be taken as that A^α has an inexact value which is not restricted to the set Γ. One can read this negation is two ways: as that A^α has an inexact value restricted to a set Δ different to Γ, or as that A^α has an inexact value restricted to the set $\mathbb{R} - \Gamma$. Here I adopt the second reading because if A^α has an exact value (which still can be the case if, for instance, A^α has a discrete spectral resolution and if $[E_A^\alpha(a_j)] = \delta_{jk}$), then $[E_A^\alpha(\Gamma)] = 0$ is equivalent to A^α having a value restricted to $\mathbb{R} - \Gamma$. Let's write this negation as $[A^\alpha] \notin^* \Gamma$. The generalised link then becomes

$$
\left.
\begin{array}{lll}
[A^\alpha] \in^* \Gamma & \textit{if and only if} & [E_A^\alpha(\Gamma)] = 1, \\
[A^\alpha] \notin^* \Gamma & \textit{if and only if} & [E_A^\alpha(\Gamma)] = 0, \\[4pt]
\text{it is indefinite whether or} & & \\
\text{not } [A^\alpha] \text{ is restricted to } \Gamma & \multicolumn{2}{l}{\textit{if and only if } E_A^\alpha(\Gamma) \textit{ is indefinite.}}
\end{array}
\right\} \quad (5.23)
$$

This link can now be consistently applied to magnitudes represented by operators with a discrete and with a continuous spectral resolution. The resulting value assignment satisfies the following relations. From the full property ascription (5.12) and the properties (5.18) of the spectral family $\{E^\alpha_A(\Gamma)\}_\Gamma$ one can derive that:[63]

$$
\left.
\begin{aligned}
&[A^\alpha] \notin^* \varnothing, \\
&[A^\alpha] \in^* \mathbb{R}, \\
&[A^\alpha] \in^* \Gamma \Rightarrow [A^\alpha] \notin^* \mathbb{R} - \Gamma, \\
&[A^\alpha] \notin^* \Gamma \Rightarrow [A^\alpha] \in^* \mathbb{R} - \Gamma.
\end{aligned}
\right\} \tag{5.24}
$$

A second series of relations is:

$$
\left.
\begin{aligned}
&[A^\alpha] \in^* \Gamma \Rightarrow [A^\alpha] \in^* \Delta \quad \text{for every } \Delta \text{ with } \Delta \supseteq \Gamma, \\
&[A^\alpha] \in^* \Gamma \Rightarrow [A^\alpha] \notin^* \Delta \quad \text{for every } \Delta \text{ with } \Delta \cap \Gamma = \varnothing, \\
&[A^\alpha] \notin^* \Gamma \Rightarrow [A^\alpha] \notin^* \Delta \quad \text{for every } \Delta \text{ with } \Delta \subseteq \Gamma, \\
&[A^\alpha] \notin^* \Gamma \Rightarrow [A^\alpha] \in^* \Delta \quad \text{for every } \Delta \text{ with } \Delta \cup \Gamma = \mathbb{R}.
\end{aligned}
\right\} \tag{5.25}
$$

A final series concerns the propositional logic of the value assignment:

$$
\left.
\begin{aligned}
&\text{not } [A^\alpha] \in^* \Gamma &&\Leftrightarrow\; [A^\alpha] \in^* \mathbb{R} - \Gamma, \\
&[A^\alpha] \in^* \Gamma \text{ and } [A^\alpha] \in^* \Delta &&\Leftrightarrow\; [A^\alpha] \in^* \Gamma \cap \Delta, \\
&[A^\alpha] \in^* \Gamma \text{ or } [A^\alpha] \in^* \Delta &&\Rightarrow\; [A^\alpha] \in^* \Gamma \cup \Delta,
\end{aligned}
\right\} \tag{5.26}
$$

(where 'not $[A^\alpha] \in^* \Gamma$' is defined as $[A^\alpha] \notin^* \Gamma$). One relation is missing in this last list and that is

$$
[A^\alpha] \in^* \Gamma \cup \Delta \;\Rightarrow\; [A^\alpha] \in^* \Gamma \text{ or } [A^\alpha] \in^* \Delta. \tag{5.27}
$$

This relation doesn't hold in general, proving once again that $[A^\alpha] \in^* \Gamma$ is a genuinely inexact value assignment. If A^α has an exact value (that is, if $[E^\alpha_A(a_j)] = \delta_{jk}$), then this missing relation holds.

   These three sets of relations give the rules obeyed by the generalised value assignment (5.23) to magnitudes and fix in that sense partly what it means to assign an inexact value to such magnitudes. It is much harder, however, to directly describe what it means, physically speaking, to assign an inexact value to a magnitude. Two answers can be given but neither yields a positive description. The first answer is that $[A^\alpha] \in^* \Gamma$ means that A^α does not have an exact value x which falls outside the set Γ (which is a straightforward consequence of the second relation of (5.25)). However, this answer only gives a negative description; it tells you what is not the case if $[A^\alpha] \in^* \Gamma$. A second answer is given by Healey (1989) when he writes about his own construction of a value assignment to magnitudes (his construction differs in a number of points from the one presented here, but his remarks still

---

[63] See also Healey (1989, Sect. 2) for a discussion of this generalised value assignment.

apply) that 'The content of the claim [that $[E_A^\alpha(\Gamma)] = 1$] is best unpacked by examining the inferences which may be drawn from it. One such inference is that if one were to observe whether or not α has $E_A^\alpha(\Gamma)$, one would find that it does: And if one were to conduct a less precise observation as to whether or not α has $E_A^\alpha(\Delta)$ (with $\Gamma \subset \Delta$), one would find that it does. On the other hand, there is no *maximally* precise observation of A^α which would locate its value with maximal precision [...].'[64] However, this description of the physical meaning of $[E_A^\alpha(\Gamma)] = 1$ or, equivalently, of $[A^\alpha] \in^* \Gamma$, is formulated in terms of possible measurement outcomes. And this seems to be against the spirit of interpretations of quantum mechanics: the aim of an interpretation is to describe reality and thus to give a meaning to quantum mechanical assertions which goes beyond the realm of measurement outcomes.

A positive description of the meaning of the assignment $[A^\alpha] \in^* \Gamma$ seems to be given by the generalised link (5.23) itself, namely that $[A^\alpha] \in^* \Gamma$ implies that α possesses the property represented by the projection $E_A^\alpha(\Gamma)$. However, if this is the only positive description of the physical meaning of $[A^\alpha] \in^* \Gamma$, the gain of the generalised link seems to be of minimal value because then linking the property ascription $[E_A^\alpha(\Gamma)] = 1$ with $[A^\alpha] \in^* \Gamma$ does not tell much about the value of A^α.

MATHEMATICS

All the relations in (5.24), (5.25) and (5.26) can be transposed into relations between the values of the projections $E_A^\alpha(\Gamma)$. These latter relations can then be proved by means of the full property ascription (5.12) and the properties (5.18).

The first two relations in (5.24) are easily proved. Both $\mathbb{O}^\alpha$ and $\mathbb{I}^\alpha$ are, according to (5.12), members of every set $\mathscr{D}\mathscr{P}_k$ and are always assigned the values 0 and 1, respectively. So, it is with probability 1 the case that $[\mathbb{O}^\alpha] = 0$ and $[\mathbb{I}^\alpha] = 1$. Also by using the properties (5.18), it follows that $[A^\alpha] \in^* \mathbb{R}$ and $[A^\alpha] \notin^* \varnothing$ with probability 1.

The third relation in (5.24) is equivalent to

$$[E_A^\alpha(\Gamma)] = 1 \;\Rightarrow\; [E_A^\alpha(\mathbb{R} - \Gamma)] = 0 \tag{5.28}$$

and can be proved because the full property ascription (5.12) satisfies Closure. It thus follows that if $[E_A^\alpha(\Gamma)] = 1$, then the negation of $E_A^\alpha(\Gamma)$ simultaneously has the value 0. This negation $\mathbb{I}^\alpha - E_A^\alpha(\Gamma)$ is according to (5.18) equal to $E_A^\alpha(\mathbb{R} - \Gamma)$ so if $[E_A^\alpha(\Gamma)] = 1$, then $[E_A^\alpha(\mathbb{R} - \Gamma)] = 0$. The fourth relation in (5.24) can be proved analogously.

---

[64] Quotation from Healey (1989, page 75) with his italics and with the notation adjusted to the present notation.

The first relation in (5.25) is equivalent to

$$[E_A^\alpha(\Gamma)] = 1 \Rightarrow [E_A^\alpha(\Delta)] = 1 \quad \text{for every } \Delta \text{ with } \Delta \supseteq \Gamma. \tag{5.29}$$

Since $\Delta \supseteq \Gamma$, it follows that $\Gamma \cap \Delta = \Gamma$. Using the third property of (5.18), the above relation holds if

$$[E_A^\alpha(\Gamma)] = 1 \Rightarrow [E_A^\alpha(\Delta)] = 1 \text{ for every } E_A^\alpha(\Delta) \text{ with } E_A^\alpha(\Delta)E_A^\alpha(\Gamma) = E_A^\alpha(\Gamma).$$
$$\tag{5.30}$$

This last relation holds because the full property ascription (5.12) satisfies Weakening.

The second relation in (5.25) can be proved analogously by using that the full property ascription (5.12) satisfies Exclusion. The third relation in (5.25) follows by the fourth relation in (5.24) and the second of (5.25). The fourth relation in (5.25) follows by the third relation in (5.24) and the first of (5.25).

The first relation of (5.26) holds by definition. The $\Leftarrow$ part of the second relation of (5.26) is a consequence of the first relation of (5.25) since $\Gamma \cap \Delta$ is a subset of both Γ and Δ. The $\Rightarrow$ part of the second relation of (5.26) is proved as follows:

If for a system α the projections $E_A^\alpha(\Gamma)$ and $E_A^\alpha(\Delta)$ simultaneously have the value 1, then they are simultaneously members of the actual property ascription $\mathscr{DP}_k$ to α. According to (5.12), one can conclude (using decomposition (5.4)) that a member of $\mathscr{DP}_k$ with the value 1 is a sum of the core projection C_k^α and some projection orthogonal to C_k^α. Hence, it holds that $E_A^\alpha(\Gamma)C_k^\alpha = C_k^\alpha$ and $E_A^\alpha(\Delta)C_k^\alpha = C_k^\alpha$. Using the properties (5.18), it follows that $E_A^\alpha(\Gamma \cap \Delta)C_k^\alpha = C_k^\alpha$. Hence, $E_A^\alpha(\Gamma \cap \Delta)$ is simultaneously with $E_A^\alpha(\Gamma)$ and $E_A^\alpha(\Delta)$ a member of $\mathscr{DP}_k$ with the value 1.

The proof of the third relation of (5.26) is left to the reader.

6

Joint property ascriptions

The previous chapter dealt with the property ascription to single systems. In this chapter I consider correlations between the properties ascribed to different systems by the bi, spectral and atomic modal interpretations. These correlations are captured by means of a **joint property ascription**, that is, by means of joint probabilities $p(C_a^\alpha, C_b^\beta, C_c^\gamma, \dots)$ that the core property ascriptions to the systems α, β, γ, ..., are simultaneously $[C_a^\alpha] = 1$, $[C_b^\beta] = 1$, $[C_c^\gamma] = 1$, ..., respectively. I start by listing the existing results and then discuss the possibilities and impossibilities of extending these results to general joint property ascriptions.

6.1 A survey

According to the rules of the bi modal interpretation, the joint property ascription to two disjoint systems α and β is given by

$$p(P_a^\alpha, P_b^\beta) = \mathrm{Tr}^\omega(|\Psi^\omega\rangle\langle\Psi^\omega| \, [P_a^\alpha \otimes P_b^\beta]), \tag{6.1}$$

if the composite $\omega = \alpha\beta$ has a pure state $|\Psi^\omega\rangle$. If one accepts perspectivalism (discussed in Section 4.3), this is all one needs to know about joint property ascriptions. One can, of course, bisect a composite ω in more than one way, say, firstly into α and β and secondly into γ and δ. But these different bisections correspond to different perspectives and, because it makes sense only to simultaneously consider properties of systems defined from one perspective, any attempt to find a joint property ascription to α, β, γ and δ is superfluous. If, however, one rejects perspectivalism, the search for such a joint property ascription does make sense. Unfortunately, a no-go theorem, presented in Section 6.3, proves that if one accepts Instantaneous Autonomy, Dynamical Autonomy for measurements, and Empirical Adequacy (see Section 3.3), the

---

Joint property ascription

A joint property ascription to the systems $\{\alpha, \beta, \gamma, \ldots\}$ consists of the joint probabilities $p(C_a^\alpha, C_b^\beta, C_c^\gamma, \ldots)$ that the core property ascriptions to these systems are simultaneously $[C_a^\alpha] = 1$, $[C_b^\beta] = 1$, $[C_c^\gamma] = 1$, etc.

---

bi modal interpretation cannot be supplemented by general joint property ascriptions.

According to the rules of the spectral modal interpretation, the joint property ascription to the disjoint systems α, β, γ, ... is given by

$$p(P_a^\alpha, P_b^\beta, P_c^\gamma, \ldots) = \mathrm{Tr}^\omega (W^\omega [P_a^\alpha \otimes P_b^\beta \otimes P_c^\gamma \otimes \cdots]), \tag{6.2}$$

with ω equal to the composite of the systems α, β, γ, etc.

Again, if one accepts perspectivalism, this is the end of the story because then one only simultaneously considers the properties of sets of disjoint systems. However, if one denies perspectivalism, one can continue to try to also fix joint property ascriptions to non-disjoint systems. In the next section I show that in special cases one can indeed give such ascriptions. However, the no-go theorem of Section 6.3 proves that joint property ascriptions to non-disjoint systems do not exist in general.

In the atomic modal interpretation joint property ascriptions are given for collections containing a molecule β and the atoms α_q, α_r, α_s, ... in that molecule:

$$p(P_a^{\alpha_q}, P_b^{\alpha_r}, \ldots, P_{a'b'c'\ldots}^\beta) = \delta_{aa'} \delta_{bb'} \cdots \mathrm{Tr}^\beta (W^\beta P_{abc\ldots}^\beta), \tag{6.3}$$

where $P_{abc\ldots}^\beta := P_a^{\alpha_q} \otimes P_b^{\alpha_r} \otimes P_c^{\alpha_s} \otimes \cdots$. Joint property ascriptions for general collections of atoms and molecules exist and are derived in Section 6.4.

6.2 Snoopers

In the spectral modal interpretation there exist in special cases joint property ascriptions to non-disjoint systems. Take a collection $\{\omega, \alpha, \beta, \gamma, \ldots\}$ of a number of disjoint systems α, β, γ, ..., and their composite $\omega = \alpha\beta\gamma\cdots$. This collection contains non-disjoint systems (α and ω, for instance) and one can derive the joint probabilities

$$p(P_a^\alpha, P_b^\beta, \ldots, P_d^\omega) \tag{6.4}$$

for a joint property ascription to this collection (Vermaas 1996, App. A). The ingredients of this derivation are the assumption of Instantaneous Autonomy and the joint probabilities (6.2) for disjoint systems.

At this point, the reader might want to argue that the above joint probabilities cannot exist, for if they did exist, then one could construct joint probabilities for sets of *non-commuting* operators. And notably Fine (1982) has proved that these latter joint probabilities cannot exist. To develop this argument, define a mapping which associates the operator P_a^α with $P_a^\alpha \otimes \mathbb{I}^{\omega/\alpha}$, which associates P_b^β with $P_b^\beta \otimes \mathbb{I}^{\omega/\beta}$, and so on. Under this mapping (6.4) becomes a joint probability

$$p(P_a^\alpha \otimes \mathbb{I}^{\omega/\alpha}, P_b^\beta \otimes \mathbb{I}^{\omega/\beta}, \dots, P_d^\omega) \tag{6.5}$$

over projections defined on $\mathcal{H}^\omega$ which, in general, do not commute.[65] (Note that (6.5) is not a joint probability for some property ascription to ω; it is solely a mathematical object constructed by means of (6.4).) Furthermore, since the joint probabilities (6.4) need to be consistent with the joint probabilities (6.2) (that is, the marginal $\sum_d p(P_a^\alpha, P_b^\beta, \dots, P_d^\omega)$ should return (6.2)), the joint probabilities (6.5) should satisfy

$$\sum_d p(P_a^\alpha \otimes \mathbb{I}^{\omega/\alpha}, P_b^\beta \otimes \mathbb{I}^{\omega/\beta}, \dots, P_d^\omega) = \mathrm{Tr}^\omega(W^\omega [P_a^\alpha \otimes \mathbb{I}^{\omega/\alpha}][P_b^\beta \otimes \mathbb{I}^{\omega/\beta}] \cdots).$$

$$\tag{6.6}$$

Hence, if the joint probabilities (6.4) exist, one can enrich quantum mechanics with joint probabilities for non-commuting projections consistent with the joint Born probabilities for commuting projections (the right-hand side of (6.6)). And Theorem 7 in Fine (1982, page 1309), in particular, seems to rule out such an enrichment.

The strongest rebuttal of this argument is, however, an explicit derivation of (6.4).[66] To do so, suppose that there exists a system σ disjoint to ω and that the composite $\sigma\omega$ has a state $W^{\sigma\omega}$ which satisfies the condition

$$\forall d, e : \qquad d \neq e \ \Rightarrow \ \mathrm{Tr}^{\sigma\omega}(W^{\sigma\omega} [P_e^\sigma \otimes P_d^\omega]) = 0. \tag{6.7}$$

This assumption does not constrain the state of ω: for any possible state W^ω with spectral resolution $W^\omega = \sum_j w_j^\omega P_j^\omega$, the state $W^{\sigma\omega}$ can be

---

[65] Let, for instance, the state of ω be $|\psi^\omega\rangle = \frac{\sqrt{3}}{2}|e_1^\alpha\rangle \otimes |e_1^\beta\rangle + \frac{1}{2}|e_2^\alpha\rangle \otimes |e_2^\beta\rangle$ with $\langle e_1^\alpha | e_2^\alpha \rangle = \langle e_1^\beta | e_2^\beta \rangle = 0$. Then the projections $P_1^\omega = |\psi^\omega\rangle\langle\psi^\omega|$ and $P_1^\alpha \otimes \mathbb{I}^\beta = |e_1^\alpha\rangle\langle e_1^\alpha| \otimes \mathbb{I}^\beta$ do not commute.

[66] Fine's (1982) Theorem 7 rules out joint probabilities for non-commuting projections which are consistent with the joint Born probabilities *and* which are (among other things) definable for all possible states W^ω of ω. I here escape Fine's theorem because I don't require that the joint probabilities (6.4) are, given a fixed set of projections $P_a^\alpha, P_b^\beta, \dots, P_d^\omega$, definable for all possible states W^ω; I only need that the joint probabilities are definable for those states W^ω which generate $\{P_a^\alpha\}_a$, $\{P_b^\beta\}_b, \dots, \{P_d^\omega\}_d$ as the eigenprojections of its partial traces $W^\alpha, W^\beta, \dots$, and as the eigenprojections of W^ω itself, respectively.

Snooper

A system σ is called a snooper for a system α if the joint property ascription to α and σ satisfies $p(C_j^\alpha, C_k^\sigma) = 0$ for all $j \neq k$. The actually possessed core property of the snooper thus reveals with probability 1 the actually possessed core property of α (σ 'snoops into' α).

---

chosen as

$$W^{\sigma\omega} = \sum_j w_j^\omega P_j^\sigma \otimes P_j^\omega, \tag{6.8}$$

with $\{P_j^\sigma\}_j$ a set of orthogonal projections.

Due to the above condition there exist strict correlations between the core projections to σ and ω: the joint probabilities (6.2) for disjoint systems yield with probability 1 that $[P_d^\omega] = 1$ if and only if $[P_d^\sigma] = 1$. The system σ thus acts like an indicator or snooper which records the actual core property ascription to ω. For this reason I call σ a **snooper** system for ω.

From the strict correlations between the core property ascriptions to σ and ω, it follows that (6.4) is equal to the joint probability $p(P_a^\alpha, P_b^\beta, \ldots, P_d^\sigma)$ for the joint property ascription to α, β, ... and σ. This latter joint probability can be determined with (6.2) because α, β, ..., σ are all mutually disjoint systems. Given the condition (6.7), one can thus derive that

$$p(P_a^\alpha, P_b^\beta, \ldots, P_d^\omega) = \text{Tr}^{\sigma\omega}(W^{\sigma\omega} [P_d^\sigma \otimes P_a^\alpha \otimes P_b^\beta \otimes \cdots]) \tag{6.9}$$

and rewrite this (see the MATHEMATICS) as

$$p(P_a^\alpha, P_b^\beta, \ldots, P_d^\omega) = \text{Tr}^\omega(W^\omega P_d^\omega [P_a^\alpha \otimes P_b^\beta \otimes \cdots]). \tag{6.10}$$

This second result is still only proved in the special case that there exists a snooper σ for ω. By now invoking Instantaneous Autonomy one can turn this result into a generally valid one.

Consider any system ω with a fixed state W^ω. If there exists a snooper σ for ω, the joint property ascription to $\{\omega, \alpha, \beta, \ldots\}$ is given by (6.10). If there does not exist a snooper for ω, this joint property ascription is unknown. Instantaneous Autonomy demands that in both cases the property ascription is equal since the state of ω is in both cases equal. Hence, also if there does not exist a snooper for ω, the joint probabilities are given by (6.10).[67]

---

[67] Note that according to (6.10) the joint property ascription to $\{\alpha, \beta, \ldots, \omega\}$ is only a function of the state of ω as demanded by the necessary condition of Instantaneous Autonomy. If this joint property ascription were a function of the state $W^{\sigma\omega}$ of ω plus snooper, as it appears to be according to formula (6.9), this necessary condition would be violated and application of Instantaneous Autonomy would not be possible.

MATHEMATICS

To prove from condition (6.7) that

$$\text{Tr}^{\sigma\omega}(W^{\sigma\omega}\,[P_d^{\sigma}\otimes P_a^{\alpha}\otimes P_b^{\beta}\otimes\cdots]) = \text{Tr}^{\omega}(W^{\omega}\,P_d^{\omega}\,[P_a^{\alpha}\otimes P_b^{\beta}\otimes\cdots]) \qquad (6.11)$$

I give a lemma and a theorem:

Lemma 6.1

For each density operator W and each orthonormal set $\{|e_j\rangle\}_j$ of vectors, the following holds

$$\langle e_j|W|e_j\rangle = 0 \qquad \Leftrightarrow \qquad \forall k: \qquad \langle e_j|W|e_k\rangle = \langle e_k|W|e_j\rangle = 0. \qquad (6.12)$$

Proof: The $\Leftarrow$ part is trivial. To prove the $\Rightarrow$ part, consider a density operator W. Because W is self-adjoint and positive, it follows that $\langle\psi|W|\psi\rangle \geq 0$ for every vector $|\psi\rangle$. Now take $\langle e_j|W|e_j\rangle = 0$ and assume that $\langle e_j|W|e_k\rangle \neq 0$. Define the vector $|\psi\rangle = \lambda|e_j\rangle + |e_k\rangle$. Then

$$\langle\psi|W|\psi\rangle = \langle e_k|W|e_k\rangle + 2\,\text{Re}[\bar{\lambda}\,\langle e_j|W|e_k\rangle]. \qquad (6.13)$$

If $\langle e_k|W|e_k\rangle = 0$, choose $\lambda = -1/\langle e_k|W|e_j\rangle$, if $\langle e_k|W|e_k\rangle \neq 0$ (W is positive, so $\langle e_k|W|e_k\rangle \geq 0$), choose $\lambda = -\langle e_k|W|e_k\rangle/\langle e_k|W|e_j\rangle$. In both cases $\langle\psi|W|\psi\rangle$ is negative, contradicting that W is positive. By *reductio ad absurdum* it follows that if $\langle e_j|W|e_j\rangle = 0$, then $\langle e_j|W|e_k\rangle = 0$. The complex conjugate of this result yields $\langle e_k|W|e_j\rangle = 0$. $\qquad\square$

Theorem 6.1

If $\{Q_q^{\kappa}\}_q$ is an orthogonal set of projections on $\mathscr{H}^{\kappa}$ with $\sum_q Q_q^{\kappa} = \mathbb{I}^{\kappa}$, and if $\{Q_r^{\lambda}\}_r$ is an orthogonal set of projections on $\mathscr{H}^{\lambda}$ with $\sum_r Q_r^{\lambda} = \mathbb{I}^{\lambda}$, and $W^{\kappa\lambda}$ is a density operator with partial trace $\text{Tr}^{\kappa}(W^{\kappa\lambda}) = W^{\lambda}$, then

$$\forall q, r: \qquad \text{Tr}^{\kappa\lambda}(W^{\kappa\lambda}[Q_q^{\kappa}\otimes Q_r^{\lambda}]) = \delta_{qr}\,\text{Tr}^{\kappa\lambda}(W^{\kappa\lambda}[Q_q^{\kappa}\otimes Q_q^{\lambda}]) \qquad (6.14)$$

if and only if

$$\forall s: \qquad \text{Tr}^{\kappa}(W^{\kappa\lambda}[Q_s^{\kappa}\otimes\mathbb{I}^{\lambda}]) = Q_s^{\lambda}W^{\lambda} = W^{\lambda}Q_s^{\lambda}. \qquad (6.15)$$

Proof: To prove the 'if' part of the theorem, consider a density operator $W^{\kappa\lambda}$ that obeys (6.15). Then, for all q and r:

$$\text{Tr}^{\kappa\lambda}(W^{\kappa\lambda}[Q_q^{\kappa}\otimes Q_r^{\lambda}]) = \text{Tr}^{\lambda}(\text{Tr}^{\kappa}(W^{\kappa\lambda}[Q_q^{\kappa}\otimes\mathbb{I}^{\lambda}])\,Q_r^{\lambda}) = \text{Tr}^{\lambda}(W^{\lambda}Q_q^{\lambda}Q_r^{\lambda})$$
$$= \delta_{qr}\,\text{Tr}^{\lambda}(W^{\lambda}Q_q^{\lambda}Q_q^{\lambda}) = \delta_{qr}\,\text{Tr}^{\kappa\lambda}(W^{\kappa\lambda}[Q_q^{\kappa}\otimes Q_q^{\lambda}]). \qquad (6.16)$$

To prove the 'only if' part, take a density operator $W^{\kappa\lambda}$ that obeys (6.14). Define orthonormal bases $\{|\psi_{a,b}^{\kappa}\rangle\}_{a,b}$ for $\mathscr{H}^{\kappa}$ and $\{|\psi_{c,d}^{\lambda}\rangle\}_{c,d}$ for $\mathscr{H}^{\lambda}$ such that all Q_q^κs and Q_r^λs can be expanded as, respectively, $Q_q^\kappa = \sum_b |\psi_{q,b}^\kappa\rangle\langle\psi_{q,b}^\kappa|$ and $Q_r^\lambda = \sum_d |\psi_{r,d}^\lambda\rangle\langle\psi_{r,d}^\lambda|$. An orthonormal basis for $\mathcal{H}^{\kappa\lambda}$ is then $\{|\psi_{a,b}^\kappa\rangle \otimes |\psi_{c,d}^\lambda\rangle\}_{a,b,c,d}$. If one performs the trace in (6.14) with respect to this basis, one obtains

$$a \neq c \quad \Rightarrow \quad \sum_{b,d} \langle\psi_{a,b}^\kappa|\langle\psi_{c,d}^\lambda|W^{\kappa\lambda}|\psi_{a,b}^\kappa\rangle|\psi_{c,d}^\lambda\rangle = 0. \tag{6.17}$$

Because the density operator $W^{\kappa\lambda}$ is a positive operator, each term in this summation must be positive or equal to 0. It thus follows that each term is 0 and from Lemma 6.1 one can conclude that

$$a \neq c \quad \text{or} \quad e \neq g \quad \Rightarrow \quad \langle\psi_{a,b}^\kappa|\langle\psi_{c,d}^\lambda|W^{\kappa\lambda}|\psi_{e,f}^\kappa\rangle|\psi_{g,h}^\lambda\rangle = 0. \tag{6.18}$$

It can be shown that (6.15) holds because the matrix elements of the three operators in (6.15) with respect to the basis $\{|\psi_{c,d}^\lambda\rangle\}_{c,d}$, coincide. Consider the matrix elements of the operator $\mathrm{Tr}^\kappa(W^{\kappa\lambda}[Q_s^\kappa \otimes \mathbb{I}^\lambda])$ with s arbitrary. From (6.18) it follows that

$$\langle\psi_{c,d}^\lambda|\mathrm{Tr}^\kappa(W^{\kappa\lambda}[Q_s^\kappa \otimes \mathbb{I}^\lambda])|\psi_{g,h}^\lambda\rangle = \sum_{a,b}\langle\psi_{a,b}^\kappa|\langle\psi_{c,d}^\lambda|(W^{\kappa\lambda}[Q_s^\kappa \otimes \mathbb{I}^\lambda])|\psi_{a,b}^\kappa\rangle|\psi_{g,h}^\lambda\rangle$$

$$= \sum_b \langle\psi_{s,b}^\kappa|\langle\psi_{c,d}^\lambda|W^{\kappa\lambda}|\psi_{s,b}^\kappa\rangle|\psi_{g,h}^\lambda\rangle$$

$$= \sum_b \delta_{cs}\,\delta_{gs}\,\langle\psi_{s,b}^\kappa|\langle\psi_{s,d}^\lambda|W^{\kappa\lambda}|\psi_{s,b}^\kappa\rangle|\psi_{s,h}^\lambda\rangle. \tag{6.19}$$

The matrix elements of $Q_s^\lambda W^\lambda$ are according to (6.18) equal to

$$\langle\psi_{c,d}^\lambda|Q_s^\lambda\,\mathrm{Tr}^\kappa(W^{\kappa\lambda})|\psi_{g,h}^\lambda\rangle = \sum_{a,b}\delta_{cs}\,\langle\psi_{a,b}^\kappa|\langle\psi_{s,d}^\lambda|W^{\kappa\lambda}|\psi_{a,b}^\kappa\rangle|\psi_{g,h}^\lambda\rangle$$

$$= \sum_b \delta_{cs}\,\delta_{gs}\,\langle\psi_{s,b}^\kappa|\langle\psi_{s,d}^\lambda|W^{\kappa\lambda}|\psi_{s,b}^\kappa\rangle|\psi_{s,h}^\lambda\rangle. \tag{6.20}$$

The same expression can be derived for the matrix elements of $W^\lambda Q_s^\lambda$. These matrix elements all coincide so (6.15) holds. □

The equality (6.11) can now be proved by first noting that by definition

$$\mathrm{Tr}^{\sigma\omega}(W^{\sigma\omega}\,[P_d^\sigma \otimes P_a^\alpha \otimes P_b^\beta \otimes \cdots])$$

$$= \mathrm{Tr}^\omega\left(\mathrm{Tr}^\sigma(W^{\sigma\omega}[P_d^\sigma \otimes \mathbb{I}^\omega])\,[P_a^\alpha \otimes P_b^\beta \otimes \cdots]\right). \tag{6.21}$$

The state $W^{\sigma\omega}$ obeys condition (6.7) and Theorem 6.1 yields that $\mathrm{Tr}^\sigma(W^{\sigma\omega}\,[P_d^\sigma \otimes \mathbb{I}^\omega])$ equals $W^\omega P_d^\omega$ for all d. Substitution of this into (6.21) gives (6.11).

6.3 A no-go theorem

In special cases one can thus give joint property ascriptions to collections of non-disjoint systems in the spectral modal interpretation. In this section I prove that such property ascriptions cannot be given in general. Firstly, I determine a necessary condition for the existence of general joint property ascriptions. Then, I establish a no-go theorem by proving that the spectral modal interpretation cannot always meet this necessary condition (Vermaas 1997).[68]

Consider a collection $\{\kappa, \lambda, \nu, \ldots\}$ of subsystems of a large system ω. The states of κ, λ ... are then partial traces of the state W^ω. If a joint property ascription to $\{\kappa, \lambda, \nu, \ldots\}$ exists, there exist probabilities $p(P_a^\kappa, P_b^\lambda, \ldots)$ which necessarily generate a *classical* probability space $\langle E, \mathscr{A}, p \rangle$ obeying Kolmogorov's axioms: take for the set E of elementary events the set $\{P_k^\kappa\}_k \times \{P_l^\lambda\}_l \times \cdots$ with $\times$ the Cartesian product for sets; let $\mathscr{A}$ be the σ-algebra generated by the members of E, and let the probability measure p be equal to $p(P_a^\kappa, P_b^\lambda, \ldots)$ itself. Furthermore, the *marginals* of $p(P_a^\kappa, P_b^\lambda, \ldots)$ necessarily give the probabilities for the (joint) property ascriptions to subsets of $\{\kappa, \lambda, \nu, \ldots\}$. For instance, the probabilities $p(P_b^\lambda, P_c^\nu, \ldots)$ for the joint property ascription to $\{\lambda, \nu, \ldots\}$ must be the marginal $\sum_a p(P_a^\kappa, P_b^\lambda, P_c^\nu, \ldots)$. Hence, joint property ascriptions exist only if the following condition is satisfied:

Necessary condition for joint property ascriptions

For every collection $\{\kappa, \lambda, \ldots\}$ of subsystems of a system ω and for every state W^ω, there exist classical probabilities $p(P_a^\kappa, P_b^\lambda, \ldots)$ which are consistent with the probabilities for the property ascriptions to subsets of $\{\kappa, \lambda, \ldots\}$.

This condition can easily be satisfied: just let $p(P_a^\kappa, P_b^\lambda, \ldots)$ be the product of its one-slotted marginals $p(P_a^\kappa)$, $p(P_b^\lambda)$, etc. However, in the spectral modal interpretation some joint probabilities are already fixed, namely those for disjoint systems and those derived in the last section. And these joint probabilities are not, in general, products of their one-slotted marginals. Hence, the question becomes whether the necessary condition can be satisfied given the constraint that some joint probabilities are already defined by the spectral modal interpretation. The answer is no: it can be proved that the joint probabilities (6.2) for disjoint systems make it impossible for the above necessary condition to be satisfied.

---

[68] The no-go theorem in Vermaas (1997) is the third in a series which started with no-go theorems by Bacciagaluppi (1995) and by Clifton (1996). These earlier theorems, however, do not pertain to the spectral modal interpretation but disprove the possibility of enriching it with a property ascription rule called Property Composition (see Section 13.2 and Bacciagaluppi and Vermaas (1999)).

In order to sketch this no-go proof, consider a composite ω of four disjoint systems α, β, γ and δ. Assume that the state of ω is pure and consider the systems $\{\alpha, \alpha\beta, \alpha\beta\gamma\}$. If there exists a joint property ascription to these systems, then, according to the above necessary condition, the corresponding joint probabilities $p(P_a^\alpha, P_b^{\alpha\beta}, P_c^{\alpha\beta\gamma})$ should be classical probabilities and be consistent with the (joint) property ascriptions to subsets of $\{\alpha, \alpha\beta, \alpha\beta\gamma\}$.

Let, for a moment, $p(a, b, c)$ be shorthand notation for $p(P_a^\alpha, P_b^{\alpha\beta}, P_c^{\alpha\beta\gamma})$. A general result is that a set of values $\{p(a, b, c)\}_{a,b,c}$ defines classical probabilities that satisfy

$$p(a, b, c) \geq 0, \qquad \sum_{a,b,c} p(a, b, c) = 1, \qquad (6.22)$$

if and only if the marginals $p(a)$, $p(b)$, $p(c)$, $p(a, b)$, $p(a, c)$ and $p(b, c)$ of these values (which are defined by $p(a) = \sum_{b,c} p(a, b, c)$ and $p(a, b) = \sum_c p(a, b, c)$, etc.) satisfy the so-called Bell–Wigner inequalities:[69]

$$
\left.
\begin{aligned}
&0 \leq p(x, y) \leq p(x) \leq 1, && \forall x, y \in \{a, b, c\} \text{ with } x \neq y, \\
&p(x) + p(y) - p(x, y) \leq 1, && \forall x, y \in \{a, b, c\} \text{ with } x \neq y, \\
&p(a) + p(b) + p(c) - p(a, b) - p(a, c) - p(b, c) \leq 1, \\
&p(a) - p(a, b) - p(a, c) + p(b, c) \geq 0, \\
&p(b) - p(a, b) - p(b, c) + p(a, c) \geq 0, \\
&p(c) - p(a, c) - p(b, c) + p(a, b) \geq 0.
\end{aligned}
\right\}
\qquad (6.23)
$$

Thus the fact that the probabilities $p(P_a^\alpha, P_b^{\alpha\beta}, P_c^{\alpha\beta\gamma})$ should be classical probabilities implies that their marginals should satisfy these Bell–Wigner inequalities. So, $p(P_a^\alpha, P_b^{\alpha\beta}, P_c^{\alpha\beta\gamma})$ should satisfy, say, the fifth inequality, which reads

$$p(P_b^{\alpha\beta}) - p(P_a^\alpha, P_b^{\alpha\beta}) - p(P_b^{\alpha\beta}, P_c^{\alpha\beta\gamma}) + p(P_a^\alpha, P_c^{\alpha\beta\gamma}) \geq 0. \qquad (6.24)$$

And the fact that the probabilities $p(P_a^\alpha, P_b^{\alpha\beta}, P_c^{\alpha\beta\gamma})$ should be consistent with the property ascriptions to subsets of $\{\alpha, \alpha\beta, \alpha\beta\gamma\}$, implies that the (joint) probabilities in this fifth inequality may be calculated by directly applying the spectral modal interpretation to these subsets. Hence, one can check this inequality without being forced to determine $p(P_a^\alpha, P_b^{\alpha\beta}, P_c^{\alpha\beta\gamma})$ itself.

The probabilities in the fifth Bell–Wigner inequality (6.24) can all be calculated by the spectral modal interpretation. The first probability is clearly

$$p(P_b^{\alpha\beta}) = \text{Tr}^{\alpha\beta}(P_b^{\alpha\beta} W^{\alpha\beta}) = \text{Tr}^\omega(W^\omega [P_b^{\alpha\beta} \otimes \mathbb{I}^{\gamma\delta}]), \qquad (6.25)$$

---

[69] See Pitowsky (1989, Sect. 2.4) and Beltrametti and Maczynski (1993, Sect. V).

and the joint probabilities $p(P_a^\alpha, P_b^{\alpha\beta})$, $p(P_a^\alpha, P_c^{\alpha\beta\gamma})$ and $p(P_b^{\alpha\beta}, P_c^{\alpha\beta\gamma})$ can be determined with the joint probabilities (6.2) for disjoint systems. Consider, for instance, $p(P_a^\alpha, P_b^{\alpha\beta})$. Since α and $\alpha\beta$ are not mutually disjoint, one cannot calculate $p(P_a^\alpha, P_b^{\alpha\beta})$ directly with (6.2). But because ω has a pure state, it follows from (6.2) that the core properties of $\alpha\beta$ and of $\gamma\delta$ are one-to-one correlated ($\gamma\delta$ thus acts like a snooper for $\alpha\beta$; see the MATHEMATICS). Hence, $p(P_a^\alpha, P_b^{\alpha\beta})$ is equal to $p(P_a^\alpha, P_b^{\gamma\delta})$ and this latter joint probability can be calculated with (6.2). Exploiting also the one-to-one correlations between the core property ascriptions to δ and $\alpha\beta\gamma$, it follows that

$$
\left.
\begin{aligned}
p(P_a^\alpha, P_b^{\alpha\beta}) &= p(P_a^\alpha, P_b^{\gamma\delta}) = \mathrm{Tr}^\omega(W^\omega \, [P_a^\alpha \otimes \mathbb{I}^\beta \otimes P_b^{\gamma\delta}]), \\
p(P_a^\alpha, P_c^{\alpha\beta\gamma}) &= p(P_a^\alpha, P_c^{\delta}) = \mathrm{Tr}^\omega(W^\omega \, [P_a^\alpha \otimes \mathbb{I}^{\beta\gamma} \otimes P_c^{\delta}]), \\
p(P_b^{\alpha\beta}, P_c^{\alpha\beta\gamma}) &= p(P_b^{\alpha\beta}, P_c^{\delta}) = \mathrm{Tr}^\omega(W^\omega \, [P_b^{\alpha\beta} \otimes \mathbb{I}^\gamma \otimes P_c^{\delta}]).
\end{aligned}
\right\} \quad (6.26)
$$

To sum up, if a joint property ascription to $\{\alpha, \alpha\beta, \alpha\beta\gamma\}$ exists, the spectral modal interpretation satisfies the above necessary condition. And if this necessary condition is satified, then the spectral modal interpretation satisfies the fifth Bell–Wigner inequality (6.24). It is, however, possible to choose the state of ω such that this Bell–Wigner inequality is violated (see the MATHEMATICS). Hence, it follows by *reductio ad absurdum* that there does not always exist a joint property ascription to $\{\alpha, \alpha\beta, \alpha\beta\gamma\}$ in the spectral modal interpretation.

This no-go theorem proves that the spectral modal interpretation cannot be supplemented by general joint property ascriptions to non-disjoint systems. Thus there exist only general correlations between disjoint systems, supporting the view that if one accepts the spectral modal interpretation, one has to accept perspectivalism as well.

The no-go theorem also proves that the bi modal interpretation cannot be supplemented by joint property ascriptions if one accepts Instantaneous Autonomy, Dynamical Autonomy for measurements, and Empirical Adequacy. The bi modal interpretation leads in this case to the spectral modal interpretation (as proved in Appendix A) such that joint property ascriptions also have to obey (6.2) and the above necessary condition.

MATHEMATICS

I firstly prove a lemma for the spectral modal interpretation which says that if two disjoint systems σ and τ have a pure composite state $|\Psi^{\sigma\tau}\rangle$, then the properties ascribed to σ and to τ are one-to-one correlated. (This lemma essentially says that if two systems have a pure composite state, then the spectral modal interpretation yields the same core property ascription as the bi modal interpretation.)

Lemma 6.2
If a composite $\sigma\tau$ has a pure state $|\Psi^{\sigma\tau}\rangle$, the eigenvalues of W^σ and W^τ can be labelled such that $w_j^\sigma = w_j^\tau$ for all j. Given this labelling, the core property ascriptions to σ and τ are one-to-one correlated, that is, $p(P_j^\sigma, P_k^\tau) = 0$ for all $j \neq k$.

Proof: The biorthogonal decomposition of $|\Psi^{\sigma\tau}\rangle$ yields that the eigenvalues of both W^σ and W^τ are $\{\lambda_j\}_j$. Now label the eigenprojections of W^σ and W^τ such that $P_j^\sigma = P^\sigma(\lambda_j)$ and $P_j^\tau = P^\tau(\lambda_j)$ for all j. Substitution of the biorthogonal decomposition into the joint probabilities (6.2) for $\{\sigma, \tau\}$ gives $p(P_j^\sigma, P_k^\tau) = n[I_j]\lambda_j \delta_{jk}$.   □

Now take $\omega = \alpha\beta\gamma\delta$ and assume that ω has a pure state. From the lemma it then follows that the properties of $\alpha\beta$ are one-to-one correlated with the properties of $\gamma\delta$, and that the properties of $\alpha\beta\gamma$ are one-to-one correlated with the properties of δ. Hence, $\gamma\delta$ is a snooper for $\alpha\beta$ and δ is a snooper for $\alpha\beta\gamma$, proving the equalities (6.26).

I secondly prove that one can give a pure state for ω such that the fifth Bell–Wigner inequality (6.24) is violated. This inequality gives with the equalities (6.26)

$$p(P_b^{\alpha\beta}) - p(P_a^\alpha, P_b^{\gamma\delta}) - p(P_b^{\alpha\beta}, P_c^\delta) + p(P_a^\alpha, P_c^\delta) \geq 0. \tag{6.27}$$

Let the Hilbert spaces $\mathcal{H}^\alpha$, $\mathcal{H}^\beta$, $\mathcal{H}^\gamma$ and $\mathcal{H}^\delta$ all be two-dimensional and let, respectively, $\{|e_j^\alpha\rangle\}_{j=1}^2$, $\{|e_j^\beta\rangle\}_{j=1}^2$, $\{|e_j^\gamma\rangle\}_{j=1}^2$ and $\{|e_j^\delta\rangle\}_{j=1}^2$ be orthonormal bases for these Hilbert spaces. With these bases one can construct orthonormal bases for all tensor product Hilbert spaces. For instance, a basis for $\mathcal{H}^\alpha \otimes \mathcal{H}^\beta$ associated with $\alpha\beta$ is given by $\{|e_j^\alpha\rangle \otimes |e_k^\beta\rangle\}_{j,k=1}^2$. Choose $|\Psi^\omega\rangle$ to be

$$|\Psi^\omega\rangle = \sqrt{\tfrac{2}{3}} |\psi_1^{\alpha\beta}\rangle \otimes |\phi_1^{\gamma\delta}\rangle + \sqrt{\tfrac{1}{3}} |\psi_2^{\alpha\beta}\rangle \otimes |\phi_2^{\gamma\delta}\rangle \tag{6.28}$$

with $|\psi_1^{\alpha\beta}\rangle = \tfrac{1}{10}\sqrt{10}\,(2, 2, -1, 1)$ and $|\psi_2^{\alpha\beta}\rangle = \tfrac{1}{5}\sqrt{5}\,(-1, 0, 0, 2)$ with respect to the basis $\{|e_1^\alpha\rangle \otimes |e_1^\beta\rangle, |e_1^\alpha\rangle \otimes |e_2^\beta\rangle, |e_2^\alpha\rangle \otimes |e_1^\beta\rangle, |e_2^\alpha\rangle \otimes |e_2^\beta\rangle\}$ and with $|\phi_1^{\gamma\delta}\rangle = \tfrac{1}{5}\sqrt{5}\,(2, 0, 0, -1)$ and $|\phi_2^{\gamma\delta}\rangle = \tfrac{1}{5}\sqrt{5}\,(1, 0, 0, 2)$ with respect to the basis $\{|e_1^\gamma\rangle \otimes |e_1^\delta\rangle, |e_1^\gamma\rangle \otimes |e_2^\delta\rangle, |e_2^\gamma\rangle \otimes |e_1^\delta\rangle, |e_2^\gamma\rangle \otimes |e_2^\delta\rangle\}$.

In order to evaluate the inequality (6.27) one has to determine the eigenprojections of the states of $\alpha\beta$, $\gamma\delta$, α and δ. The state $|\Psi^\omega\rangle$ is chosen such that (6.27) is violated for the eigenprojections corresponding to the largest eigenvalue of all the states. The states of $\alpha\beta$ and $\gamma\delta$ are

$$
\left.
\begin{aligned}
W^{\alpha\beta} &= \tfrac{2}{3} |\psi_1^{\alpha\beta}\rangle\langle\psi_1^{\alpha\beta}| + \tfrac{1}{3} |\psi_2^{\alpha\beta}\rangle\langle\psi_2^{\alpha\beta}|, \\
W^{\gamma\delta} &= \tfrac{2}{3} |\phi_1^{\gamma\delta}\rangle\langle\phi_1^{\gamma\delta}| + \tfrac{1}{3} |\phi_2^{\gamma\delta}\rangle\langle\phi_2^{\gamma\delta}|.
\end{aligned}
\right\}
\tag{6.29}
$$

Since $\langle\psi_1^{\alpha\beta}|\psi_2^{\alpha\beta}\rangle = 0$ and $\langle\phi_1^{\gamma\delta}|\phi_2^{\gamma\delta}\rangle = 0$, the above decompositions are spectral resolutions. Their largest eigenvalue is in both cases $\tfrac{2}{3}$, the corresponding eigenprojections are, respectively, $P_1^{\alpha\beta} = |\psi_1^{\alpha\beta}\rangle\langle\psi_1^{\alpha\beta}|$ and $P_1^{\gamma\delta} = |\phi_1^{\gamma\delta}\rangle\langle\phi_1^{\gamma\delta}|$.

The reduced states of α and δ can be determined with explicit matrix representations. For instance $W^{\alpha\beta}$ is equal to

$$
W^{\alpha\beta} = \tfrac{1}{15}
\begin{pmatrix}
5 & 4 & -2 & 0 \\
4 & 4 & -2 & 2 \\
-2 & -2 & 1 & -1 \\
0 & 2 & -1 & 5
\end{pmatrix}
\tag{6.30}
$$

with respect to the basis $\{|e_1^\alpha\rangle \otimes |e_1^\beta\rangle, |e_1^\alpha\rangle \otimes |e_2^\beta\rangle, |e_2^\alpha\rangle \otimes |e_1^\beta\rangle, |e_2^\alpha\rangle \otimes |e_2^\beta\rangle\}$. Partial tracing yields

$$
W^\alpha = \tfrac{1}{5}
\begin{pmatrix}
3 & 0 \\
0 & 2
\end{pmatrix}
\tag{6.31}
$$

with respect to the basis $\{|e_1^\alpha\rangle, |e_2^\alpha\rangle\}$. The eigenprojection corresponding to the largest eigenvalue of this state of α is given by $P_1^\alpha = |e_1^\alpha\rangle\langle e_1^\alpha|$. Analogously one can calculate that the eigenprojection corresponding to the largest eigenvalue of W^δ is $P_1^\delta = |e_1^\delta\rangle\langle e_1^\delta|$.

One can now calculate all terms in (6.27) for $a = b = c = 1$. One obtains $p(P_1^{\alpha\beta}) = \tfrac{2}{3}$, $p(P_1^\alpha, P_1^{\gamma\delta}) = \tfrac{8}{15}$, $p(P_1^{\alpha\beta}, P_1^\delta) = \tfrac{8}{15}$ and $p(P_1^\alpha, P_1^\delta) = \tfrac{1}{3}$ such that the left-hand side of (6.27) becomes $-\tfrac{1}{15}$. This inequality is thus violated.

6.4 The atomic modal interpretation

For the atomic modal interpretation it is possible to give general joint property ascriptions. The joint property ascriptions to arbitrary sets of atoms can be determined with (6.3). Take a set of atoms $\{\alpha_j\}_j$. The joint probabilities to these atoms are then

$$
p(P_{a_1}^{\alpha_1}, P_{a_2}^{\alpha_2}, P_{a_3}^{\alpha_3}, \dots) = \text{Tr}^\omega(W^\omega [P_{a_1}^{\alpha_1} \otimes P_{a_2}^{\alpha_2} \otimes P_{a_3}^{\alpha_3} \otimes \cdots])
\tag{6.32}
$$

with ω equal to the composite $\alpha_1\alpha_2\alpha_3\cdots$.

Consider then the joint property ascription to the atoms $\{\alpha_j\}_j$ plus a molecular system β consisting of the atoms $\{\alpha_k, \dots, \alpha_m\}$. The core property ascription to this molecule is according to (6.3) given by $[P_{b_k \cdots b_m}^\beta] = 1$ if and only if the core property ascriptions to its atoms are, respectively, $[P^{\alpha_k}_{b_k}] = 1$, ..., $[P^{\alpha_m}_{b_m}] = 1$. Hence,

$$p(P^{\beta}_{b_k \cdots b_m}, P^{\alpha_1}_{a_1}, P^{\alpha_2}_{a_2}, \ldots) = \delta_{b_k a_k} \cdots \delta_{b_m a_m} \operatorname{Tr}^{\omega}(W^{\omega} [P^{\alpha_1}_{a_1} \otimes P^{\alpha_2}_{a_2} \otimes \cdots]). \qquad (6.33)$$

Using that $P^{\beta}_{b_k \cdots b_m} P^{\alpha_k}_{a_k} \otimes \cdots \otimes P^{\alpha_m}_{a_m}$ is equal to $\delta_{b_k a_k} \cdots \delta_{b_m a_m} P^{\alpha_k}_{a_k} \otimes \cdots \otimes P^{\alpha_m}_{a_m}$, the above result can be rewritten as

$$p(P^{\beta}_{b_k \cdots b_m}, P^{\alpha_1}_{a_1}, P^{\alpha_2}_{a_2}, \ldots) = \operatorname{Tr}^{\omega}(W^{\omega} [P^{\beta}_{b_k \cdots b_m} \otimes \mathbb{I}^{\omega/\beta}][P^{\alpha_1}_{a_1} \otimes P^{\alpha_2}_{a_2} \otimes \cdots]). \qquad (6.34)$$

For a second molecule $\gamma = \alpha_q \cdots \alpha_s$ one can equally derive that

$$p(P^{\beta}_{b_k \cdots b_m}, P^{\gamma}_{c_q \cdots c_s}, P^{\alpha_1}_{a_1}, P^{\alpha_2}_{a_2}, \ldots) =$$
$$\operatorname{Tr}^{\omega}(W^{\omega} [P^{\beta}_{b_k \cdots b_m} \otimes \mathbb{I}^{\omega/\beta}][P^{\gamma}_{c_q \cdots c_s} \otimes \mathbb{I}^{\omega/\gamma}][P^{\alpha_1}_{a_1} \otimes P^{\alpha_2}_{a_2} \otimes \cdots]). \qquad (6.35)$$

By repeating this procedure over and over again, one can construct joint probabilities for any set of atoms and molecules.

7

Discontinuities, instabilities and other bad behaviour

In this chapter I begin the discussion of the question of how the core properties of systems evolve in time. This question breaks up in two separate subquestions. Firstly, there is the question of how *the set* of all possible core properties of a system evolves. Secondly, one has the question of how, given this dynamics of the set of possible core properties, *the actually possessed* core property of that system evolves. Here, I consider only the first question of the dynamics of the whole set of core properties. In the next chapter I focus on the evolution of the actually possessed core property.

The modal interpretations that I consider ascribe core properties to systems by means of the spectral resolutions of states. The dynamics of the set of core properties is thus determined by the dynamical behaviour of these spectral resolutions. Study of the dynamics of the spectral resolutions now reveals a number of problematic features of the modal property ascription. Firstly, the set of the eigenprojections of a state $W^\alpha(t)$ can evolve discontinuously. Secondly, this set can evolve highly unstably. Finally, the eigenprojections of a state can evolve rather deviantly when compared with the evolution of the state itself. As a result of all this, the evolution of the set of core properties of a system can be discontinuous, unstable and deviant as well.

7.1 Discontinuities

Up to now I have only considered the property ascription to systems at single instants. It is now time to set things in motion.

Consider two interacting spin $\frac{1}{2}$-particles σ and τ and assume that at $t = 0$ they are both in a pure spin up state in the $\vec{z}$ direction. So, let σ have a state $|u_{\vec{z}}^\sigma\rangle$ and let τ have a state $|u_{\vec{z}}^\tau\rangle$. Let the composite $\sigma\tau$ evolve freely and let its Hamiltonian be $H^{\sigma\tau} = -\mathrm{i}\,|u_{\vec{z}}^\sigma\rangle\langle d_{\vec{z}}^\sigma| \otimes |u_{\vec{z}}^\tau\rangle\langle d_{\vec{z}}^\tau| + \mathrm{i}\,|d_{\vec{z}}^\sigma\rangle\langle u_{\vec{z}}^\sigma| \otimes |d_{\vec{z}}^\tau\rangle\langle u_{\vec{z}}^\tau|$. Then the

Schrödinger equation yields that the evolution of the state of $\sigma\tau$ is given by

$$|\Psi^{\sigma\tau}(t)\rangle = \cos t \, |u^\sigma_{\hat{z}}\rangle \otimes |u^\tau_{\hat{z}}\rangle + \sin t \, |d^\sigma_{\hat{z}}\rangle \otimes |d^\tau_{\hat{z}}\rangle. \tag{7.1}$$

The state of one of the particles, say σ, thus evolves as

$$W^\sigma(t) = \cos^2 t \, |u^\sigma_{\hat{z}}\rangle\langle u^\sigma_{\hat{z}}| + \sin^2 t \, |d^\sigma_{\hat{z}}\rangle\langle d^\sigma_{\hat{z}}|. \tag{7.2}$$

In the bi, spectral and atomic modal interpretations the core property ascription to this particle σ is given by

$$\left\{\begin{array}{ll} [|u^\sigma_{\hat{z}}\rangle\langle u^\sigma_{\hat{z}}|] = 1 & \text{with probability } \cos^2 t \\ [|d^\sigma_{\hat{z}}\rangle\langle d^\sigma_{\hat{z}}|] = 1 & \text{with probability } \sin^2 t \end{array} \quad \text{at} \quad t \neq \pi/4 \ (\text{mod } \pi/2), \right. \\ \left. [\mathbb{I}^\sigma] \quad\quad = 1 \quad \text{with probability } 1 \quad\quad\quad \text{at} \quad t = \pi/4 \ (\text{mod } \pi/2). \right\} \tag{7.3}$$

So, the core properties evolve as follows: at $t \neq \pi/4 \ (\text{mod } \pi/2)$ the set of core properties is given by $\{C^\sigma_j\}_j = \{|u^\sigma_{\hat{z}}\rangle\langle u^\sigma_{\hat{z}}|, |d^\sigma_{\hat{z}}\rangle\langle d^\sigma_{\hat{z}}|\}$ and at $t = \pi/4 \ (\text{mod } \pi/2)$ it is given by $\{\mathbb{I}^\sigma\}$.

This evolution raises two questions. Firstly, how does the *actually* possessed core property of σ evolve? Consider, for instance, the time interval $[0, \pi/2]$. At the beginning at $t = 0$, the actual spin of σ is with certainty up in the $\hat{z}$ direction. That is, $[|u^\sigma_{\hat{z}}\rangle\langle u^\sigma_{\hat{z}}|] = 1$ with probability 1. Then the probability that the spin is actually down, that is, $[|d^\sigma_{\hat{z}}\rangle\langle d^\sigma_{\hat{z}}|] = 1$, increases. And at the end of the interval at $t = \pi/2$, the spin is down with certainty. But how does the actually possessed spin evolve in this interval? Does it oscillate randomly between up and down, before it ends being down with certainty? Or does the spin flip only once from up to down? This question of the evolution of the actually possessed core property is taken up in the next chapter.

A second question is what happens at $t = \pi/4$. At this instant the state $W^\sigma(t)$ passes through a degeneracy in its spectral resolution: at $t \neq \pi/4$ the state has two eigenvalues and thus two eigenprojections $|u^\sigma_{\hat{z}}\rangle\langle u^\sigma_{\hat{z}}|$ and $|d^\sigma_{\hat{z}}\rangle\langle d^\sigma_{\hat{z}}|$, and at $t = \pi/4$ the state has only one degenerate eigenvalue $\frac{1}{2}$ and one eigenprojection $\mathbb{I}^\sigma$. The core property ascription to σ consequently exhibits a discontinuity: the set of core projections changes at $t = \pi/4$ from the set $\{|u^\sigma_{\hat{z}}\rangle\langle u^\sigma_{\hat{z}}|, |d^\sigma_{\hat{z}}\rangle\langle d^\sigma_{\hat{z}}|\}$ to the set $\{\mathbb{I}^\sigma\}$, and back again. The question now is whether this discontinuity reveals a real discontinuity in the properties of σ or whether it proves that modal interpretations are not properly interpreting states with degenerate spectral resolutions.

This possibility that modal interpretations do not accurately interpret degenerate states becomes more pressing if one considers the two spin $\frac{1}{2}$-particles in the Einstein–Podolsky–Rosen–Bohm experiment. In this experiment the two particles are emitted in the singlet state $|\Psi^{\sigma\tau}(t)\rangle = \frac{1}{2}\sqrt{2}(|u_{\frac{1}{2}}^{\sigma}\rangle \otimes |d_{\frac{1}{2}}^{\tau}\rangle - |d_{\frac{1}{2}}^{\sigma}\rangle \otimes |u_{\frac{1}{2}}^{\tau}\rangle)$. Particle σ thus has the permanently degenerate state

$$W^{\sigma}(t) = \frac{1}{2}|u_{\frac{1}{2}}^{\sigma}\rangle\langle u_{\frac{1}{2}}^{\sigma}| + \frac{1}{2}|d_{\frac{1}{2}}^{\sigma}\rangle\langle d_{\frac{1}{2}}^{\sigma}| = \frac{1}{2}\mathbb{I}^{\sigma}. \tag{7.4}$$

The core property ascription to this particle is therefore the tautology that $[\mathbb{I}^{\sigma}] = 1$ with probability 1. So, if modal interpretations indeed properly interpret degenerate states, their property ascription can be quite trivial.

These observations led to two attempts to improve on the modal interpretation of degenerate states. Before discussing these attempts I give two definitions: let's call degeneracies of states that are restricted to single instants, like the ones in (7.2), *passing degeneracies*,[70] and let's call degeneracies that occur during whole intervals, like in (7.4), *permanent degeneracies*.

The first attempt was by Elby and Bub (1994) who tried to devise a modal interpretation which always singles out one-dimensional core projections. They considered a composite of three disjoint systems α, β and γ with a pure state $|\Psi^{\alpha\beta\gamma}\rangle$ and wrote down the so-called *triorthogonal decomposition*,

$$|\Psi^{\alpha\beta\gamma}\rangle = \sum_{j} c_j |c_j^{\alpha}\rangle \otimes |c_j^{\beta}\rangle \otimes |c_j^{\gamma}\rangle \tag{7.5}$$

(the sets $\{|c_j^{\alpha}\rangle\}_j$, $\{|c_j^{\beta}\rangle\}_j$ and $\{|c_j^{\gamma}\rangle\}_j$ are sets of pair-wise orthogonal vectors). Elby and Bub then proposed that the core property ascription to, for instance, α is $[|c_a^{\alpha}\rangle\langle c_a^{\alpha}|] = 1$ with probability $|c_a|^2$. This property ascription makes sense because they proved that if a triorthogonal decomposition of $|\Psi^{\alpha\beta\gamma}\rangle$ exists, then the projections $\{|c_j^{\alpha}\rangle\langle c_j^{\alpha}|\}_j$ are uniquely fixed by this decomposition (see also Bub (1997, Sect. 5.5) for the proof).[71]

If this 'tri modal interpretation' had been defensible, one would indeed have had a procedure for improving on the bi, spectral and atomic modal interpretations of degenerate states. To see this, consider a spin $\frac{1}{2}$-particle σ which is part of a composite $\sigma\tau\tau'$ with the pure state $|\Psi^{\sigma\tau\tau'}(t)\rangle = c_1 |u_{\frac{1}{2}}^{\sigma}\rangle \otimes |u_{\frac{1}{2}}^{\tau}\rangle \otimes |u_{\frac{1}{2}}^{\tau'}\rangle + c_2 |d_{\frac{1}{2}}^{\sigma}\rangle \otimes |d_{\frac{1}{2}}^{\tau}\rangle \otimes |d_{\frac{1}{2}}^{\tau'}\rangle$. The state of σ is then $W^{\sigma}(t) = |c_1|^2 |u_{\frac{1}{2}}^{\sigma}\rangle\langle u_{\frac{1}{2}}^{\sigma}| + |c_2|^2 |d_{\frac{1}{2}}^{\sigma}\rangle\langle d_{\frac{1}{2}}^{\sigma}|$. Thus, if $W^{\sigma}(t)$ is non-degenerate, the bi, spectral and atomic modal interpretations agree with the tri modal interpretation that the core properties of σ are given by $\{|u_{\frac{1}{2}}^{\sigma}\rangle\langle u_{\frac{1}{2}}^{\sigma}|, |d_{\frac{1}{2}}^{\sigma}\rangle\langle d_{\frac{1}{2}}^{\sigma}|\}$. But if $W^{\sigma}(t)$ is degenerate, the bi, spectral and atomic modal interpretations ascribe the

---

[70] More precisely, a degeneracy of a state $W^{\alpha}(t)$ at t_0 is called passing if there exists a punctuated interval $(t_0 - \varepsilon, t_0 + \varepsilon)/\{t_0\}$ in which $W^{\alpha}(t)$ does not have this degeneracy.

[71] The triorthogonal decomposition thus outclasses the biorthogonal decomposition: the biorthogonal decomposition only uniquely singles out the one-dimensional projections $\{|c_j^{\alpha}\rangle\langle c_j^{\alpha}|\}_j$ if the values $\{|c_j|^2\}_j$ are non-degenerate (see page 51) whereas the triorthogonal decomposition always singles out these projections, even if the values $\{|c_j|^2\}_j$ are degenerate.

core property $\mathbb{I}^\sigma$, whereas the tri modal interpretation still ascribes the core properties $\{|u_{\hat{z}}^\sigma\rangle\langle u_{\hat{z}}^\sigma|, |d_{\hat{z}}^\sigma\rangle\langle d_{\hat{z}}^\sigma|\}$. Unfortunately, the tri modal interpretation is not defensible. Clifton (1994) proved that not all possible states $|\Psi^{\alpha\beta\gamma}\rangle$ allow a triorthogonal decomposition (7.5). Hence, the property ascription by Elby and Bub is applicable only in special cases.

The second attempt was the one by Bacciagaluppi, Donald and Vermaas (1995). The aim was to remove the discontinuities in the property ascription of modal interpretations, by basing the core property ascription to a system at time t_0 on the state of the system as it evolves in an interval around t_0.

Consider again the degeneracy at $t_0 = \pi/4$ in the state (7.2) of the spin particle σ. In a small punctuated interval $(\pi/4 - \varepsilon, \pi/4 + \varepsilon)/\{\pi/4\}$ around this degeneracy, the spectral resolutions of $W^\sigma(t)$ yield for every instant t the eigenvalues $\cos^2 t$ and $\sin^2 t$ and the corresponding eigenprojections $|u_{\hat{z}}^\sigma\rangle\langle u_{\hat{z}}^\sigma|$ and $|d_{\hat{z}}^\sigma\rangle\langle d_{\hat{z}}^\sigma|$, respectively. These individual spectral resolutions of $W^\sigma(t)$ can be joined together to give what I propose to call a continuous *dynamical decomposition*. Such a dynamical decomposition has the general form $W^\alpha(t) = \sum_q f_q^\alpha(t)\, T_q^\alpha(t)$, where $\{f_q^\alpha(t)\}_q$ are continuous eigenvalue functions and where $\{T_q^\alpha(t)\}_q$ are eigenprojection functions which are continuous with regard to the trace norm $\|.\|_1$.[72] In the case of our spin particle, these functions are given by $f_1^\sigma(t) = \cos^2 t$, $f_2^\sigma(t) = \sin^2 t$, $T_1^\sigma(t) = |u_{\hat{z}}^\sigma\rangle\langle u_{\hat{z}}^\sigma|$ and $T_2^\sigma(t) = |d_{\hat{z}}^\sigma\rangle\langle d_{\hat{z}}^\sigma|$. This dynamical decomposition of $W^\sigma(t)$ can now be extended to the instant $t_0 = \pi/4$ of the degeneracy: the left-hand limits $\lim_{t\uparrow\pi/4} f_q^\sigma(t)$ and the right-hand limits $\lim_{t\downarrow\pi/4} f_q^\sigma(t)$ both yield the values $f_1^\sigma(\pi/4) = \frac{1}{2}$ and $f_2^\sigma(\pi/4) = \frac{1}{2}$, and the limits $\lim_{t\uparrow\pi/4} T_q^\sigma(t)$ and $\lim_{t\downarrow\pi/4} T_q^\sigma(t)$ both yield the projections $T_1^\sigma(\pi/4) = |u_{\hat{z}}^\sigma\rangle\langle u_{\hat{z}}^\sigma|$ and $T_2^\sigma(\pi/4) = |d_{\hat{z}}^\sigma\rangle\langle d_{\hat{z}}^\sigma|$. So, by continuing the eigenvalue and the eigenprojection functions, one can extend the dynamical decomposition of $W^\sigma(t)$ defined on $(\pi/4-\varepsilon, \pi/4+\varepsilon)/\{\pi/4\}$ to the whole interval $(\pi/4-\varepsilon, \pi/4+\varepsilon)$. This extended dynamical decomposition yields a resolution of the degenerate state $W^\sigma(\pi/4)$ which decomposes this state in more terms than the spectral resolution of $W^\sigma(\pi/4)$ does. The spectral resolution $W^\sigma(\pi/4) = \sum_j w_j^\sigma(\pi/4) P_j^\sigma(\pi/4)$ yields one eigenprojection $P_1^\sigma(\pi/4) = \mathbb{I}^\sigma$, whereas the extended dynamical decomposition $W^\sigma(\pi/4) = \sum_q f_q^\sigma(\pi/4)\, T_q^\sigma(\pi/4)$ yields two projections $\{T_1^\sigma(\pi/4), T_2^\sigma(\pi/4)\} = \{|u_{\hat{z}}^\sigma\rangle\langle u_{\hat{z}}^\sigma|, |d_{\hat{z}}^\sigma\rangle\langle d_{\hat{z}}^\sigma|\}$.

Let's call the functions $\{T_q^\alpha(t)\}_q$ the **continuous trajectories of eigenprojections** of $W^\alpha(t)$. These continuous trajectories of projections evolve (by definition) without discontinuities through the degeneracy at $t_0 = \pi/4$. The idea by Bacciagaluppi, Donald and Vermaas was now to define a modal

---

[72] See footnote 7 for the definition of the trace norm.

A continuous trajectory of eigenprojections

A projection-valued function $T_q^\alpha(t)$ is a continuous trajectory of eigenprojections of an evolving state $W^\alpha(t)$ if $T_q^\alpha(t)$ is an eigenprojection of $W^\alpha(t)$, for every t, and if $T_q^\alpha(t)$ is continuous with respect to the trace norm $\|.\|_1$.

---

interpretation, called the *extended modal interpretation*, in which the core properties of a system α at t_0 are not given by the eigenprojections $\{P_j^\alpha(t_0)\}_j$ of the state of α, but by the projections $\{T_q^\alpha(t_0)\}_q$ which lie on the continuous trajectories of eigenprojections of the state of α.

If this idea had worked in general, one would again have had a method of improving on the bi, spectral and atomic modal interpretations. Assume for a moment that every evolving state $W^\alpha(t)$ allows at all times t_0 a unique dynamical decomposition $W^\alpha(t) = \sum_q f_q^\alpha(t) \, T_q^\alpha(t)$. Then the extended modal interpretation and the bi, spectral and atomic modal interpretations all agree that if the state of an atomic system is non-degenerate in a small interval around t_0, the core properties of that system at t_0 are given by the eigenprojections of that state at t_0 (if there are no degeneracies, the projections $\{T_q^\alpha(t_0)\}_q$ are equal to the eigenprojections $\{P_j^\alpha(t_0)\}_j$). However, if a state passes through a degeneracy at t_0 (that is, $W^\alpha(t_0)$ is degenerate at t_0 but non-degenerate at a small punctuated interval around t_0), the extended modal interpretation and the bi, spectral and atomic modal interpretations differ because then the projections $\{T_q^\alpha(t_0)\}_q$ are not equal to the eigenprojections $\{P_j^\alpha(t_0)\}_j$. Moreover, at a passing degeneracy the core property ascription by the bi, spectral and atomic modal interpretations evolves discontinuously, whereas the sets of core properties ascribed by the extended modal interpretation evolves by definition continuously. The extended modal interpretation does not, however, improve on the property ascription to systems with permanently degenerate states. Consider, for instance, the state $W^\sigma(t)$ given in (7.4). Because this state is permanently degenerate, it follows that the dynamical decomposition in a small punctuated interval around any instant t_0 is given by $W^\sigma(t) = f_1^\sigma(t) \, T_1^\sigma(t)$ with $f_1^\sigma(t) = \frac{1}{2}$ and $T_1^\sigma(t) = \mathbb{I}^\sigma$. If this dynamical decomposition is extended to t_0, one obtains that the only projection lying at t_0 on a continuous trajectory of eigenprojections of $W^\sigma(t)$ is given by $T_1^\sigma(t_0) = \mathbb{I}^\sigma$. For permanently degenerate states, the core properties ascribed by the extended modal interpretation are thus equal to the core properties ascribed by the bi, spectral and atomic modal interpretations. Hence, the extended modal interpretation is only an improvement in the sense that discontinuities in the core property ascription related to passing degeneracies are removed.[73]

Unfortunately, there are indications that the extended modal interpretation also cannot be applied in general. That is, it need not be the case that in quantum mechanics states have continuous trajectories of eigenprojections. The extended modal interpretation is thus threatened by the same verdict as the tri modal interpretation of Elby and Bub: that is, it is useless because it is applicable only in special cases. But irrespectively of this possible verdict, the results of Bacciagaluppi, Donald and Vermaas proved their worth since they revealed a number of things about the evolution of the core properties ascribed by the bi, spectral and atomic modal interpretations. It was shown that, given certain assumptions, evolving states can be decomposed in terms of continuously evolving trajectories of eigenprojections. And these trajectories, when they exist, sometimes evolve in undesirable ways. Notably, small changes in the state can induce large changes in the set of eigenprojections and the evolution of the eigenprojections of a state can exhibit deviant behaviour when compared with the evolution of the state itself. The evolution of the set of core properties ascribed by the bi, spectral and atomic modal interpretations can thus be not only discontinuous, but also unstable and deviant.

In this chapter I sketch the results of Bacciagaluppi, Donald and Vermaas (1995). A rigorous review would lead to much mathematical *force of arms.* Therefore, in order to concentrate on the results themselves, I choose to sometimes give up on full mathematical precision. For the precise mathematics I refer the reader to the original paper;[74] for a concise discussion the reader may consult Donald (1998). In Sections 7.2 and 7.3 I consider the dynamics of the spectral resolutions of states by discussing the question of whether the eigenprojections of evolving states can be joined to continuous trajectories. In Section 7.4 I discuss the consequences of the dynamics of spectral resolutions, especially with regard to the instability of the modal property ascription. Readers who are allergic to continuity proofs are advised to skip the next two sections and go straight to Section 7.4 on page 127.

---

[73] Guido Bacciagaluppi (private communication, 1998) noted that the extended modal interpretation does not comply with my assumption of Instantaneous Autonomy. Compare, for instance, two spin $\frac{1}{2}$-particles σ, one with the state (7.2) and the other with the state (7.4). In the first case the extended modal interpretation ascribes the core properties $\{|u_{\frac{\sigma}{2}}\rangle\langle u_{\frac{\sigma}{2}}|, |d_{\frac{\sigma}{2}}\rangle\langle d_{\frac{\sigma}{2}}|\}$ at $t = \pi/4$ and in the second case it ascribes the core property $\mathbb{I}^{\sigma}$ at $t = \pi/4$, although in both cases the state $W^{\sigma}(\pi/4)$ of these particles is given by $\frac{1}{2}\mathbb{I}^{\sigma}$. One can save Instantaneous Autonomy by reformulating it as: *If two systems have equal states during a small time interval around t_0, then the instantaneous property ascriptions to those systems at t_0 are also equal.*

[74] See also Bacciagaluppi (2000).

7.2 Continuous trajectories of eigenprojections

A necessary condition for applying the extended modal interpretation out-lined above is that every evolving state $W^\alpha(t)$ in quantum mechanics allows at every instant t_0 a dynamical decomposition $W^\alpha(t_0) = \sum_q f_q^\alpha(t_0) \, T_q^\alpha(t_0)$ in terms of projections $\{T_q^\alpha(t_0)\}_q$ which lie on continuous trajectories $\{T_q^\alpha(t)\}_q$ of eigenprojections of $W^\alpha(t)$. Moreover, this dynamical decomposition should be (implicitly) definable (see page 42) from the state evolution $W^\alpha(t)$. For, if there does not always exist such a decomposition, the core property ascrip-tion of the extended modal interpretation is on occasions inapplicable. And, if the dynamical decomposition is not definable from the state evolution $W^\alpha(t)$, then this evolution does not provide the means to fix a dynamical decomposition.

If one compares the dynamical decomposition with the spectral resolu-tion $W^\alpha(t_0) = \sum_j w_j(t_0) \, P_j^\alpha(t_0)$, it follows that every eigenprojection $P_k^\alpha(t_0)$ that corresponds to a non-zero eigenvalue $w_k(t_0)$ is given by the sum $P_k^\alpha(t_0) = \sum_{\{q \mid f_q^\alpha(t_0) = w_k(t_0)\}} T_q^\alpha(t_0)$. Hence, if there exists a dynamical decompo-sition which is implicitly definable from $W^\alpha(t_0)$, then every eigenprojection $P_k^\alpha(t_0)$ with $w_k(t_0) \neq 0$ is a sum of projections $\{T_q^\alpha(t_0)\}_q$ on eigenprojection trajectories $\{T_q^\alpha(t)\}_q$ implicitly definable from $W^\alpha(t)$. Conversely, if every eigenprojection $P_k^\alpha(t_0)$ with $w_k(t_0) \neq 0$ is a sum of projections $\{T_q^\alpha(t_0)\}_q$ on trajectories $\{T_q^\alpha(t)\}_q$, implicitly definable from $W^\alpha(t)$, then one can construct a dynamical decomposition which is implicitly definable from $W^\alpha(t_0)$: just replace all the eigenprojections $P_k^\alpha(t_0)$ in the spectral resolution of $W^\alpha(t_0)$ by the sums over the projections $\{T_q^\alpha(t_0)\}_q$. So, the condition that there exists a dynamical decomposition of every state at every instant is equivalent to the following condition:

Dynamical Decomposition

For every state evolution $W^\alpha(t)$, for every instant t_0 and for every eigenprojection $P_k^\alpha(t_0)$ of $W^\alpha(t_0)$ with $w_k(t_0) \neq 0$, there exists a set of continuous trajectories $\{T_q^\alpha(t)\}_q$ of eigenprojections implicitly definable from $W^\alpha(t)$ on a small interval I around t_0 for which holds that $P_k^\alpha(t_0) = \sum_q T_q^\alpha(t_0)$.

One can show that the state of a freely evolving system α satisfies this condition. Take any eigenprojection $P_k^\alpha(t_0)$ of the state $W^\alpha(t_0)$ with $w_k(t_0) \neq 0$ and consider a small interval I around t_0. If $W^\alpha(t)$ evolves freely on this interval, its evolution is by the Schrödinger equation (see Section 2.1) equal to

$$W^\alpha(t) = U^\alpha(t, t_0) \, W^\alpha(t_0) \, U^\alpha(t_0, t), \tag{7.6}$$

where $U^\alpha(x, y)$ is given by $\exp([(x - y)/i\hbar] H^\alpha)$. Substitution of the spectral resolution of $W^\alpha(t_0)$ yields that $W^\alpha(t)$ can be decomposed as

$$W^\alpha(t) = \sum_j w_j(t_0)\, U^\alpha(t, t_0)\, P_j^\alpha(t_0)\, U^\alpha(t_0, t). \qquad (7.7)$$

This decomposition is a spectral resolution for all $t \in I$ because the values $\{w_j(t_0)\}_j$ are distinct and the projections $\{U^\alpha(t, t_0)\, P_j^\alpha(t_0)\, U^\alpha(t_0, t)\}_j$ are pairwise orthogonal. Hence, the eigenvalue $w_k(t_0)$ of $W^\alpha(t_0)$ is also an eigenvalue of $W^\alpha(t)$ for every instant $t \in I$ around t_0. The eigenprojection $P_k^\alpha(t)$ that corresponds to this eigenvalue $w_k(t_0)$ of $W^\alpha(t)$ is given by $U^\alpha(t, t_0)\, P_k^\alpha(t_0)\, U^\alpha(t_0, t)$ for every $t \neq t_0$. These eigenprojections $P_k^\alpha(t)$, together with $P_k^\alpha(t_0)$, can be joined on I to a trajectory

$$T_k^\alpha(t) = U^\alpha(t, t_0)\, P_k^\alpha(t_0)\, U^\alpha(t_0, t), \qquad (7.8)$$

which is continuous with regard to the trace norm.[75] This proves that $P_k^\alpha(t_0)$ is given by the 'sum' $P_k^\alpha(t_0) = T_k^\alpha(t_0)$ with $T_k^\alpha(t_0)$ lying on a continuous trajectory $T_k^\alpha(t)$ of eigenprojections of the state $W^\alpha(t)$ as it evolves around t_0.

Moreover, since each eigenprojection $U^\alpha(t, t_0)\, P_k^\alpha(t_0)\, U^\alpha(t_0, t)$ is implicitly definable from the state $W^\alpha(t)$ for all $t \in I$, the trajectory $T_k^\alpha(t)$ is implicitly definable from the state evolution of $W^\alpha(t)$ on I. Hence, any eigenprojection $P_k^\alpha(t_0)$ of a freely evolving state with $w_k(t_0) \neq 0$ satisfies the Dynamical Decomposition condition.

The question is now whether the same positive result holds for the states of interacting systems. In this and the next section, building on the results by Bacciagaluppi, Donald and Vermaas (1995), I give an answer to this question. I do so by means of four propositions about the dynamics of states (see the MATHEMATICS for proofs or references to proofs):

The first proposition yields that in quantum mechanics states evolve continuously:

Proposition 7.1
If a system α is part of a composite ω which, as a whole, evolves freely by means of the Schrödinger equation, then the state $W^\alpha(t)$ of that system evolves continuously with regard to the trace norm.

In quantum mechanics states of freely evolving composites indeed evolve by means of the Schrödinger equation. So, if one assumes that, say, the state of the whole universe evolves freely, then the states of all the systems in the universe evolve continuously with regard to the trace norm.

---

[75] Continuity of this trajectory $T_k^\alpha(t)$ can be proved with Proposition 7.3, which I give below.

The second proposition yields the continuity of the evolution of the eigenvalues of a state. Consider a state evolution $W^\alpha(t)$ and let

$$W^\alpha(t) = \sum_{i=1}^{N} r_i^\alpha(t) |r_i^\alpha(t)\rangle \langle r_i^\alpha(t)| \tag{7.9}$$

be a decomposition of this state at all times t in terms of eigenvalues $\{r_i^\alpha(t)\}_i$ and one-dimensional eigenprojections $\{|r_i^\alpha(t)\rangle\langle r_i^\alpha(t)|\}_i$ (if $W^\alpha(t)$ is degenerate at t, some of the $r_i^\alpha(t)$s are equal). Let the eigenvalues be ordered like $r_1^\alpha(t) \geq r_2^\alpha(t) \geq \ldots \geq r_N^\alpha(t)$. The number N is the dimension of $\mathscr{H}^\alpha$ and may be finite or infinite. It then holds that:

Proposition 7.2
If a state $W^\alpha(t)$ evolves continuously with regard to the trace norm, then each eigenvalue function $r_i^\alpha(t)$ evolves continuously as well.

The third proposition yields constraints on the evolution of the one-dimensional eigenprojections $\{|r_i^\alpha(t)\rangle\langle r_i^\alpha(t)|\}_i$ in the decomposition (7.9). Take an interval (a, b) with $a < b$ and consider all the one-dimensional eigenprojections $|r_i^\alpha(t)\rangle\langle r_i^\alpha(t)|$ whose eigenvalues $r_i^\alpha(t)$ are within (a, b) for a small time interval around t_0. Let $P_{(a,b)}^\alpha(t)$ be the sum of these projections $\{|r_i^\alpha(t)\rangle\langle r_i^\alpha(t)|\}_i$, so

$$P_{(a,b)}^\alpha(t) = \sum_{r_i^\alpha(t)\in(a,b)} |r_i^\alpha(t)\rangle \langle r_i^\alpha(t)|. \tag{7.10}$$

Then it holds that:

Proposition 7.3
If a state $W^\alpha(t)$ evolves continuously with regard to the trace norm, then each projection-valued function $P_{(a,b)}^\alpha(t)$, where a and b are not eigenvalues of $W^\alpha(t)$, evolves continuously as well with regard to the trace norm.

Let's return to the question of whether the states of interacting systems satisfy the Dynamical Decomposition condition and see what these three propositions yield (I present the fourth proposition in the next section). According to Proposition 7.1, in quantum mechanics all the states of interacting systems evolve continuously with regard to the trace norm. Propositions 7.2 and 7.3 thus always apply. Consider now any eigenprojection $P_k^\alpha(t_0)$ of a state $W^\alpha(t_0)$ that corresponds to a non-zero eigenvalue $w_k(t_0)$. One can distinguish five different cases. In the first case $P_k^\alpha(t_0)$ corresponds to a non-degenerate eigenvalue and in the other four cases it corresponds to a degenerate eigenvalue. Let's start with the first case.

CASE 1: $P_k^\alpha(t_0)$ corresponds to a non-degenerate eigenvalue $w_k(t_0) \neq 0$.

In this case $P_k^\alpha(t_0)$ is a one-dimensional projection and the eigenvalue $w_k(t_0)$ is different to all other eigenvalues of $W^\alpha(t_0)$. In terms of the decomposition (7.9) of $W^\alpha(t)$, this means that there exists an eigenprojection function $|r_{i'}^\alpha(t)\rangle\langle r_{i'}^\alpha(t)|$ such that $P_k^\alpha(t_0) = |r_{i'}^\alpha(t_0)\rangle\langle r_{i'}^\alpha(t_0)|$ and such that $w_k(t_0) = r_{i'}^\alpha(t_0)$. Moreover, it holds that $r_{i'-1}^\alpha(t_0) > r_{i'}^\alpha(t_0) > r_{i'+1}^\alpha(t_0)$. By Proposition 7.2 all functions $\{r_i^\alpha(t)\}_i$ evolve continuously. One can thus construct a small interval I around t_0 with $r_{i'-1}^\alpha(t) > r_{i'}^\alpha(t) > r_{i'+1}^\alpha(t)$ for all $t \in I$. And by Proposition 7.3 the eigenprojection function $|r_{i'}^\alpha(t)\rangle\langle r_{i'}^\alpha(t)|$, corresponding to the eigenvalue function $r_{i'}^\alpha(t)$, is continuous on this interval with regard to the trace norm (this continuity follows from Proposition 7.3 if one chooses the interval (a, b) such that it contains only $r_{i'}^\alpha(t)$ on I). Hence, if one defines

$$T_k^\alpha(t) := |r_{i'}^\alpha(t)\rangle\langle r_{i'}^\alpha(t)|, \tag{7.11}$$

one has a continuous trajectory of eigenprojections of $W^\alpha(t)$ on the interval I with $P_k^\alpha(t_0) = T_k^\alpha(t_0)$. Furthermore, since the value $r_{i'}^\alpha(t)$ is a non-degenerate eigenvalue of $W^\alpha(t)$ for all $t \in I$, the corresponding eigenprojection $|r_{i'}^\alpha(t)\rangle\langle r_{i'}^\alpha(t)|$ is implicitly definable from $W^\alpha(t)$ for all $t \in I$. Hence, the trajectory $T_k^\alpha(t)$ is implicitly definable from the state evolution of $W^\alpha(t)$ on I. Any eigenprojection that corresponds to a non-zero and non-degenerate eigenvalue thus satisfies the Dynamical Decomposition condition.

CASE 2: $P_k^\alpha(t_0)$ corresponds to a permanently degenerate eigenvalue $w_k(t_0) \neq 0$.

In this second case $P_k^\alpha(t_0)$ is a K-dimensional projection with $K > 1$. In terms of the decomposition (7.9) of $W^\alpha(t)$, this means that there exist K eigenprojection functions $\{|r_{i'}^\alpha(t)\rangle\langle r_{i'}^\alpha(t)|, |r_{i'+1}^\alpha(t)\rangle\langle r_{i'+1}^\alpha(t)|, \dots, |r_{i'+K-1}^\alpha(t)\rangle\langle r_{i'+K-1}^\alpha(t)|\}$ such that $P_k^\alpha(t_0) = \sum_{i=i'}^{i'+K-1} |r_i^\alpha(t_0)\rangle\langle r_i^\alpha(t_0)|$ and $w_k(t_0) = r_i^\alpha(t_0)$ for all $i \in \{i', \dots, i' + K - 1\}$. Moreover, it holds that $r_{i'-1}^\alpha(t_0) > r_{i'}^\alpha(t_0) = \dots = r_{i'+K-1}^\alpha(t_0) > r_{i'+K}^\alpha(t_0)$. Because $w_k(t_0)$ is permanently degenerate, the functions $\{r_i^\alpha(t)\}_{i=i'}^{i'+K-1}$ remain equal around t_0. By Proposition 7.2 all functions $\{r_i^\alpha(t)\}_i$ evolve continuously and one can thus construct a small interval I around t_0 with $r_{i'-1}^\alpha(t) > r_{i'}^\alpha(t) = \dots = r_{i'+K-1}^\alpha(t) > r_{i'+K}^\alpha(t)$ for all $t \in I$. And by Proposition 7.3 the projection $\sum_{i=i'}^{i'+K-1} |r_i^\alpha(t)\rangle\langle r_i^\alpha(t)|$ is continuous on this interval with regard to the trace norm (choose the interval (a, b) such that it contains only the values $r_{i'}^\alpha(t) = \dots = r_{i'+K-1}^\alpha(t)$ on I). Hence, if one defines

$$T_k^\alpha(t) := \sum_{i=i'}^{i'+K-1} |r_i^\alpha(t)\rangle\langle r_i^\alpha(t)|, \tag{7.12}$$

one obtains a continuous trajectory of eigenprojections of $W^\alpha(t)$ on the interval I with $P_k^\alpha(t_0) = T_k^\alpha(t_0)$. Furthermore, since the value $r_{i'}^\alpha(t) = \ldots = r_{i'+K-1}^\alpha(t)$ is a K-fold degenerate eigenvalue of W^α for all $t \in I$, the corresponding K-dimensional eigenprojection $\sum_{i=i'}^{i'+K-1} |r_i^\alpha(t)\rangle\langle r_i^\alpha(t)|$ is implicitly definable from $W^\alpha(t)$ for all $t \in I$. Hence, the trajectory $T_k^\alpha(t)$ is implicitly definable from the state $W^\alpha(t)$ as it evolves on I. Any eigenprojection that corresponds to a non-zero and permanently degenerate eigenvalue thus satisfies the Dynamical Decomposition condition as well.

CASE 3: $P_k^\alpha(t_0)$ corresponds to an eigenvalue $w_k(t_0) \neq 0$ which passes a degeneracy.

In this case $P_k^\alpha(t_0)$ is again K-dimensional with $K > 1$. One thus has K eigenprojection functions $\{|r_i^\alpha(t)\rangle\langle r_i^\alpha(t)|\}_{i=i'}^{i'+K-1}$ such that $P_k^\alpha(t_0) = \sum_{i=i'}^{i'+K-1} |r_i^\alpha(t_0)\rangle\langle r_i^\alpha(t_0)|$ and such that $w_k(t_0) = r_i^\alpha(t_0)$ for all $i \in \{i', \ldots, i' + K - 1\}$. Now, however, because $w_k(t_0)$ is passing a degeneracy, the functions $\{r_i^\alpha(t)\}_{i=i'}^{i'+K-1}$ are all distinct on an interval around t_0. This means that the Dynamical Decomposition condition can be violated. To see this, assume for simplicity that one is only dealing with two non-zero eigenvalue functions. So

$$W^\alpha(t) = r_1^\alpha(t) |r_1^\alpha(t)\rangle\langle r_1^\alpha(t)| + r_2^\alpha(t) |r_2^\alpha(t)\rangle\langle r_2^\alpha(t)| \tag{7.13}$$

with $r_1^\alpha(t_0) = r_2^\alpha(t_0) = \frac{1}{2}$ and with $r_1^\alpha(t) \neq r_2^\alpha(t)$ for all t in a small punctuated interval $I/\{t_0\}$ around t_0. The degenerate eigenprojection $P_k^\alpha(t_0)$ is thus $|r_1^\alpha(t_0)\rangle\langle r_1^\alpha(t_0)| + |r_2^\alpha(t_0)\rangle\langle r_2^\alpha(t_0)|$. Because the eigenvalues $r_1^\alpha(t)$ and $r_1^\alpha(t)$ are distinct on $I/\{t_0\}$, the eigenprojections $|r_1^\alpha(t)\rangle\langle r_1^\alpha(t)|$ and $|r_2^\alpha(t)\rangle\langle r_2^\alpha(t)|$ evolve by Proposition 7.3 continuously with regard to the trace norm on this punctuated interval $I/\{t_0\}$. And the sum $|r_1^\alpha(t)\rangle\langle r_1^\alpha(t)| + |r_2^\alpha(t)\rangle\langle r_2^\alpha(t)|$ of the eigenprojections evolves by Proposition 7.3 continuously with regard to the trace norm on the whole interval I. But this need not imply that the individual eigenprojections $|r_1^\alpha(t)\rangle\langle r_1^\alpha(t)|$ and $|r_2^\alpha(t)\rangle\langle r_2^\alpha(t)|$ evolve continuously at the instant t_0 of the degeneracy. There are, in fact, three possibilities and for one of them the eigenprojections $\{|r_i^\alpha(t)\rangle\langle r_i^\alpha(t)|\}_i$ do not evolve continuously at t_0.

The *first possibility* is that the eigenprojection function $|r_1^\alpha(t)\rangle\langle r_1^\alpha(t)|$ before t_0 evolves continuously to the function $|r_1^\alpha(t)\rangle\langle r_1^\alpha(t)|$ after t_0 and that $|r_2^\alpha(t)\rangle\langle r_2^\alpha(t)|$ before t_0 evolves continuously to $|r_2^\alpha(t)\rangle\langle r_2^\alpha(t)|$ after t_0. One can then define

$$T_1^\alpha(t) = |r_1^\alpha(t)\rangle\langle r_1^\alpha(t)|, \qquad T_2^\alpha(t) = |r_2^\alpha(t)\rangle\langle r_2^\alpha(t)|, \tag{7.14}$$

and write $P_k^\alpha(t_0)$ as $T_1^\alpha(t_0) + T_2^\alpha(t_0)$. The Dynamical Decomposition condition is thus satisfied. (Since the eigenvalues $r_1^\alpha(t)$ and $r_2^\alpha(t)$ are non-degenerate on $I/\{t_0\}$, the eigenprojections $|r_1^\alpha(t)\rangle\langle r_1^\alpha(t)|$ and $|r_2^\alpha(t)\rangle\langle r_2^\alpha(t)|$ are implicitly definable from $W^\alpha(t)$ for all $t \in I/\{t_0\}$. The trajectories $T_1^\alpha(t)$ and $T_2^\alpha(t)$ are thus implicitly definable from $W^\alpha(t)$ on $I/\{t_0\}$ and since these trajectories on $I/\{t_0\}$ extend uniquely to t_0, $T_1^\alpha(t)$ and $T_2^\alpha(t)$ are also implicitly definable from $W^\alpha(t)$ on the whole interval I.) If this first possibility applies, the passing degeneracy can be understood as that the eigenvalue functions $r_1^\alpha(t)$ and $r_2^\alpha(t)$ touch at the time of the degeneracy. This is illustrated by the passing degeneracy at $t_0 = 0$ in the state evolution

$$W^\sigma(t) = (1 - \tfrac{1}{2}\cos^2 t)\,|u_{\hat{z}}^\sigma\rangle\langle u_{\hat{z}}^\sigma| + \tfrac{1}{2}\cos^2 t\,|d_{\hat{z}}^\sigma\rangle\langle d_{\hat{z}}^\sigma|. \tag{7.15}$$

The *second possibility* is that the eigenprojection function $|r_1^\alpha(t)\rangle\langle r_1^\alpha(t)|$ before t_0 evolves continuously to the function $|r_2^\alpha(t)\rangle\langle r_2^\alpha(t)|$ after t_0 and that $|r_2^\alpha(t)\rangle\langle r_2^\alpha(t)|$ before t_0 evolves continuously to $|r_1^\alpha(t)\rangle\langle r_1^\alpha(t)|$ after t_0. One can then define

$$T_1^\alpha(t) = \begin{cases} |r_1^\alpha(t)\rangle\langle r_1^\alpha(t)| & \text{if } t \le t_0, \\ |r_2^\alpha(t)\rangle\langle r_2^\alpha(t)| & \text{if } t > t_0, \end{cases} \qquad T_2^\alpha(t) = \begin{cases} |r_2^\alpha(t)\rangle\langle r_2^\alpha(t)| & \text{if } t \le t_0, \\ |r_1^\alpha(t)\rangle\langle r_1^\alpha(t)| & \text{if } t > t_0. \end{cases} \tag{7.16}$$

One again has $P_k^\alpha(t_0) = T_1^\alpha(t_0) + T_2^\alpha(t_0)$ and the Dynamical Decomposition condition is satisfied. (The implicit definability of these trajectories follows by an argument similar to the argument given for the first possibility.) If this second possibility applies, the passing degeneracy can be understood as that the eigenvalue functions $r_1^\alpha(t)$ and $r_2^\alpha(t)$ cross, as is illustrated by the degeneracy at $t_0 = \pi/4$ in the state evolution (7.2).

The *third possibility* is, however, that the eigenprojection functions do not evolve continuously through t_0. An illustration of this is

$$W^\sigma(t) = \begin{cases} \cos^2 t\,|u_{\hat{z}}^\sigma\rangle\langle u_{\hat{z}}^\sigma| + \sin^2 t\,|d_{\hat{z}}^\sigma\rangle\langle d_{\hat{z}}^\sigma| & \text{if } t \le \pi/4, \\ \cos^2 t\,|u_{\hat{y}}^\sigma\rangle\langle u_{\hat{y}}^\sigma| + \sin^2 t\,|d_{\hat{y}}^\sigma\rangle\langle d_{\hat{y}}^\sigma| & \text{if } t > \pi/4. \end{cases} \tag{7.17}$$

One can still define continuous trajectories $\{T_q^\sigma(t)\}_q$ for this state evolution: they are $\{|u_{\hat{z}}^\sigma\rangle\langle u_{\hat{z}}^\sigma|, |d_{\hat{z}}^\sigma\rangle\langle d_{\hat{z}}^\sigma|\}$ before $t_0 = \pi/4$ and they are $\{|u_{\hat{y}}^\sigma\rangle\langle u_{\hat{y}}^\sigma|, |d_{\hat{y}}^\sigma\rangle\langle d_{\hat{y}}^\sigma|\}$ after $t_0 = \pi/4$. But these trajectories can never be joined to continuous trajectories defined on an interval I around $t_0 = \pi/4$. Hence, the Dynamical Decomposition condition is violated.

CASE 4: $P_k^\alpha(t_0)$ corresponds to an eigenvalue $w_k(t_0) \ne 0$ which is partly passing and partly permanent.

In this case there again exist $K > 1$ eigenprojection functions $\{|r_i^\alpha(t)\rangle\langle r_i^\alpha(t)|\}_{i=i'}^{i'+K-1}$ such that $P_k^\alpha(t_0) = \sum_{i=i'}^{i'+K-1} |r_i^\alpha(t_0)\rangle\langle r_i^\alpha(t_0)|$ and $w_k(t_0) = r_i^\alpha(t_0)$ for all $i \in \{i',\ldots,i'+K-1\}$. Now, however, K_1 functions $\{r_i^\alpha(t)\}_{i=i'}^{i'+K_1-1}$ are distinct around t_0 and $K_2 = K - K_1$ functions $\{r_i^\alpha(t)\}_{i=i'+K_1}^{i'+K-1}$ remain equal around t_0. In this fourth case the Dynamical Decomposition condition can again be violated because the K_1 eigenprojection functions $\{|r_i^\alpha(t)\rangle\langle r_i^\alpha(t)|\}_{i=i'+K_1}^{i'+K_2-1}$ need not evolve continuously through t_0, as was the shown by the third possibility of CASE 3.

CASE 5: $P_k^\alpha(t_0)$ corresponds to a degenerate eigenvalue $w_k(t_0) \neq 0$ which is neither passing nor permanent.

In this final case one also has $K > 1$ eigenprojection functions $\{|r_i^\alpha(t)\rangle\langle r_i^\alpha(t)|\}_{i=i'}^{i'+K-1}$ such that $P_k^\alpha(t_0) = \sum_{i=i'}^{i'+K-1} |r_i^\alpha(t_0)\rangle\langle r_i^\alpha(t_0)|$ and $w_k(t_0) = r_i^\alpha(t_0)$ for all $i \in \{i',\dots,i'+K-1\}$. Now, however, the K functions $\{r_i^\alpha(t)\}_{i=i'}^{i'+K-1}$ cannot be divided in K_1 functions which are distinct around t_0 and $K_2 = K - K_1$ functions which remain equal around t_0. Instead K_3 functions are distinct from one another before t_0 and $K_4 = K - K_3$ functions remain equal before t_0. And $K_5 \neq K_3$ functions are distinct after t_0 and $K_6 = K - K_5$ functions remain equal after t_0. A simple example of this final case is given by the state evolution

$$W^\sigma(t) = \begin{cases} \cos^2 t \,|u_{\hat{z}}^\sigma\rangle\langle u_{\hat{z}}^\sigma| + \sin^2 t \,|d_{\hat{z}}^\sigma\rangle\langle d_{\hat{z}}^\sigma| & \text{if } t \leq \pi/4, \\ \frac{1}{2}\,\mathbb{I}^\sigma & \text{if } t > \pi/4. \end{cases} \tag{7.18}$$

This state is degenerate at $t_0 = \pi/4$ with $P_1^\sigma(t_0) = \mathbb{I}^\sigma$. The eigenvalue functions are

$$r_1^\sigma(t) = \begin{cases} \cos^2 t & \text{if } t \leq \pi/4, \\ \frac{1}{2} & \text{if } t > \pi/4, \end{cases} \qquad r_2^\sigma(t) = \begin{cases} \sin^2 t & \text{if } t \leq \pi/4, \\ \frac{1}{2} & \text{if } t > \pi/4. \end{cases} \tag{7.19}$$

These functions are distinct before t_0 (so $K_3 = 2$) and equal after t_0 (so $K_5 = 0 \neq K_3$). The two initially non-degenerate eigenvalue functions thus merge to one permanently degenerate eigenvalue function. A second illustration can be obtained by reversing the time in the state evolution (7.18). One then has $K_3 = 0$ and $K_5 = 2$ and one can speak about one permanently degenerate eigenvalue function which splits into two distinct eigenvalue functions. More generally, one can understand $P_k^\alpha(t_0)$ in this fifth case as a degenerate eigenprojection which corresponds to eigenvalue functions which both merge and split at t_0.

This final case also violates the Dynamical Decomposition condition. Consider the state evolution (7.18). Before t_0 the eigenvalue functions are distinct. Hence, analogously to CASE 1, one can construct the trajectories $T_1^\sigma(t) = |u_{\hat{z}}^\sigma\rangle\langle u_{\hat{z}}^\sigma|$ and $T_2^\sigma(t) = |d_{\hat{z}}^\sigma\rangle\langle d_{\hat{z}}^\sigma|$ for all $t < \pi/4$ and these trajectories are implicitly definable from the state evolution $W^\sigma(t)$ before t_0. After t_0 the eigenvalues are equal to one another. And, analogously to CASE 2, one can construct the trajectory $T_1^\sigma(t) = \mathbb{I}^\sigma$ for all $t > \pi/4$ and this trajectory is also implicitly definable from the state evolution $W^\sigma(t)$ after t_0. These trajectories do not, however, connect continuously at $t_0 = \pi/4$. Hence, there do not exist trajectories $\{T_q^\sigma(t)\}_q$ of eigenprojections of $W^\alpha(t)$ which are continuous around t_0 with regard to the trace norm, such that the Dynamical Decomposition condition cannot be satisfied.

It is, of course, possible to construct trajectories which are continuous around t_0: take, for instance, $T_1^\sigma(t) = |u_{\hat{z}}^\sigma\rangle\langle u_{\hat{z}}^\sigma|$ and $T_2^\sigma(t) = |d_{\hat{z}}^\sigma\rangle\langle d_{\hat{z}}^\sigma|$. But one does not meet the Dynamical Decomposition condition by means of these trajectories because they are not implicitly definable from the state $W^\sigma(t)$. After $t_0 = \pi/4$, the state is degenerate. The only projection which is implicitly definable from this state is thus the degenerate eigenprojection $\mathbb{I}^\sigma$ and there are no means to define the trajectories $T_1^\sigma(t)$ and $T_2^\sigma(t)$ from this degenerate eigenprojection.

To sum up, Propositions 7.1, 7.2 and 7.3 restrict the discontinuous dynamics of the spectral resolutions of states in quantum mechanics. The eigenvalues always evolve along continuous functions. And the eigenprojections that correspond to non-degenerate eigenvalue functions or to permanently degenerate eigenvalue functions evolve along trajectories which are continuously with regard to the trace norm. However, if two or more eigenvalue functions cross or touch such that one passes a degeneracy, or if two or more eigenvalue functions merge or split, the corresponding eigenprojections can evolve discontinuously. Propositions 7.1, 7.2 and 7.3 therefore do not warrant that the Dynamical Decomposition condition holds for all possible states in quantum mechanics. In the next section I present a fourth proposition which yields that, given certain assumptions, state evolutions which violate Dynamical Decomposition do not occur.

MATHEMATICS

I start by giving a lemma:

Lemma 7.1
If $A^{\alpha\beta}$ is a trace class operator of $\mathcal{H}^{\alpha\beta}$, then $A^\alpha := \mathrm{Tr}^\beta(A^{\alpha\beta})$ is a trace class operator of $\mathcal{H}^\alpha$ and $\|A^\alpha\|_1 \leq \|A^{\alpha\beta}\|_1$.

This lemma is proved as Lemma 4.7 in Bacciagaluppi, Donald and Vermaas (1995). Proposition 7.1 can be derived by means of this lemma.

Proof of Proposition 7.1: Consider, firstly, a freely evolving system ω with a pure state $|\psi^\omega(t)\rangle$ and with a Hamiltonian H^ω. The Schrödinger evolution is then given by $|\psi^\omega(t)\rangle = \exp([(t-s)/i\hbar] H^\omega)|\psi^\omega(s)\rangle$.

Calculation yields that the operator $|a\rangle\langle a| - |b\rangle\langle b|$ has two non-zero eigenvalues equal to $\pm\sqrt{1 - |\langle b|a\rangle|^2}$. The difference $|\psi^\omega(t)\rangle\langle\psi^\omega(t)| - |\psi^\omega(s)\rangle\langle\psi^\omega(s)|$ is thus a trace class operator with

$$\| \, |\psi^\omega(t)\rangle\langle\psi^\omega(t)| - |\psi^\omega(s)\rangle\langle\psi^\omega(s)| \, \|_1 = 2\sqrt{1 - |\langle\psi^\omega(s)|\psi^\omega(t)\rangle|^2}. \qquad (7.20)$$

The inner product $\langle\psi^\omega(s)|\psi^\omega(t)\rangle$ is equal to $\int_{E^\omega=-\infty}^{+\infty} \exp([(t-s)/i\hbar]\, E^\omega)$ $|\langle E^\omega|\psi^\omega(t)\rangle|^2 dE^\omega$ and this function is according to Kawata (1972, Theorem 3.5.1) continuous in t. Since $\lim_{t\to s}\langle\psi^\omega(s)|\psi^\omega(t)\rangle = 1$, it follows that (7.20) is a continuous function of t equal to 0 in the limit $t \to s$. Hence, $|\psi^\omega(t)\rangle\langle\psi^\omega(t)|$ evolves continuous with regard to the trace norm.

Consider, secondly, a freely evolving system ω with a non-pure state $W^\omega(t)$ and a Hamiltonian H^ω. One can always construct a (hypothetical) freely evolving composite $\omega\sigma$ with a pure state $|\psi^{\omega\sigma}(t)\rangle$ such that the Schrödinger evolution of ω is equal to the evolution of the partial trace of the Schrödinger evolution of the state of $\omega\sigma$, that is $W^\omega(t) = \mathrm{Tr}^\sigma(|\psi^{\omega\sigma}(t)\rangle\langle\psi^{\omega\sigma}(t)|)$. Let the Hamiltonian of $\omega\sigma$ be $H^{\omega\sigma} = H^\omega \otimes \mathbb{I}^\sigma + \mathbb{I}^\omega \otimes H^\sigma$ and let $|\psi^{\omega\sigma}(0)\rangle = \sum_{j,k} c_j |p_{jk}^\omega\rangle \otimes |E_{jk}^\sigma\rangle$, where $\{c_j\}_j$ are values with $|c_j|^2 = w_j^\omega(0)$, where $\{|p_{jk}^\omega\rangle\}_{j,k}$ are pair-wise orthogonal vectors with $\sum_k |p_{jk}^\omega\rangle\langle p_{jk}^\omega| = P_j^\omega(0)$ and where $\{|E_{jk}^\sigma\rangle\}_{j,k}$ are pair-wise orthogonal eigenvectors of the Hamiltonian H^σ. Lemma 7.1 then implies that

$$\| W^\omega(t) - W^\omega(s)\|_1 \leq \| \, |\psi^{\omega\sigma}(t)\rangle\langle\psi^{\omega\sigma}(t)| - |\psi^{\omega\sigma}(s)\rangle\langle\psi^{\omega\sigma}(s)| \, \|_1. \qquad (7.21)$$

Hence, continuity of the evolution of $|\psi^{\omega\sigma}(t)\rangle\langle\psi^{\omega\sigma}(t)|$ with regard to the trace norm implies continuity of the evolution of $W^\omega(t)$ with regard to the trace norm.

Consider, finally, a system α which is part of a composite ω which, as a whole evolves freely by means of the Schrödinger equation. The state $W^\omega(t)$ of this composite then evolves continuously with regard to the trace norm. The state $W^\alpha(t)$ is a partial trace of $W^\omega(t)$ and by Lemma 7.1 it follows that $W^\alpha(t)$ evolves continuously as well with regard to the trace norm. $\qquad\square$

Proposition 7.2 is a consequence of:

Lemma 7.2
Let W^α, $\widehat{W}^\alpha$ be density operators on a Hilbert space $\mathscr{H}^\alpha$ and let $\{r_i^\alpha\}_{i=1}^N$, $\{\hat{r}_i^\alpha\}_{i=1}^N$ be the corresponding sequences of ordered eigenvalues, as in (7.9). Then $|r_i^\alpha - \hat{r}_i^\alpha| \leq \| W^\alpha - \widehat{W}^\alpha\|_1$, for $i = 1,\dots,N$.

This lemma is proved as Lemma 2.1 in Bacciagaluppi, Donald and Vermaas (1995).

Proof of Proposition 7.2: Take a state $W^\alpha(t)$ which evolves continuously with regard to the trace norm such that $\lim_{t \to s} \| W^\alpha(t) - W^\alpha(s) \|_1$ converges to 0. By Lemma 7.2 it then follows that $\lim_{t \to s} |r_i^\alpha(t) - r_i^\alpha(s)|$ also converges to 0, so the eigenvalue $r_i^\alpha(t)$ evolves continuously as well.   □

I give a third lemma to prove Proposition 7.3:

Lemma 7.3
Let W^α be a density operator on $\mathcal{H}^\alpha$. Suppose that $a < b$ and that a and b are not eigenvalues of W^α. If $\dim(\mathcal{H}^\alpha) = \infty$, then suppose also that $a, b \neq 0$. Choose $\varepsilon \in (0, \frac{1}{2})$ such that $\inf\{|r_i^\alpha - a|\} > \varepsilon$ and $\inf\{|r_i^\alpha - b|\} > \varepsilon$ (where the $r_i^\alpha s$ are the eigenvalues of W^α which lie in (a, b)). Then, for any density operator $\widehat{W}^\alpha$ such that $\| \widehat{W}^\alpha - W^\alpha \|_1 < \frac{4}{9}\varepsilon^2$, $\mathrm{Tr}^\alpha[(\widehat{P}_{(a,b)}^\alpha - P_{(a,b)}^\alpha)^2] < \varepsilon$.

Here, $P_{(a,b)}^\alpha$ and $\widehat{P}_{(a,b)}^\alpha$ are the sums of the one-dimensional eigenprojections of W^α and $\widehat{W}^\alpha$, respectively, as in (7.10). This lemma is proved as Lemma 2.2 in Bacciagaluppi, Donald and Vermaas (1995).

Proof of Proposition 7.3: Take a state $W^\alpha(t)$ which evolves continuously with regard to the trace norm and consider a projection $P_{(a,b)}^\alpha(t) \neq \mathbb{O}^\alpha$, where a and b are not eigenvalues of $W^\alpha(t)$. I start by proving that this projection evolves continuously with regard to the trace norm in the case that the interval (a, b) does not contain the value 0, that is, if $a \geq 0$.

The projection $P_{(a,b)}^\alpha(t)$ is the sum of all the eigenprojections $\{|r_i^\alpha(t)\rangle\langle r_i^\alpha(t)|\}$ of $W^\alpha(t)$ for which the corresponding eigenvalues lie in the interval (a, b). And since the eigenvalues are ordered, these eigenvalues are given by a sequence $b > r_k^\alpha(t) \geq r_{k+1}^\alpha(t) \geq \ldots \geq r_l^\alpha(t) > a \geq 0$. One thus has $\inf\{|r_i^\alpha(t) - a|\} = r_l^\alpha(t) - a > 0$ and $\inf\{|r_i^\alpha(t) - b|\} = b - r_k^\alpha(t) > 0$.

Consider all the values $\varepsilon > 0$ which are strictly smaller than $\min\{\frac{1}{2}, r_l^\alpha(t) - a, b - r_k^\alpha(t)\}$. There indeed exist such values ε because $\frac{1}{2}$, $r_l^\alpha(t) - a$ and $b - r_k^\alpha(t)$ are all strictly larger than 0. For these values ε one has that $\varepsilon \in (0, \frac{1}{2})$, that $\inf\{|r_i^\alpha(t) - a|\} > \varepsilon$ and that $\inf\{|r_i^\alpha(t) - b|\} > \varepsilon$. Hence, by Lemma 7.3, one obtains that

$$\| W^\alpha(t) - W^\alpha(s) \|_1 < \tfrac{4}{9}\varepsilon^2 \quad \Longrightarrow \quad \mathrm{Tr}^\alpha[(P_{(a,b)}^\alpha(t) - P_{(a,b)}^\alpha(s))^2] < \varepsilon. \tag{7.22}$$

The projection $P_{(a,b)}^\alpha(t)$ projects on a finite-dimensional subspace of $\mathcal{H}^\alpha$, for if it projected on an infinite-dimensional subspace, there would exist an infinite number of eigenprojections $|r_i^\alpha(t)\rangle\langle r_i^\alpha(t)|$ with $r_i^\alpha(t) > a \geq 0$ which contradicts the fact that the trace $\mathrm{Tr}^\alpha(W^\alpha(t)) = \sum_i r_i^\alpha(t)$ is equal to 1. It follows analogously that $P_{(a,b)}^\alpha(s)$ also projects on a finite-dimensional subspace.

Consider now the operator $P^\alpha_{(a,b)}(t) - P^\alpha_{(a,b)}(s)$. This operator is self-adjoint and has a finite-dimensional support. (The support is equal to or smaller than the join of the subspaces onto which $P^\alpha_{(a,b)}(t)$ and $P^\alpha_{(a,b)}(s)$ project. These latter subspaces are both finite-dimensional, so the support is finite-dimensional as well). The operator $P^\alpha_{(a,b)}(t) - P^\alpha_{(a,b)}(s)$ can thus be taken as defined on a finite-dimensional Hilbert space (that is, the Hilbert space given by all the vectors in the join of the subspaces onto which $P^\alpha_{(a,b)}(t)$ and $P^\alpha_{(a,b)}(s)$ project) and therefore allows a discrete spectral resolution $\sum^n_{j=1} x_j |x^\alpha_j\rangle\langle x^\alpha_j|$ with a finite number of terms (so $n < \infty$).

Substitution of this spectral resolution in the right-hand inequality $\text{Tr}^\alpha[(P^\alpha_{(a,b)}(t) - P^\alpha_{(a,b)}(s))^2] < \varepsilon$ of (7.22) yields $\sum^n_{j=1} x^2_j < \varepsilon$. So, one obtains for all j that $x^2_j < \varepsilon$ and $|x_j| < \sqrt{\varepsilon}$. Substitution of the spectral resolution in $\|P^\alpha_{(a,b)}(t) - P^\alpha_{(a,b)}(s)\|_1$ gives $\sum^n_{j=1} |x_j|$. Hence, $\|P^\alpha_{(a,b)}(t) - P^\alpha_{(a,b)}(s)\|_1 < n\sqrt{\varepsilon}$. One can therefore rephrase (7.22) as

$$\|W^\alpha(t) - W^\alpha(s)\|_1 < \tfrac{4}{9}\varepsilon^2 \implies \|P^\alpha_{(a,b)}(t) - P^\alpha_{(a,b)}(s)\|_1 < n\sqrt{\varepsilon}. \qquad (7.23)$$

The projection $P^\alpha_{(a,b)}(t)$ evolves continuously with regard to the trace norm if for all (small) $\tilde{\varepsilon} > 0$ there exists a $\delta > 0$ such that

$$\|W^\alpha(t) - W^\alpha(s)\|_1 < \delta \implies \|P^\alpha_{(a,b)}(t) - P^\alpha_{(a,b)}(s)\|_1 < \tilde{\varepsilon}. \qquad (7.24)$$

If one takes $\tilde{\varepsilon} = n\sqrt{\varepsilon}$, (7.23) is equivalent to

$$\|W^\alpha(t) - W^\alpha(s)\|_1 < [4/9n^2]\,\tilde{\varepsilon}^4 \implies \|P^\alpha_{(a,b)}(t) - P^\alpha_{(a,b)}(s)\|_1 < \tilde{\varepsilon} \qquad (7.25)$$

for all $0 < \tilde{\varepsilon} < n\min\{\frac{1}{\sqrt{2}}, \sqrt{r^\alpha_i(t) - a}, \sqrt{b - r^\alpha_k(t)}\}$. So, if one chooses $\delta = [4/9n^2]\,\tilde{\varepsilon}^4$, then (7.24) holds. Hence, in the case that (a,b) does not contain the value 0, $P^\alpha_{(a,b)}(t)$ evolves continuously with regard to the trace norm.

Consider, secondly, $P^\alpha_{(a,b)}(t)$ in the case that (a,b) does contain the value 0, that is, if $a < 0$. In this case $P^\alpha_{(a,b)}(t)$ is the sum of all the eigenprojections $\{|r^\alpha_i(t)\rangle\langle r^\alpha_i(t)|\}$ of $W^\alpha(t)$ for which the corresponding eigenvalues are smaller than b (all the eigenvalues of a state $W^\alpha(t)$ lie in the interval $[0,1]$; there thus do not exist eigenvalues of $W^\alpha(t)$ which are smaller than a). These eigenvalues are given by a sequence $b > r^\alpha_k(t) \geq r^\alpha_{k+1}(t) \geq \ldots \geq r^\alpha_N(t) \geq 0$, where N can be infinite. Because I only consider non-trivial projections $P^\alpha_{(a,b)}(t) \neq \mathbb{O}^\alpha$, the interval (a,b) contains at least one eigenvalue and, in order that this is possible, b is strictly larger than 0.

Consider now the projection $P^\alpha_{(b,1]}(t)$. Because the eigenvalues of $W^\alpha(t)$ are not strictly larger than 1, it follows that all the eigenvalues $\{r^\alpha_i(t)\}^{k-1}_{i=1}$ of $W^\alpha(t)$ which do not lie in the interval (a,b), lie in the interval $(b,1]$. Hence, $P^\alpha_{(b,1]}(t)$ is the sum of all the eigenprojection $\{|r^\alpha_i(t)\rangle\langle r^\alpha_i(t)|\}^{k-1}_{i=1}$ which are not part of $P^\alpha_{(a,b)}(t)$, so $P^\alpha_{(a,b)}(t) = \mathbb{I}^\alpha - P^\alpha_{(b,1]}(t)$. Using this relation, one can conclude that

$$\|P^\alpha_{(a,b)}(t) - P^\alpha_{(a,b)}(s)\|_1 = \|P^\alpha_{(b,1]}(s) - P^\alpha_{(b,1]}(t)\|_1. \qquad (7.26)$$

Since b is strictly larger than 0, the interval $(b, 1]$ does not contain the value 0. So, according to the above proof, $P^\alpha_{(b,1]}(t)$ evolves continuously with regard to the trace norm. Hence, $\lim_{t \to s} \|P^\alpha_{(b,1]}(s) - P^\alpha_{(b,1]}(t)\|_1$ converges to 0, which yields that $\lim_{t \to s} \|P^\alpha_{(a,b)}(t) - P^\alpha_{(a,b)}(s)\|_1$ converges to 0. Hence, if (a, b) does contain the value 0, $P^\alpha_{(a,b)}(t)$ also evolves continuously with regard to the trace norm. $\qquad\square$

7.3 Analytic trajectories of eigenprojections

The fourth proposition gives stronger continuity results for the spectral dynamics of a state $W^\alpha(t)$ than Propositions 7.2 and 7.3. The applicability of this fourth proposition is, on the other hand, not secured by the continuous evolution of $W^\alpha(t)$ with regard to the trace norm. It reads:

Proposition 7.4
Let a state $W^\alpha(t)$ evolve analytically on an interval I around t_0. Let $w^\alpha_k(t_0)$ be an eigenvalue of $W^\alpha(t_0)$ with a corresponding K-dimensional eigenprojection $P^\alpha_k(t_0)$, where $K < \infty$. Then there is an open interval $I_k \subset I$ around t_0 on which there exist K (not necessarily distinct) analytic functions $\{\tilde{r}^\alpha_i(t)\}^K_{i=1}$ and K projection-valued functions $\{|\tilde{r}^\alpha_i(t)\rangle\langle\tilde{r}^\alpha_i(t)|\}^K_{i=1}$ which are analytic with regard to the trace norm. It holds that $\tilde{r}^\alpha_i(t_0) = w^\alpha_k(t_0)$ for all $i \in \{1, 2, \dots, K\}$ and that $\{\tilde{r}^\alpha_i(t)\}^K_{i=1}$ is a set of eigenvalues of $W^\alpha(t)$ for each $t \in I_k$. It also holds that $\sum^K_{i=1} |\tilde{r}^\alpha_i(t_0)\rangle\langle\tilde{r}^\alpha_i(t_0)| = P^\alpha_k(t_0)$ and that $\{|\tilde{r}^\alpha_i(t)\rangle\langle\tilde{r}^\alpha_i(t)|\}^K_{i=1}$ is a set of pairwise orthogonal eigenprojections of $W^\alpha(t)$ for each $t \in I_k$, where $|\tilde{r}^\alpha_i(t)\rangle\langle\tilde{r}^\alpha_i(t)|$ corresponds to the eigenvalue $\tilde{r}^\alpha_i(t)$.

(Note that the eigenvalues $\{\tilde{r}^\alpha_i(t)\}_i$ and the corresponding one-dimensional eigenprojections $\{|\tilde{r}^\alpha_i(t)\rangle\langle\tilde{r}^\alpha_i(t)|\}_i$ need not satisfy the convention $\tilde{r}^\alpha_1(t) \geq \tilde{r}^\alpha_2(t) \geq$ etc. They thus need not be equal to the eigenvalues $\{r^\alpha_i(t)\}_i$ and the eigenprojections $\{|r^\alpha_i(t)\rangle\langle r^\alpha_i(t)|\}_i$, respectively, defined by the decomposition (7.9).)

To unpack the meaning of Proposition 7.4, I start by discussing the notion of analytic evolution.

A complex-valued function $a(z)$ is called analytic on a region D of the complex plane $\mathbb{C}$ if the limit of $[a(z_0 + h) - a(z_0)]/h$ exists for all $z_0 \in D$ as h goes to zero in $\mathbb{C}$. This notion of analyticity can be generalised to operator-valued functions.[76] Consider an operator-valued function $A^\alpha(z) : z \in D \mapsto$

---

[76] Reed and Simon (1972, Sect. VI.3) and, more extensively, Bacciagaluppi (1996b, Sect 6.4).

$A^\alpha(z)$ which is a map from a region D in the complex plane to the set of operators on $\mathcal{H}^\alpha$ which are bounded with regard to the operator norm.[77] This function is called analytic on D if the limit of $[A^\alpha(z_0 + h) - A^\alpha(z_0)]/h$ exists with regard to the operator norm for all $z_0 \in D$ as h goes to zero in $\mathbb{C}$.

A state evolution $W^\alpha(t)$ is now taken as analytic on a time interval I if there exists a region D in the complex plane which is an extension of I, and if there exists a function $A^\alpha(z)$ which is analytic on D and which yields $W^\alpha(t)$ when restricted to I, that is, $A^\alpha(t) = W^\alpha(t)$. In order to obtain a straightforward criterion for the analyticity of a state evolution, one can prove the following

Sufficient condition for analytic state evolution

A state $W^\alpha(t)$ evolves analytically on an interval I if there exists a decomposition $W^\alpha(t) = \sum_{j,k=1}^{N<\infty} a_{jk}(t)\,|e_j^\alpha\rangle\langle e_k^\alpha|$ which (A) contains a finite number of terms, where (B) the set $\{|e_j^\alpha\rangle\}_{j=1}^N$ is a fixed orthonormal basis for $\mathcal{H}^\alpha$, and where (C) the coefficients $\{a_{jk}(t)\}_{j,k=1}^N$ can be extended to functions $\{a_{jk}(z)\}_{j,k=1}^N$ which are analytic on a complex region D which includes I.

Take now a state $W^\alpha(t)$ which evolves analytically in an interval I by considering, for instance, a state that satisfies the above sufficient condition. Then Proposition 7.4 applies to any non-zero eigenvalue of the state at any instant $t_0 \in I$. To see this, note that any eigenprojection $P_k^\alpha(t_0)$ corresponding to a non-zero eigenvalue $w_k(t_0)$ is finite-dimensional, that is, $\mathrm{Tr}^\alpha(P_k^\alpha(t_0)) < \infty$. If $P_k^\alpha(t_0)$ were infinite-dimensional, then the trace $\mathrm{Tr}^\alpha(W^\alpha(t_0)) = \sum_j w_j(t_0)\,\mathrm{Tr}^\alpha(P_j^\alpha(t_0))$ would be infinite (every eigenvalue $w_j(t_0)$ is positive or equal to zero). The trace $\mathrm{Tr}^\alpha(W^\alpha(t_0))$ is, however, equal to 1, so $P_k^\alpha(t_0)$ is finite-dimensional.

Proposition 7.4 yields strong continuity results for such analytically evolving states. A first result is that any non-zero eigenvalue $w_k^\alpha(t_0)$ corresponding to a K-dimensional eigenprojection $P_k^\alpha(t_0)$ lies on K eigenvalue functions $\{\tilde{r}_i^\alpha(t)\}_{i=1}^K$ which are analytic on an interval I_k around t_0. A property of functions that are analytic on an interval I is that if they have the same values on a subinterval of I, they have the same values on the whole interval I. Two eigenvalue functions $\tilde{r}_i^\alpha(t)$ and $\tilde{r}_{i'}^\alpha(t)$ are thus either equal to one another at single instants in I_k or equal on the whole interval I_k. The eigenvalue functions $\{\tilde{r}_i^\alpha(t)\}$ can therefore touch or cross in I_k. However, they cannot merge or split in I_k for then two or more eigenvalue functions $\{\tilde{r}_i^\alpha(t)\}$ must be equal on one subinterval of I_k and be distinct on another.

---

[77] The operator norm $\|.\|$ of A is defined by $\|A\| := \sup_{|\psi\rangle \in \mathcal{H}} \|A|\psi\rangle\| / \| |\psi\rangle \|$.

A second result by Proposition 7.4 is that any K-dimensional eigenprojection $P_k^\alpha(t_0)$ corresponding to a non-zero eigenvalue $w_k^\alpha(t_0)$ is equal to the sum of K projections $\{|\widetilde{r}_i^\alpha(t_0)\rangle\langle\widetilde{r}_i^\alpha(t_0)|\}_{i=1}^K$ which lie on K eigenprojection functions $\{|\widetilde{r}_i^\alpha(t)\rangle\langle\widetilde{r}_i^\alpha(t)|\}_{i=1}^K$ which are analytic with regard to the trace norm on an interval I_k around t_0. By this analyticity, these eigenprojection functions $\{|\widetilde{r}_i^\alpha(t)\rangle\langle\widetilde{r}_i^\alpha(t)|\}_{i=1}^K$ form trajectories of eigenprojections which are continuous with regard to the trace norm.

With these results one can prove that any state $W^\alpha(t)$ which evolves analytically and by means of the Schrödinger equation meets the Dynamical Decomposition condition. That is, any eigenprojection $P_k^\alpha(t_0)$ of such a state with $w_k(t_0) \neq 0$, is a sum $P_k^\alpha(t_0) = \sum_q T_q^\alpha(t_0)$ of projections on continuous trajectories $\{T_q^\alpha(t)\}_q$ of $W^\alpha(t)$. To prove this, I return to the five possible cases which I discussed in the previous section.

For CASES 1 and 2 (see page 108) of an eigenprojection $P_k^\alpha(t_0)$ corresponding to an eigenvalue $w_k(t_0) \neq 0$ which is non-degenerate or permanently degenerate, respectively, it has already been proved that the Dynamical Decomposition condition is satisfied. This followed from the Propositions 7.1, 7.2 and 7.3 and from the fact that states evolve by the Schrödinger equation.

So let's consider CASE 3 (page 109) of an eigenprojection $P_k^\alpha(t_0)$ corresponding to an eigenvalue $w_k(t_0) \neq 0$ that passes a degeneracy. Proposition 7.4 yields that there exist K eigenvalue functions $\{\widetilde{r}_i^\alpha(t)\}_{i=1}^K$ and K eigenprojection functions $\{|\widetilde{r}_i^\alpha(t)\rangle\langle\widetilde{r}_i^\alpha(t)|\}_{i=1}^K$ of $W^\alpha(t)$ such that $P_k^\alpha(t_0) = \sum_{i=1}^K |\widetilde{r}_i^\alpha(t_0)\rangle\langle\widetilde{r}_i^\alpha(t_0)|$ and such that $w_k(t_0) = \widetilde{r}_i^\alpha(t_0)$ for all $i \in \{1,\ldots,K\}$. The eigenvalue functions $\{\widetilde{r}_i^\alpha(t)\}_{i=1}^K$ are analytic on an interval I_k around t_0 and, because $w_k(t_0)$ is a passing degeneracy, this interval I_k can be taken such that all $\{\widetilde{r}_i^\alpha(t)\}_{i=1}^K$ are distinct on the punctuated interval $I_k/\{t_0\}$. So, for every $t \in I_k/\{t_0\}$ these values $\{\widetilde{r}_i^\alpha(t)\}_{i=1}^K$ are all non-degenerate eigenvalues of $W^\alpha(t)$. This implies that for every $t \in I_k/\{t_0\}$ the corresponding eigenprojections $\{|\widetilde{r}_i^\alpha(t)\rangle\langle\widetilde{r}_i^\alpha(t)|\}_{i=1}^K$ are implicitly definable from $W^\alpha(t)$. These eigenprojections can now be joined to continuous trajectories of $W^\alpha(t)$ on the punctuated interval $I_k/\{t_0\}$ by taking

$$T_{ki}^\alpha(t) = |\widetilde{r}_i^\alpha(t)\rangle\langle\widetilde{r}_i^\alpha(t)| \tag{7.27}$$

for all $i \in \{1,\ldots,K\}$, because, as I said before, the analytic eigenvalue functions $\{|\widetilde{r}_i^\alpha(t)\rangle\langle\widetilde{r}_i^\alpha(t)|\}_{i=1}^K$ are continuous with regard to the trace norm. Moreover, these trajectories can be uniquely extended to the instant t_0 because, again by the continuity of the functions $\{|\widetilde{r}_i^\alpha(t)\rangle\langle\widetilde{r}_i^\alpha(t)|\}_{i=1}^K$, $\lim_{t\to t_0} T_{ki}^\alpha(t)$ exists and is equal to $|\widetilde{r}_i^\alpha(t_0)\rangle\langle\widetilde{r}_i^\alpha(t_0)|$. The trajectories (7.27) are thus continuous on the whole interval I_k. The trajectories $\{T_{ki}^\alpha(t)\}_i$ are, as I said, implicitly definable from the state evolution $W^\alpha(t)$ on the interval $I_k/\{t_0\}$. Because these trajectories extend uniquely to t_0, they are implicitly definable from $W^\alpha(t)$ on the whole of I_k. Hence, for any eigenprojection $P_k^\alpha(t_0)$ with $w_k(t_0) \neq 0$ that passes a degeneracy, there exists a set of continuous trajectories $\{T_{ki}^\alpha(t)\}_{i=1}^K$ of eigenprojections implicitly definable from $W^\alpha(t)$ such that $P_k^\alpha(t_0) = \sum_i T_{ki}^\alpha(t_0)$. (Proposition 7.4 thus rules out that analytically evolving states $W^\alpha(t)$ have eigenprojection trajectories which evolve discontinuously through t_0. The third possibility of CASE 3, illustrated by the state evolution (7.17), therefore does not occur if one considers only analytic state evolutions.)

For CASE 4 (page 110) one can also prove that the Dynamical Decomposition condition is satisfied. In this case Proposition 7.4 yields that there are K eigenvalue functions $\{\tilde{r}_i^\alpha(t)\}_{i=1}^K$ and K eigenprojection functions $\{|\tilde{r}_i^\alpha(t)\rangle\langle\tilde{r}_i^\alpha(t)|\}_{i=1}^K$ such that $P_k^\alpha(t_0) = \sum_{i=1}^K |\tilde{r}_i^\alpha(t_0)\rangle\langle\tilde{r}_i^\alpha(t_0)|$ and such that $w_k(t_0) = \tilde{r}_i^\alpha(t_0)$ for all $i \in \{1,\dots,K\}$. However, now, because $w_k(t_0)$ is partly a passing and partly a permanent degeneracy, there exists an interval I_k around t_0 on which K_1 functions in the set $\{\tilde{r}_i^\alpha(t)\}_{i=1}^K$ are distinct from one another before t_0 and on which $K_2 = K - K_1$ functions are equal to one another before t_0. And K_1 functions in the set $\{\tilde{r}_i^\alpha(t)\}_{i=1}^K$ are distinct after t_0 and K_2 functions are equal after t_0. A consequence of Proposition 7.4 is that the K_2 functions which are equal to one another before t_0 are also the K_2 functions which are equal to one another after t_0 (this follows because the function $\{\tilde{r}_i^\alpha(t)\}_{i=1}^K$ are all analytic on I_k, and functions, which are analytic on an interval I and are equal to one another on a subinterval of I, are equal on the whole of I). Hence, one can order the eigenvalue functions $\{\tilde{r}_i^\alpha(t)\}_{i=1}^K$ such that the first K_1 functions $\{\tilde{r}_i^\alpha(t)\}_{i=1}^{K_1}$ are the functions which are distinct from one another on the punctuated interval $I_k/\{t_0\}$ and such that the remaining K_2 functions $\{\tilde{r}_i^\alpha(t)\}_{i=K_1+1}^K$ are equal on that interval. The rest of the proof is a variation of the proof for CASE 3. For every $t \in I_k/\{t_0\}$, the eigenprojections $\{|\tilde{r}_i^\alpha(t_0)\rangle\langle\tilde{r}_i^\alpha(t_0)|\}_{i=1}^{K_1}$ and $\sum_{i=K_1+1}^K |\tilde{r}_i^\alpha(t_0)\rangle\langle\tilde{r}_i^\alpha(t_0)|$ are implicitly definable from $W^\alpha(t)$ (the eigenprojections $\{|\tilde{r}_i^\alpha(t_0)\rangle\langle\tilde{r}_i^\alpha(t_0)|\}_{i=1}^{K_1}$ correspond to the non-degenerate eigenvalues, the eigenprojection $\sum_{i=K_1+1}^K |\tilde{r}_i^\alpha(t_0)\rangle\langle\tilde{r}_i^\alpha(t_0)|$ corresponds to the permanent degenerate eigenvalue). These eigenprojections can be joined to the trajectories

$$
\left.
\begin{aligned}
T_{ki}^\alpha(t) &= |\tilde{r}_i^\alpha(t)\rangle\langle\tilde{r}_i^\alpha(t)|, &&\text{for all } i \in \{1,\dots,K_1\}, \\
T_{kK_1+1}^\alpha(t) &= \sum_{i=K_1+1}^K |\tilde{r}_i^\alpha(t)\rangle\langle\tilde{r}_i^\alpha(t)|.
\end{aligned}
\right\}
\tag{7.28}
$$

These trajectories extend uniquely to continuous trajectories $\{T_{ki}^\alpha(t)\}_{i=1}^{K_1+1}$ of eigenprojections of $W^\alpha(t)$ on the whole interval I_k by taking $T_{ki}^\alpha(t_0) = |\tilde{r}_i^\alpha(t_0)\rangle\langle\tilde{r}_i^\alpha(t_0)|$ for all $i \in \{1,\dots,K_1\}$ and by taking $T_{kK_1+1}^\alpha(t_0) = \sum_{i=K_1+1}^K |\tilde{r}_i^\alpha(t_0)\rangle\langle\tilde{r}_i^\alpha(t_0)|$. These extended trajectories are therefore also implicitly definable from the state evolution $W^\alpha(t)$. Finally, one has that $P_k^\alpha(t_0) = \sum_{i=1}^{K_1+1} T_{ki}^\alpha(t_0)$, which proves the Dynamical Decomposition condition for this fourth case.[78]

Consider finally CASE 5 (page 111). This case is analogous to CASE 4, but now there exists an interval I_k around t_0 on which K_3 functions in the set $\{\tilde{r}_i^\alpha(t)\}_{i=1}^K$ are distinct to one another before t_0 and on which $K_4 = K - K_3$ functions are equal to one another before t_0. And K_5 functions in the set $\{\tilde{r}_i^\alpha(t)\}_{i=1}^K$ are distinct after t_0 and $K_6 = K - K_5$ functions are equal after t_0. As was noted in the proof for CASE 4, a consequence of Proposition 7.4 is that the K_4 functions which are equal to one another before t_0 are by their analyticity on I_k also equal to one another after t_0. Conversely, the K_6 functions which are equal to one another after t_0 are also equal to one another before t_0. It thus follows that the number of functions which are equal to one another before t_0 is equal to the number of functions which are equal after t_0, so $K_4 = K_6$. This implies that CASE 5 is not possible for an analytically evolving state $W^\alpha(t)$ because in that case it holds that $K_4 \neq K_6$. (Proposition 7.4 thus rules out that analytically evolving states $W^\alpha(t)$ have eigenvalue functions which merge or split at t_0. The state evolution (7.18) does not therefore occur if one considers only analytic state evolutions.)

To conclude, any state $W^\alpha(t)$ which evolves analytically and by means of the Schrödinger equation meets the Dynamical Decomposition condition. That is, any possible case of an eigenprojection $P_k^\alpha(t_0)$ of such a state with $w_k(t_0) \neq 0$, is a sum $P_k^\alpha(t_0) = \sum_q T_q^\alpha(t_0)$ of projections on continuous trajectories $\{T_q^\alpha(t)\}_q$ of $W^\alpha(t)$.

Proposition 7.4 thus again restricts discontinuous dynamics of spectral resolutions of states in quantum mechanics. If $W^\alpha(t)$ evolves analytically, its non-zero eigenvalues always evolve along analytic functions and the corresponding eigenprojections always evolve along analytic trajectories. Two or more non-zero eigenvalue functions can cross or touch at an instant, giving rise to a passing degeneracy in the spectral resolution of $W^\alpha(t)$, but the corresponding trajectories of eigenprojections pass continuously through this degeneracy. Moreover, non-zero eigenvalue functions can neither merge nor split.

The evolution of the spectral resolution of states can be captured more fully if one also takes into account the eigenprojection $P_k^\alpha(t_0)$ which corresponds to the zero eigenvalue $w_k(t_0) = 0$. Let's consider, firstly, the case in which the Hilbert space $\mathcal{H}^\alpha$ of a system α is finite-dimensional and the state

---

[78] Note that this proof of the Dynamical Decomposition condition can easily be generalised to eigenprojections $P_k^\alpha(t_0)$ corresponding to eigenvalues $w_k(t_0) \neq 0$ which are partly a passing degeneracy and partly a sum of N permanent degeneracies.

$W^\alpha(t)$ of that system evolves analytically during a (large) interval $I = (t_1, t_2)$. In this case the eigenprojection $P_k^\alpha(t_0)$ with $w_k(t_0) = 0$ is finite-dimensional as well (one can derive that $\text{Tr}^\alpha(P_k^\alpha(t_0)) \leq \text{Tr}^\alpha(\mathbb{I}^\alpha) < \infty$). Hence, Proposition 7.4 also applies to this zero eigenprojection.

Consider any instant $t_0 \in I$. Because the Dynamical Decomposition condition holds for every eigenprojection of $W^\alpha(t_0)$, the spectral resolution of $W^\alpha(t_0)$ can be rewritten as a dynamical decomposition $W^\alpha(t_0) = \sum_{k,i} f_{ki}^\alpha(t_0) T_{ki}^\alpha(t_0)$, where the trajectories $\{T_{ki}^\alpha(t)\}_i$ associated with the eigenprojection $P_k^\alpha(t_0)$ evolve analytically on an open interval I_k around t_0. The interval on which all the trajectories $\{T_{ki}^\alpha(t)\}_{k,i}$ evolve analytically is given by the meet $\cap_k I_k$. Since $W^\alpha(t)$ has a finite number of eigenprojections (if there are an infinite number of eigenprojections, one has $\infty = \text{Tr}^\alpha(\sum_k P_k^\alpha(t_0)) = \text{Tr}^\alpha(\mathbb{I}^\alpha)$, which contradicts that $\text{Tr}^\alpha(\mathbb{I}^\alpha) < \infty$ for finite-dimensional Hilbert spaces), this meet is an open interval around t_0 as well. One can thus extend the dynamical decomposition defined at t_0 to a dynamical decomposition $W^\alpha(t) = \sum_{k,i} f_{ki}^\alpha(t) T_{ki}^\alpha(t)$ which holds for every t in the open interval $\cap_k I_k$ around t_0.

Let the interval $\cap_k I_k$ be given by $(\tilde{t}, \hat{t})$ and consider the instant $\hat{t}$. One can construct a dynamical decomposition $W^\alpha(t) = \sum_{\hat{k},\hat{i}} \widehat{f}_{\hat{k}\hat{i}}^\alpha(t) \widehat{T}_{\hat{k}\hat{i}}^\alpha(t)$ on an open interval $\cap_{\hat{k}} \widehat{I}_{\hat{k}}$ around $\hat{t}$. The two intervals $\cap_k I_k$ and $\cap_{\hat{k}} \widehat{I}_{\hat{k}}$ have a non-trivial meet $[\cap_k I_k] \cap [\cap_{\hat{k}} \widehat{I}_{\hat{k}}]$ and on this meet the eigenvalue functions of $W^\alpha(t)$ are given by both $\{f_{ki}^\alpha(t)\}_{k,i}$ and $\{\widehat{f}_{\hat{k}\hat{i}}^\alpha(t)\}_{\hat{k},\hat{i}}$. And the trajectories of eigenprojections of $W^\alpha(t)$ are given by both $\{T_{ki}^\alpha(t)\}_{k,i}$ and $\{\widehat{T}_{\hat{k}\hat{i}}^\alpha(t)\}_{\hat{k},\hat{i}}$. One can thus identify these eigenvalue functions and trajectories, respectively. And if one relabels them as $\{f_q^\alpha(t)\}_q$ and $\{T_q^\alpha(t)\}_q$, one obtains eigenvalue functions and eigenprojection trajectories of $W^\alpha(t)$ that are analytic on the join $[\cap_k I_k] \cup [\cap_{\hat{k}} \widehat{I}_{\hat{k}}]$. The dynamical decompositions on $\cap_k I_k$ and on $\cap_{\hat{k}} \widehat{I}_{\hat{k}}$, can thus be extended to a dynamical decomposition

$$W^\alpha(t) = \sum_q f_q^\alpha(t) T_q^\alpha(t) \tag{7.29}$$

on this join $[\cap_k I_k] \cup [\cap_{\hat{k}} \widehat{I}_{\hat{k}}]$. One can now do the same for $\tilde{t}$ such that this decomposition extends to the join $[\cap_{\tilde{k}} \widetilde{I}_{\tilde{k}}] \cup [\cap_k I_k] \cup [\cap_{\hat{k}} \widehat{I}_{\hat{k}}]$. And by repeating this procedure over and over again (so by writing $[\cap_{\tilde{k}} \widetilde{I}_{\tilde{k}}] \cup [\cap_k I_k] \cup [\cap_{\hat{k}} \widehat{I}_{\hat{k}}]$ as $(\tilde{t}, \hat{t})$ and then again considering the dynamical decompositions at $\tilde{t}$ and at $\hat{t}$) one obtains that $W^\alpha(t)$ has the dynamical decomposition (7.29) on the whole interval I on which it evolves analytically. It follows that if a state $W^\alpha(t)$ defined on a finite-dimensional Hilbert space evolves analytically on an interval I, then all eigenvalues (zero and non-zero) evolve along analytic functions on I and all corresponding eigenprojections evolve along analytic trajectories on I.

Consider, secondly, the case that the Hilbert space $\mathcal{H}^\alpha$ of a system α is infinite-dimensional and that the state $W^\alpha(t)$ evolves analytically during an interval $I = (t_1, t_2)$. The construction of the global dynamical decomposition (7.29) is in this case in general not possible. Firstly, the zero eigenprojection $P_k^\alpha(t_0)$ with $w_k(t_0) = 0$ may be infinite-dimensional. Hence, Proposition 7.4 need not apply to the zero eigenprojection such that may evolve discontinuously. Secondly, there may be an infinite number of eigenprojections $P_k^\alpha(t_0)$ with $w_k(t_0) \neq 0$. This implies that the meet $\cap_k I_k$ of the intervals on which all the trajectories $\{T_{ki}^\alpha(t)\}_{k,i}$ associated with these eigenprojections $P_k^\alpha(t_0)$ evolve analytically need not be an open interval around t_0. This meet $\cap_k I_k$ can instead converge to only $\{t_0\}$.[79]

An analytically evolving state $W^\alpha(t)$ defined on an infinite-dimensional Hilbert space can thus, in general, only be decomposed at each individual instant t_0 as $W^\alpha(t_0) = \sum_q f_q^\alpha(t_0) T_q^\alpha(t_0)$ with $f_q^\alpha(t_0) > 0$. Each eigenvalue function $f_q^\alpha(t)$ and corresponding trajectory $T_q^\alpha(t)$ can be analytically continued around t_0 as long as $f_q^\alpha(t)$ remains non-zero. However, there need not exist an interval around t_0 on which all functions $\{f_q^\alpha(t)\}_q$ and trajectories $\{T_q^\alpha(t)\}_q$ can be simultaneously continued.

The following picture seems appropriate. At t_{birth} a new term $f_q^\alpha(t) T_q^\alpha(t)$ in the spectral resolution of $W^\alpha(t)$ comes into existence. That is, at t_{birth} the eigenvalue function $f_q^\alpha(t)$ is born by becoming non-zero and a corresponding trajectory $T_q^\alpha(t)$ of eigenprojections breaks away from the null eigenspace of $W^\alpha(t)$. Then, after t_{birth} this term evolves analytically as long as $f_q^\alpha(t)$ is non-zero. Finally, at t_{death}, the eigenvalue function $f_q^\alpha(t)$ dies by becoming zero again and the trajectory $T_q^\alpha(t)$ dissolves in the null eigenspace. The dynamics of the spectral resolution of $W^\alpha(t)$ is a sum of such temporal living terms $f_q^\alpha(t) T_q^\alpha(t)$. At each instant t_0 one can decompose $W^\alpha(t)$ by means of these terms, so $W^\alpha(t_0) = \sum_q f_q(t_0) T_q^\alpha(t_0)$, but it need not be the case that this decomposition can be extended to an interval around t_0.

To sum up: If $W^\alpha(t)$ evolves analytically and $\dim(\mathcal{H}^\alpha)$ is finite, the spectral dynamics of $W^\alpha(t)$ is given by the global dynamical decomposition (7.29) consisting of permanently existing terms $f_q^\alpha(t) T_q^\alpha(t)$ of analytic eigenvalue functions $f_q^\alpha(t)$ and analytic trajectories $T_q^\alpha(t)$ of eigenprojections. If $\dim(\mathcal{H}^\alpha)$ is infinite, the spectral dynamics of $W^\alpha(t)$ is given by sums of temporary existing terms $f_q^\alpha(t) T_q^\alpha(t)$ of non-zero analytic eigenvalue functions $f_q^\alpha(t)$ and analytic trajectories $T_q^\alpha(t)$ of eigenprojections.

---

[79] See Example 5.5 in Bacciagaluppi, Donald and Vermaas (1995).

Two questions still need attention: Do all states evolve analytically in quantum mechanics? And what can happen if states do not evolve analytically?

Let's consider the first question. Take any system α whose state evolves by means of the Schrödinger equation. This system α evolves freely or interacts with other systems. In both cases one can construct a composite ω that contains α as a subsystem, which evolves freely by means of the Schrödinger equation and which has a pure state $|\Psi^\omega(t)\rangle$.[80] Hence, the state evolution of any system α can always be taken as a partial trace $W^\alpha(t) = \text{Tr}^{\omega/\alpha}(|\Psi^\omega(t)\rangle\langle\Psi^\omega(t)|)$ with $|\Psi^\omega(t)\rangle$ evolving by the Schrödinger equation.

It can be proved that if the Schrödinger evolution of a state vector $|\Psi^\omega(t)\rangle$ is analytic with regard to the Hilbert space norm $\||\Psi^\omega(t)\rangle\|$, then any state $W^\alpha(t)$ obtained as a partial trace $\text{Tr}^{\omega/\alpha}(|\Psi^\omega(t)\rangle\langle\Psi^\omega(t)|)$ evolves analytically as well.[81] So, the question of whether the state of a system α evolves analytically, reduces to the question of whether one can construct a system ω that contains α as a subsystem, and that has a state vector $|\Psi^\omega(t)\rangle$ which evolves analytically by the Schrödinger equation.

In order to answer this latter question, consider the complex extension $|\Psi^\omega(z)\rangle = \exp([z/i\hbar]\,H^\omega)\,|\Psi^\omega(0)\rangle$ of the Schrödinger evolution of $|\Psi^\omega(0)\rangle$. The vector $|\Psi^\omega(z)\rangle$ is by definition equal to

$$|\Psi^\omega(z)\rangle = \sum_{n=0}^{\infty} \frac{z^n}{(i)^n n!}\,[H^\omega/\hbar]^n\,|\Psi^\omega(0)\rangle. \tag{7.30}$$

This vector $|\Psi^\omega(z)\rangle$ exists if all the vectors $\{[H^\omega/\hbar]^n|\Psi^\omega(0)\rangle\}_{n=0}^{\infty}$ exist and if $|\Psi^\omega(z)\rangle$ is an element of $\mathcal{H}^\omega$, that is, if $|\Psi^\omega(z)\rangle$ has a finite Hilbert space

---

[80] The system ω may be an existing system or a non-existing hypothetical system. The point is that any state evolution $W^\alpha(t)$ can mathematically be taken as if it is a partial trace $\text{Tr}^{\omega/\alpha}(|\Psi^\omega(t)\rangle\langle\Psi^\omega(t)|)$ with $|\Psi^\omega(t)\rangle$ a pure state of a composite ω.

The construction of this composite ω goes as follows. If α evolves freely and has a state $W^\alpha(t)$ and a Hamiltonian H^α, then take as ω the composite $\alpha\sigma$. Let the Hamiltonian H^ω be equal to $H^\alpha \otimes \mathbb{I}^\sigma + \mathbb{I}^\alpha \otimes H^\sigma$. Let the state of ω be $|\Psi^\omega(0)\rangle = \sum_{jk} c_j\,|p_{jk}^\alpha\rangle \otimes |E_{jk}^\sigma\rangle$, where $\{c_j\}_j$ are values with $|c_j|^2 = w_j^\alpha(0)$, where $\{|p_{jk}^\alpha\rangle\}_{jk}$ are pair-wise orthogonal vectors with $\sum_k |p_{jk}^\alpha\rangle\langle p_{jk}^\alpha| = P_j^\alpha(0)$, and where $\{|E_{jk}^\sigma\rangle\}_{jk}$ are pair-wise orthogonal eigenvectors of the Hamiltonian H^σ. The partial trace $\text{Tr}^\sigma(|\Psi^\omega(t)\rangle\langle\Psi^\omega(t)|)$ then yields the free state evolution of α.

If α interacts with other systems, there exists a composite $\alpha\beta$ with a state $W^{\alpha\beta}(t)$ and a Hamiltonian $H^{\alpha\beta}$ such that $W^\alpha(t) = \text{Tr}^\beta(W^{\alpha\beta}(t))$. Construct then ω as $\alpha\beta\sigma$ in a way similar to that described in the previous paragraph, but now with $\alpha\beta$ in the rôle of α. One then obtains a system ω with a pure state which satisfies $W^{\alpha\beta}(t) = \text{Tr}^\sigma(|\Psi^\omega(t)\rangle\langle\Psi^\omega(t)|)$ and thus satisfies $W^\alpha(t) = \text{Tr}^\beta(W^{\alpha\beta}(t)) = \text{Tr}^{\beta\sigma}(|\Psi^\omega(t)\rangle\langle\Psi^\omega(t)|)$.

[81] See Theorem 4.10 in Bacciagaluppi, Donald and Vermaas (1995).

norm. This norm satifies

$$\||\Psi^{\omega}(z)\rangle\| \leq \sum_{n=0}^{\infty} \left\| \frac{z^n}{(i)^n n!} \, [H^{\omega}/\hbar]^n \, |\Psi^{\omega}(0)\rangle \right\| = \sum_{n=0}^{\infty} \frac{|z|^n}{n!} \, \| [H^{\omega}/\hbar]^n \, |\Psi^{\omega}(0)\rangle \|.$$

(7.31)

Hence, $|\Psi^{\omega}(z)\rangle$ exists if all the vectors $\{[H^{\omega}/\hbar]^n |\Psi^{\omega}\rangle\}_{n=0}^{\infty}$ exist and if this last summation is finite.

Define now an *analytic vector* for the operator $H^{\omega}/\hbar$ with regard to a complex region $D = \{z\,|\,|z| < \varepsilon\}$ around $z = 0$ as follows. For $\varepsilon > 0$, a vector $|\Psi^{\omega}\rangle \in \mathcal{H}^{\omega}$ is an analytic vector for $H^{\omega}/\hbar$ in a region $D = \{z\,|\,|z| < \varepsilon\}$, if $|\Psi^{\omega}\rangle$ is in the domain of $[H^{\omega}/\hbar]^n$ for all n, and if $\sum_{n=0}^{\infty}(|z|^n/n!)\,\|[H^{\omega}/\hbar]^n\,|\Psi^{\omega}\rangle\| < \infty$.[82] It thus follows that if $|\Psi^{\omega}(0)\rangle$ is an analytic vector for $H^{\omega}/\hbar$ for some $\varepsilon > 0$, then the vector-valued function $|\Psi^{\omega}(z)\rangle$ is defined on $D = \{z\,|\,|z| < \varepsilon\}$. This function is now analytic because the limit of $[|\Psi^{\omega}(z_0+h)\rangle - |\Psi^{\omega}(z_0)\rangle]/h$ exists for all $z_0 \in D$ as h goes to 0 in $\mathbb{C}$. Hence, if $|\Psi^{\omega}(0)\rangle$ is an analytic vector for $H^{\omega}/\hbar$ for some $\varepsilon > 0$, then the Schrödinger evolution $|\Psi^{\alpha}(t)\rangle = \exp([t/i\hbar]\,H^{\omega})\,|\Psi^{\omega}(0)\rangle$ is analytic in the time interval $I = (-\varepsilon, \varepsilon)$.

One thus has the following criterion for analytic state evolution. The evolution of the state of a system α is analytic if one can construct a system ω that has α as a subsystem and that has a state $|\Psi^{\omega}(t)\rangle$ which is an analytic vector for $H^{\omega}/\hbar$.

By means of this criterion one can deduce that if (A) the dimension of $\mathcal{H}^{\omega}$ is finite, or (B) the Hamiltonian H^{ω} is bounded with regard to the operator norm, then any possible state of a subsystem of ω evolves analytically. Condition (A) implies condition (B) because if $\dim[\mathcal{H}^{\omega}]$ is finite, every operator defined on $\mathcal{H}^{\omega}$, including $H^{\omega}/\hbar$, is bounded with regard to the operator norm. Condition (B) (and condition (A) via condition (B)) implies that the state $|\Psi^{\omega}(t)\rangle$ is an analytic vector for $H^{\omega}/\hbar$ because if an operator is bounded with regard to the operator norm, every vector in $\mathcal{H}^{\omega}$ is an analytic vector for that operator.

By condition (A) one can conclude that if a freely evolving system α is defined on a finite-dimensional Hilbert space $\mathcal{H}^{\alpha}$ or if an interacting system α is part of a freely evolving composite $\alpha\beta$, defined on a finite-dimensional Hilbert space $\mathcal{H}^{\alpha\beta}$, then the state $W^{\alpha}(t)$ always evolves analytically. This follows because in this case one can define the system ω on a finite-dimensional Hilbert space $\mathcal{H}^{\omega}$ as well (see footnote 80).

To illustrate this conclusion, take a freely evolving composite $\alpha\beta$ defined on a finite-dimensional Hilbert space $\mathcal{H}^{\alpha\beta}$. The Hamiltonian of $\alpha\beta$ has a

---

[82] See the first definition in Section X.6 of Reed and Simon (1975).

discrete spectral resolution $H^{\alpha\beta} = \sum_{j=1}^{N} E_j |E_j^{\alpha\beta}\rangle\langle E_j^{\alpha\beta}|$, where N is $\dim(\mathscr{H}^{\alpha\beta})$ and $\{|E_j^{\alpha\beta}\rangle\}_{j=1}^{N}$ is an orthonormal basis for $\mathscr{H}^{\alpha\beta}$. The state $W^{\alpha\beta}(t)$ can thus be decomposed as

$$W^{\alpha\beta}(t) = \sum_{j,k=1}^{N} \langle E_j^{\alpha\beta}|W^{\alpha\beta}(0)|E_k^{\alpha\beta}\rangle \, e^{\frac{(E_j-E_k)t}{i\hbar}} \, |E_j^{\alpha\beta}\rangle\langle E_k^{\alpha\beta}|. \tag{7.32}$$

And since the functions $\{\langle E_j^{\alpha\beta}|W^{\alpha\beta}(0)|E_k^{\alpha\beta}\rangle \exp[(E_j-E_k)z/i\hbar]\}_{j,k}$ are analytic functions and since the summation contains a finite number of terms, the state $W^{\alpha\beta}(t)$ evolves analytically according to the sufficient condition given on page 117.

Consider then the subsystem α of this composite. Let $\{|e_a^\alpha\rangle\}_a$ and $\{|f_c^\beta\rangle\}_c$ be orthonormal bases for $\mathscr{H}^\alpha$ and $\mathscr{H}^\beta$, respectively. Partial tracing of $W^{\alpha\beta}(t)$ yields that the state of α is given by $W^\alpha(t) = \sum_{a,b} a_{ab}(t)|e_a^\alpha\rangle\langle e_b^\alpha|$ with

$$a_{ab}(t) = \sum_{c,j,k} \left[\langle e_a^\alpha|\otimes\langle f_c^\beta|\right]|E_j^{\alpha\beta}\rangle\,\langle E_j^{\alpha\beta}|W^{\alpha\beta}(0)|E_k^{\alpha\beta}\rangle\,\langle E_k^{\alpha\beta}|\left[|e_b^\alpha\rangle\otimes|f_c^\beta\rangle\right] e^{\frac{(E_j-E_k)t}{i\hbar}}. \tag{7.33}$$

The functions $\{a_{ab}(z)\}_{a,b}$ are again analytic functions and the summation contains a finite number of terms, so $W^\alpha(t)$ also evolves analytically.

By condition (B) one can conclude that if a freely evolving system α has a bounded Hamiltonian H^α, or if an interacting system α is part of a freely evolving composite $\alpha\beta$ with a bounded Hamiltonian $H^{\alpha\beta}$, the state $W^\alpha(t)$ also evolves analytically. This follows because in this second case one can define the Hamiltonian of the system ω such that it is also bounded with regard to the operator norm (see footnote 80).

However, if a freely evolving systems α is defined on an infinite-dimensional Hilbert space and has an unbounded Hamiltonian, or if an interacting system α is part of a freely evolving composite $\alpha\beta$ defined on an infinite-dimensional Hilbert space and with an unbounded Hamiltonian $H^{\alpha\beta}$, it is unclear whether or not $W^\alpha(t)$ evolves analytically. Any system ω that contains α or $\alpha\beta$ is in this case also defined on an infinite-dimensional Hilbert space and also has an unbounded Hamiltonian. Hence, by our criterion, $W^\alpha(t)$ evolves analytically only if one can construct a system ω with a state $|\Psi^\omega(t)\rangle$ that satifies $W^\alpha(t) = \mathrm{Tr}^{\omega/\alpha}(|\Psi^\omega(t)\rangle\langle\Psi^\omega(t)|)$ and which is an analytic vector for $H^\omega/\hbar$. If this construction is impossible, $W^\alpha(t)$ need not evolve analytically.

The question thus becomes a question of whether one can construct for every system α a composite ω with a state $|\Psi^\omega(t)\rangle$ which is an analytic vector for $H^\omega/\hbar$. The answer is probably negative. The set of analytic vectors for

$H^\omega/\hbar$ is a dense set in $\mathscr{H}^\omega$. All the vectors in the subspace $\mathscr{E}_{[-S,S]}$ of $\mathscr{H}^\omega$ spanned by the eigenvectors $\{|E_j^\omega\rangle\}_j$ of H^ω with $-S \le E_j^\omega \le S$, where S is finite, are analytic vectors for $H^\omega/\hbar$.[83] But there does not (yet) exist an argument why $|\Psi^\omega(t)\rangle$ needs to be a member of this dense set. (If one assumes that the system ω is the whole universe, then the condition that the universe contains a finite amount of energy yields the constraint $\langle\Psi^\omega(t)|H^\omega|\Psi^\omega(t)\rangle < \infty$ on the state vector of ω. But this constraint does not yield that $|\Psi^\omega(t)\rangle$ is in the set $\mathscr{E}_{[-S,S]}$.)

So, to conclude, not every Schrödinger-like state evolution needs to be analytic: if the universe is defined on an infinite-dimensional Hilbert space, if it has an unbounded Hamiltonian with regard to the operator norm and if its state vector is not an analytic vector for $H^\omega/\hbar$, then the states of systems in the universe can evolve non-analytically.

The last question to address is that of what can happen if the Schrödinger evolution of a state is not analytic. In this case the dynamics of the spectral resolution of the state is not constrained by Proposition 7.4. The continuous eigenvalue functions can thus merge and split, and the trajectories of eigen-projections can evolve discontinuously at passing degeneracies. Matthew Donald has designed an explicit example[84] which proves that these discontinuities in the eigenprojection trajectories can indeed occur. Donald considers an interacting system α defined on a two-dimensional Hilbert space and shows that its state evolves (in terms of matrices) as

$$W^\alpha(t) = \begin{cases} w_1^\alpha(t)\begin{pmatrix}\frac{1}{2}&\frac{1}{2}\\\frac{1}{2}&\frac{1}{2}\end{pmatrix} + w_2^\alpha(t)\begin{pmatrix}\frac{1}{2}&-\frac{1}{2}\\-\frac{1}{2}&\frac{1}{2}\end{pmatrix} & \text{if } t \le 0, \\[3ex] w_1^\alpha(t)\begin{pmatrix}1&0\\0&0\end{pmatrix} + w_2^\alpha(t)\begin{pmatrix}0&0\\0&1\end{pmatrix} & \text{if } t > 0, \end{cases} \tag{7.34}$$

with $w_1^\alpha(0) = w_2^\alpha(0) = \frac{1}{2}$. The above decompositions of $W^\alpha(t)$ are spectral resolutions, revealing a true discontinuity in the evolution of the eigenprojections at $t = 0$.

MATHEMATICS

The Sufficient condition for analytic state evolution on page 117 is proved as follows:

---

[83] See Section X.6 of Reed and Simon (1975).
[84] See Example 5.6 in Bacciagaluppi, Donald and Vermaas (1995).

Proof: Assume that one can decompose a state as $W^\alpha(t) = \sum_{j,k=1}^{N<\infty} a_{jk}(t)|e_j^\alpha\rangle\langle e_k^\alpha|$, where the summation contains a finite number of terms, the set $\{|e_j^\alpha\rangle\}_{j=1}^N$ is a fixed orthonormal basis for $\mathscr{H}^\alpha$ and the coefficients $\{a_{jk}(t)\}_{j,k=1}^N$ can be extended to functions $\{a_{jk}(z)\}_{j,k=1}^N$ which are analytic on a complex region D which includes I.

Now define the operator-valued function $A^\alpha(z) = \sum_{j,k=1}^{N<\infty} a_{jk}(z)|e_j^\alpha\rangle\langle e_k^\alpha|$ on D. Then $\lim_{h\to 0}[A^\alpha(z_0+h)-A^\alpha(z_0)]/h$ is $\sum_{j,k=1}^{N<\infty}\{\lim_{h\to 0}[a_{jk}(z_0+h)-a_{jk}(z_0)]/h\}|e_j^\alpha\rangle\langle e_k^\alpha|$ for all $z_0 \in D$ (one may interchange the limit and the summation because the summation contains a finite number of terms). The limit $[a_{jk}(z_0+h)-a_{jk}(z_0)]/h$ exists for all $z_0 \in D$ as h goes to zero in $\mathbb{C}$. Hence, $A^\alpha(z)$ is analytic. And because $A^\alpha(z)$ is equal to $W^\alpha(t)$ when restricted to I, $W^\alpha(t)$ evolves analytically. $\qquad\square$

Proposition 7.4 can be proved by a theorem by Rellich (1969) and by the fact that if a vector-valued function $|\psi^\alpha(t)\rangle$ is analytic with regard to the Hilbert space norm, the projection-valued function $|\psi^\alpha(t)\rangle\langle\psi^\alpha(t)|$ is analytic with regard to the trace norm. Rellich's theorem reads:

Theorem 7.1
Let $W^\alpha(t)$ be analytic on I and suppose that $t_0 \in I$. Let $w_k^\alpha(t_0)$ be an isolated eigenvalue of $W^\alpha(t_0)$ with a corresponding K-dimensional eigenprojection $P_k^\alpha(t_0)$, where $K < \infty$. Then there is an open interval $I_k \subset I$ with $t_k \in I_0$ on which there exist K (not necessarily distinct) numerical analytic functions $\{\widetilde{r}_i^\alpha(t)\}_{i=1}^K$ and K vector-valued analytic functions $\{|\widetilde{r}_i^\alpha(t)\rangle\}_{i=1}^K$. It holds that $\widetilde{r}_i^\alpha(t_0) = w_k^\alpha(t_0)$ and, for each $t \in I_k$, $\{|\widetilde{r}_i^\alpha(t)\rangle\}_{i=1}^K$ is a set of pair-wise orthogonal eigenvectors of $W^\alpha(t)$ with $|\widetilde{r}_i^\alpha(t)\rangle$ corresponding to the eigenvalue $\widetilde{r}_i^\alpha(t)$.

This theorem is proved in Rellich (1969, Sects. 1.1 and 2.2) and in Kato (1976, Sects. II.1, II.4, II.6, VII.1 and VII.3). The proof that if the function $|\psi^\alpha(t)\rangle$ is analytic, the function $|\psi^\alpha(t)\rangle\langle\psi^\alpha(t)|$ is analytic as well can be found in, for instance, Bacciagaluppi (1996b, the proof of Theorem 6.11 on the pages 234–7).

7.4 Instabilities and other bad behaviour

The results presented in the last two sections prove that the spectral resolutions of the states of systems evolve in many cases in a very continuous way. If the states evolve by means of the Schrödinger equation, the eigenvalues of the states always evolve along continuous functions. And if the states evolve analytically, the eigenprojections of the states that correspond to non-zero eigenvalues always evolve along continuous trajectories. However, these results also prove that in some cases the spectral resolutions can evolve discontinuously. If the states evolve by means of the Schrödinger equation but not analytically, the eigenvalues of the states evolve along functions that can cross, touch, merge and split at specific instants. The corresponding eigenprojections evolve at those instants not along continuous trajectories but exhibit genuine discontinuities, as is pointedly illustrated by Matthew Donald's example (7.34).

Let's now return to modal interpretations and see what conclusions can be drawn. Take, firstly, the extended modal interpretation. In this interpretation the core property ascription is defined only for systems whose states have at all times continuous trajectories of eigenprojections. Since the results of the last two sections prove that such continuous trajectories need not always exist, a first conclusion is that the extended modal interpretation is not always applicable. This interpretation can therefore not be taken as a general interpretation of quantum mechanics.

Consider, secondly, the bi, spectral and atomic modal interpretations. The core properties of these interpretations evolve discontinuously as was shown in Section 7.1. Moreover, the attempts to (partly) remove these discontinuities by Elby and Bub (1994) and by Bacciagaluppi, Donald and Vermaas (1995) both failed. Hence, a second conclusion is that the property ascription to a system by the bi, spectral and atomic modal interpretations is unresolvable discontinuous at those instants at which the continuous eigenvalues of the state of that system cross, touch, merge or split.

Unfortunately one can arrive by means of the results of the last two sections at further negative conclusions about the bi, spectral and atomic modal interpretations. Take an atomic system α defined on an infinite-dimensional Hilbert space and with a Hamiltonian which is bounded with regard to the operator norm. The state of that system then evolves analytically. And in those time intervals in which the non-zero analytic eigenvalue functions $\{f_q^\alpha(t)\}_q$ are not passing or touching, the bi, spectral and atomic core properties of α evolve along the analytic trajectories $\{T_q^\alpha(t)\}_q$. The results of Section 7.3 revealed that these analytic trajectories can emerge at an instant t_{birth} and vanish at another instant t_{death}. This phenomenon of emerging and vanishing trajectories of eigenprojections may seem at first sight rather harmless. Because a trajectory $T_q^\alpha(t)$ emerges or vanishes only when the corresponding eigenvalue function $f_q^\alpha(t)$ is zero, the probability $\text{Tr}^\alpha(W^\alpha(t)\,T_q^\alpha(t)) = f_q^\alpha(t)\,\text{Tr}^\alpha(T_q^\alpha(t))$ with which the bi, spectral and atomic modal interpretations ascribe the core property $T_q^\alpha(t)$, is also zero. The core properties which are born are thus at their birth actually possessed with probability zero; they are not instantaneously launched with large probabilities. And the core properties which die, do not drop dead in broad daylight but rather fade away. The phenomenon of emerging and vanishing trajectories of eigenprojections thus does not lead to a wild and irregular dynamics of the set of core properties.

However, a number of examples by Matthew Donald prove that this dynamics of emerging and vanishing trajectories of eigenprojections need not harmonise with the dynamics of the state of the system. For instance, a state which evolves periodically with a period $2\pi/\theta$ can have trajectories of eigenprojections which exist from $t_{\text{birth}} = -\infty$ onwards but disappear at the finite times $t_{\text{death}} = m\pi/\theta$, $(m = 1, 2, \dots)$.[85] The vanishing trajectories thus do not straightforwardly reflect the periodicity of $W^\alpha(t)$. Hence, a third conclusion is that the dynamics of the set of core properties ascribed by the bi, spectral and atomic modal interpretations to a system can deviate from the dynamics of the state of that system.

A final conclusion about the property ascription of modal interpretations is that the dynamics of the set of core properties can be highly unstable if the spectral resolution of a state is close to a degeneracy. This instability has already been noted by, for instance, Albert and Loewer (1990, 1993). I now end by giving some explicit examples of state evolutions where this instability is clearly present. I start by reconsidering the problem of discontinuities.

Consider the set of states defined on a two-dimensional Hilbert space and represented by two-by-two density matrices $\{W\}$. This set can be parameterised by means of three real parameters v_1, v_2 and v_3 and the Pauli matrices if one takes W equal to

$$W = \tfrac{1}{2}(\mathbb{I} + \vec{v} \cdot \vec{\sigma}) \tag{7.35}$$

with[86]

$$\mathbb{I} = \begin{pmatrix} 1 & 0 \\ 0 & 1 \end{pmatrix}, \quad \sigma_1 = \begin{pmatrix} 0 & 1 \\ 1 & 0 \end{pmatrix}, \quad \sigma_2 = \begin{pmatrix} 0 & -i \\ i & 0 \end{pmatrix}, \quad \sigma_3 = \begin{pmatrix} 1 & 0 \\ 0 & -1 \end{pmatrix}. \tag{7.36}$$

Hence, the set of two-by-two density matrices has three (real) dimensions. Furthermore, the set of degenerate two-by-two density matrices is zero-dimensional since only one exists, namely $W = \tfrac{1}{2}\mathbb{I}$.

Given this difference in the dimensionality of these sets of states and degenerate states, one now can argue that it is quite rare that states actually hit degeneracies. It is of course imaginable that a state hits $\tfrac{1}{2}\mathbb{I}$, for instance, by writing down evolutions like (7.2). However, in the real world, state

---

[85] See Example 5.1 in Bacciagaluppi, Donald, and Vermaas (1995).

[86] This parameterisation defines a map from the unit ball in $\mathbb{R}^3$, called the Bloch sphere, to the set of two-by-two density matrices (see, for instance, Donald (1998)).

evolutions do not follow such well-defined and straight paths as given by (7.2). Environmental influences perturb the evolution of states, generating irregular paths containing random fluctuations. And since for any sensible measure on the set of two-by-two matrices the subset $\{\frac{1}{2}\mathbb{I}\}$ of degenerate states has measure zero, one can conclude that a generally evolving state has zero probability of ever hitting a degeneracy. Hence, even if modal property ascriptions are discontinuous at passing degeneracies, the whole problem is not worth being bothered with.

This argument that randomly fluctuating states hit degeneracies only with probability zero has been made rigorous by Matthew Donald who proved for finite-dimensional Hilbert spaces that the space of degenerate states has always co-dimension 3. That is, if the space of all states has a real dimension of N, the set of degenerate states has a real dimension of $N - 3$.[87]

If one accepts this argument, further nice results follow from the Propositions 7.1, 7.2 and 7.3. For if the states defined on a finite-dimensional Hilbert space are degenerate with probability zero, their continuous eigenvalue functions $\{f_q^{\alpha}(t)\}_q$ cross, touch, merge or split with probability zero. Hence, with probability 1 one can order these functions like $f_1^{\alpha}(t) > f_2^{\alpha}(t) > \ldots$ for all t. Proposition 7.3 then yields that the corresponding trajectories $\{T_q^{\alpha}(t)\}_q$ of eigenprojections evolve continuously. So, with probability 1, the core properties $\{T_q^{\alpha}(t)\}_q$ ascribed by the bi, spectral and atomic modal interpretations evolve continuously.

I do not like this argument. For all practical purposes it may be that states do not hit degeneracies but we are not yet discussing practical purposes. We are trying to find out whether modal interpretations give a well-developed description of reality. So, if the discontinuities related to degenerate states give rise to a problematic description of reality, one either proves that it is strictly impossible for states to hit those degeneracies, or, if such a proof is impossible, one tries to make sense of these discontinuities. And in this last case it is irrelevant whether or not it is probable that such discontinuities actually occur.

However, even if one accepts the argument, it can be shown that the possibility that a state fluctuates randomly around a degeneracy is quite problematic in itself within modal interpretations. Let's see what happens.

The idea is that a state which comes near a degeneracy does not hit the degeneracy but rather grazes it due to environmental perturbations. Let's therefore take a state evolution which hits a degeneracy with certainty and construct a family of perturbed copies of this evolution. I work on a

---

[87] See Proposition 3.2 in Bacciagaluppi, Donald and Vermaas (1995).

two-dimensional Hilbert space and represent the states by the two-by-two matrices given in (7.35). Let the unperturbed evolution be

$$W(t) = \tfrac{1}{2}\left(\mathbb{I} + g(t)\,\sigma_1\right) = \tfrac{1}{2}\begin{pmatrix} 1 & g(t) \\ g(t) & 1 \end{pmatrix},\tag{7.37}$$

where $g(t)$ is a real-valued function which is equal to zero at $t = 0$, which is monotonously and strictly increasing on an interval around $t = 0$ and which satisfies $|g(t)| < 1$ ($W(t)$ is then indeed a density matrix). This unperturbed state evolution $W(t)$ passes through the degeneracy at $t = 0$. Now assume that the environmental influences add a small term $\tfrac{1}{2}\varepsilon(t)\,\sigma_2$ to $W(t)$ with $\varepsilon(t)$ again real-valued and with $|\varepsilon(t)| \ll 1$. The family of perturbed evolutions is then given by

$$W(t, \varepsilon(t)) = \tfrac{1}{2}\begin{pmatrix} 1 & g(t) - i\varepsilon(t) \\ g(t) + i\varepsilon(t) & 1 \end{pmatrix}.\tag{7.38}$$

A member $W(t, \varepsilon(t))$ of this family hits the degeneracy at $t = 0$ if and only if $\varepsilon(0) = 0$.

I show three things by means of this family of perturbed evolutions. Firstly, an arbitrarily small perturbation of a state which hits a degeneracy can maximally change its set of eigenprojections in the limit to the degeneracy. Secondly, two arbitrarily close evolutions can have maximally different sets of eigenprojections near a degeneracy. Thirdly, the set of eigenprojections of a state which comes near a degeneracy can change maximally in an arbitrarily small time interval.

I start by giving some results. The distance $\| W(t) - W(t, \varepsilon(t))\|_1$ between the unperturbed evolution and a perturbed evolution with respect to the trace norm, is at any time t equal to $|\varepsilon(t)|$. So, by taking $\varepsilon(t)$ arbitrarily small, the perturbed evolution $W(t, \varepsilon(t))$ is arbitrarily close to $W(t)$. The distance between two perturbed evolutions $W(t, \varepsilon_1(t))$ and $W(t, \varepsilon_2(t))$ is analogously at any time equal to $|\varepsilon_1(t) - \varepsilon_2(t)|$. Hence, two perturbed evolutions can also lie arbitrarily close.

The spectral resolution of the unperturbed evolution is at all times (except at $t = 0$) given by $W(t) = w_1(t)\,P_1(t) + w_2(t)\,P_2(t)$, where $w_1(t) = (1 + g(t))/2$, where $w_2(t) = (1 - g(t))/2$ and where

$$P_1(t) = \tfrac{1}{2}\begin{pmatrix} 1 & 1 \\ 1 & 1 \end{pmatrix}, \quad P_2(t) = \tfrac{1}{2}\begin{pmatrix} 1 & -1 \\ -1 & 1 \end{pmatrix}.\tag{7.39}$$

The spectral resolution of a perturbed evolution is at all times (except at $t = 0$ if $\varepsilon(0) = 0$) given by $W(t, \varepsilon(t)) = w_1(t, \varepsilon(t))\,P_1(t, \varepsilon(t)) + w_2(t, \varepsilon(t))\,P_2(t, \varepsilon(t))$, where $w_1(t, \varepsilon(t)) = (1 + r(t))/2$, where $w_2(t, \varepsilon(t)) = (1 - r(t))/2$ and where

$$P_1(t, \varepsilon(t)) = \tfrac{1}{2} \begin{pmatrix} 1 & e^{-i\varphi(t)} \\ e^{i\varphi(t)} & 1 \end{pmatrix}, \qquad P_2(t, \varepsilon(t)) = \tfrac{1}{2} \begin{pmatrix} 1 & -e^{-i\varphi(t)} \\ -e^{i\varphi(t)} & 1 \end{pmatrix}. \qquad (7.40)$$

Here, $r(t)$ is the modulus of $g(t) + i\varepsilon(t)$ and $\varphi(t)$ is its phase.

In order to compare the sets of eigenprojections of $W(t)$ and $W(t, \varepsilon(t))$, I need a measure to capture their closeness. A convenient measure for the closeness of two projections $P = |\psi\rangle\langle\psi|$ and $\widehat{P} = |\phi\rangle\langle\phi|$ which project onto the rays $|\psi\rangle$ and $|\phi\rangle$, respectively, is given by $\mathrm{Tr}(P\widehat{P}) = |\langle\psi|\phi\rangle|^2$. If P and $\widehat{P}$ project onto the same ray, then $|\langle\psi|\phi\rangle| = 1$ and $\mathrm{Tr}(P\widehat{P}) = 1$. If P and $\widehat{P}$ project onto different rays, $\mathrm{Tr}(P\widehat{P})$ decreases if the angle $\cos^{-1}(|\langle\psi|\phi\rangle|)$ between $|\psi\rangle$ and $|\phi\rangle$ increases. So, the smaller $\mathrm{Tr}(P\widehat{P})$ is, the more different P and $\widehat{P}$ are. Now let the one-dimensional eigenprojections of two states W and $\widehat{W}$ be given by the sets $\{P_1, P_2\}$ and $\{\widehat{P}_1, \widehat{P}_2\}$, respectively. The projection $\widehat{P}_1$ is maximally different to both P_1 and P_2 if the values $\mathrm{Tr}(P_1\widehat{P}_1)$ and $\mathrm{Tr}(P_2\widehat{P}_1)$ are both as small as possible. Since the sum $P_1 + P_2$ is equal to the unit operator $\mathbb{I}$ (I am still working on a two-dimensional Hilbert space), it holds that $\mathrm{Tr}(P_1\widehat{P}_1) + \mathrm{Tr}(P_2\widehat{P}_1) = \mathrm{Tr}(\widehat{P}_1) = 1$. Hence, $\mathrm{Tr}(P_1\widehat{P}_1)$ and $\mathrm{Tr}(P_2\widehat{P}_1)$ have simultaneously their smallest value if one of them, say $\mathrm{Tr}(P_1\widehat{P}_1)$, has value $\tfrac{1}{2}$. And if $\mathrm{Tr}(P_1\widehat{P}_1) = \tfrac{1}{2}$, such that $\mathrm{Tr}(P_2\widehat{P}_1) = \tfrac{1}{2}$, one can derive that also $\mathrm{Tr}(P_1\widehat{P}_2) = \tfrac{1}{2}$ and $\mathrm{Tr}(P_2\widehat{P}_2) = \tfrac{1}{2}$ (using that $\widehat{P}_1 + \widehat{P}_2 = \mathbb{I}$). So, a good criterion for $\{P_1, P_2\}$ and $\{\widehat{P}_1, \widehat{P}_2\}$ to be maximally different is that $\mathrm{Tr}(P_1\widehat{P}_1) = \tfrac{1}{2}$.

Now, let's consider how arbitrarily small perturbations of evolving states can affect the eigenprojections of those states near degeneracies. Consider, firstly, how a perturbation of $W(t)$ given by $W(t, \varepsilon(t))$ affects the eigenprojections $\{P_1(t), P_2(t)\}$ of $W(t)$ around $t = 0$. The distance between $W(t)$ and $W(t, \varepsilon(t))$ is $|\varepsilon(t)|$ so the perturbation of $W(t)$ can be taken to be arbitrarily small by taking for all t the perturbation $\varepsilon(t)$ arbitrarily small. Calculation of $\mathrm{Tr}[P_1(t)\, P_1(t, \varepsilon(t))]$ yields

$$\mathrm{Tr}[P_1(t)\, P_1(t, \varepsilon(t))] = \mathrm{Tr}\left[\tfrac{1}{2}\begin{pmatrix} 1 & 1 \\ 1 & 1 \end{pmatrix} \tfrac{1}{2}\begin{pmatrix} 1 & e^{-i\varphi(t)} \\ e^{i\varphi(t)} & 1 \end{pmatrix}\right]$$
$$= \tfrac{1}{2} + \tfrac{1}{2}\cos\varphi(t) = \tfrac{1}{2} + \tfrac{1}{2}\cos(\tan^{-1}[\varepsilon(t)/g(t)]). \qquad (7.41)$$

So, irrespective of how small the perturbation $\varepsilon(t)$ is, if $\varepsilon(0) \neq 0$ (and if $\varepsilon(t)$ is sufficiently smooth around $t = 0$), the limit $t \to 0$ of $\mathrm{Tr}[P_1(t)\, P_1(t, \varepsilon(t))]$ goes to $\tfrac{1}{2}$ since the limit $t \to 0$ of $\varepsilon(t)/g(t)$ goes to $\pm\infty$. Hence, for arbitrarily small perturbations, the eigenprojections of $W(t)$ and $W(t, \varepsilon(t))$ can become maximally different if one approaches the degeneracy at $t = 0$.

Compare, secondly, two perturbed evolutions $W(t, -\varepsilon)$ and $W(t, \varepsilon)$ with ε

a constant larger than zero. The distance between these evolutions is 2ε. So, by taking ε arbitrarily small, they are arbitrarily close to one another and they are arbitrarily near the degeneracy $W(0) = \frac{1}{2}\mathbb{I}$ at $t = 0$. Calculation of $\mathrm{Tr}[P_1(t, -\varepsilon) P_1(t, \varepsilon)]$ yields

$$\mathrm{Tr}[P_1(t, -\varepsilon) P_1(t, \varepsilon)] = \tfrac{1}{2} + \tfrac{1}{2}\cos(2\varphi(t)) = \tfrac{1}{2} + \tfrac{1}{2}\cos(2\tan^{-1}[\varepsilon/g(t)]). \quad (7.42)$$

Now take an instant $t' < 0$. The value $g(t')$ is then strictly smaller than zero since $g(t)$ is monotonously and strictly increasing. Choose ε such that $0 < \varepsilon < -g(t')$ (ε can thus be arbitrarily small). Because $g(t)$ is increasing there is a second instant t'' that is closer to $t = 0$ than t' is (so $t' < t'' < 0$) and for which it holds that $g(t'') = -\varepsilon$. At this instant $\mathrm{Tr}[P_1(t, -\varepsilon) P_1(t, \varepsilon)]$ is $\frac{1}{2}$. So, two arbitrarily close evolutions can have maximally different sets of eigenprojections near a degeneracy.

Consider, thirdly, a single state evolution $W(t, \varepsilon)$ that comes arbitrarily close to the degeneracy $W(0) = \frac{1}{2}\mathbb{I}$ but does not hit it. Let t_1 be an instant before $t = 0$, let t_2 be one after $t = 0$ and compare the eigenprojections of $W(t, \varepsilon)$ at those two instants. The value $\mathrm{Tr}[P_1(t_1, \varepsilon) P_1(t_2, \varepsilon)]$ is

$$\mathrm{Tr}[P_1(t_1, \varepsilon) P_1(t_2, \varepsilon)] = \tfrac{1}{2} + \tfrac{1}{2}\cos(\varphi(t_1) - \varphi(t_2)). \quad (7.43)$$

For any t_1 before $t = 0$ one can choose $\varepsilon = -g(t_1)$ and $t_2 = g^{-1}(\varepsilon)$ (g is monotonously and strictly increasing, so g^{-1} exists). The phases $\varphi(t_1) = \tan^{-1}[\varepsilon/g(t_1)]$ and $\varphi(t_2) = \tan^{-1}[\varepsilon/g(t_2)]$ are then $-\pi/4$ and $\pi/4$, respectively, and $\mathrm{Tr}[P_1(t_1, \varepsilon) P_1(t_2, \varepsilon)]$ is equal to $\frac{1}{2}$. One can now take the limit $t_1 \uparrow 0$ and in this limit ε and t_2 also go to zero. If one does so, it follows that the set of eigenprojections of a state that comes arbitrarily close to a degeneracy can change maximally in an arbitrarily small time interval $t_2 - t_1$.

These results prove that whenever a state is approaching a degeneracy, it is much wiser if it accepts its fate, takes a deep breath and dives along a straight line through the degeneracy. In the above case, the unperturbed state $W(t)$ has during its dive at all times $t \neq 0$ the eigenprojections $P_1(t)$ and $P_2(t)$ given in (7.39). The set of core properties generated by $W(t)$ is thus constant and goes awry only at the isolated instant $t = 0$. If, however, the state grazes the degeneracy or fluctuates around it, the set of core properties can change rapidly, resulting in an unstable property ascription during a finite time interval.

This possibility that modal property ascriptions can be unstable due to rapidly changing eigenprojections is, of course, not limited only to randomly fluctuating states. For instance, the smoothly evolving state $W(t, \varepsilon)$ has also been proved to exhibit maximal changes in its set of eigenprojections during small time intervals.

To conclude, the study of the dynamics of the spectral resolution of evolving states not only proves that modal property ascriptions can be discontinuous and that the core projections can evolve along (deviantly) emerging and vanishing trajectories. It also proves that modal property ascriptions can be unstable.

A final remark concerns yet another source for incorrect property ascriptions. In this book I always assume that one can precisely identify the systems α, β, γ, etc. Consequently, one can also precisely identify the composites of these systems. If, however, one proceeds the other way round and starts with a set of composites, one has to answer the question of how exactly to factor these composites into disjoint subsystems. In Bacciagaluppi, Donald and Vermaas (1995, Example 7.3) it is proved that the property ascription to a subsystem can depend with high sensitivity on the precise identification of that subsystem.

8

Transition probabilities

Having dealt with the evolution of the set of core properties of a system, I continue with the evolution of the actually possessed core property.

8.1 Introduction

Consider again the spin $\frac{1}{2}$-particle σ described at the beginning of Section 7.1. Its state is $W^\sigma(t) = \cos^2 t\,|u_{\hat{z}}^\sigma\rangle\langle u_{\hat{z}}^\sigma| + \sin^2 t\,|d_{\hat{z}}^\sigma\rangle\langle d_{\hat{z}}^\sigma|$ and its actually possessed core property at time t is $|u_{\hat{z}}^\sigma\rangle\langle u_{\hat{z}}^\sigma|$ with probability $\cos^2 t$ and $|d_{\hat{z}}^\sigma\rangle\langle d_{\hat{z}}^\sigma|$ with probability $\sin^2 t$. So, at $t = 0$ the actual spin of σ is up with probability 1 and after $t = 0$ the probability that the spin is actually down becomes non-zero and increases in time. The question is now how does this actually possessed spin of σ evolve. Does the actual spin remain up as long as possible, allowing only one jump from up to down? Or does it randomly flip up and down with an increasing bias towards being down?

In this chapter I address this question of the evolution of the actually possessed core properties. More precisely, following early proposals by Dieks (1993, 1994a,b), I determine this evolution by means of the **transition probabilities** $p(C_b^\alpha(t)/C_a^\alpha(s))$, which are the conditional probabilities that the actually possessed core property of a system α is $C_b^\alpha(t)$ at time t, given that this property is $C_a^\alpha(s)$ at time $s < t$. With these transition probabilities one cannot, in general, derive the probabilities which should be assigned to whole histories of actually possessed core properties. However, transition probabilities already constrain such histories enough to derive some interesting results.

One can easily see that there exist many candidate expressions for these transition probabilities $p(C_b^\alpha(t)/C_a^\alpha(s))$. Let $p(C_a^\alpha(s), C_b^\alpha(t))$ be the joint probability that the actual core properties of α are sequentially $C_a^\alpha(s)$ at s and

135

Transition probabilities

Let the core projections of a system α be $\{C_j^\alpha(s)\}_j$ at time s and $\{C_k^\alpha(t)\}_k$ at time t (with $s < t$). The evolution of the actually possessed core property of α from s to t is then described by transition probabilities $p(C_b^\alpha(t)/C_a^\alpha(s))$ that are the conditional probabilities that $[C_b^\alpha(t)] = 1$ at t, given that $[C_a^\alpha(s)] = 1$ at s.

$C_b^\alpha(t)$ at t. One then has

$$p(C_b^\alpha(t)/C_a^\alpha(s)) = \frac{p(C_a^\alpha(s), C_b^\alpha(t))}{\sum_b p(C_a^\alpha(s), C_b^\alpha(t))}. \tag{8.1}$$

The joint probability $p(C_a^\alpha(s), C_b^\alpha(t))$ must be consistent with the core property ascriptions to α at s and at t. So, its marginals must yield the single time probabilities assigned to $C_a^\alpha(s)$ and $C_b^\alpha(t)$. That is,

$$\sum_a p(C_a^\alpha(s), C_b^\alpha(t)) = p(C_b^\alpha(t)), \qquad \sum_b p(C_a^\alpha(s), C_b^\alpha(t)) = p(C_a^\alpha(s)). \tag{8.2}$$

It is a standard result that, in general, the marginals of a joint probability do not uniquely determine that joint probability. Hence, modal interpretations, which fix the marginals $p(C_a^\alpha(s))$ and $p(C_b^\alpha(t))$, do not uniquely determine the joint probability $p(C_a^\alpha(s), C_b^\alpha(t))$ and thus do not uniquely determine the transition probabilities $p(C_b^\alpha(t)/C_a^\alpha(s))$. Therefore, since, in principle, there exist many candidate expressions, the task of giving transition probabilities becomes not so much a matter of finding a specific candidate expression, but rather one of arguing why a specific candidate is the correct one.

In the light of this underdeterminateness of transition probabilities, the results are somewhat limited. For freely evolving systems one can indeed argue for specific transition probabilities (Section 8.2). If the bi and the spectral modal interpretations satisfy the assumption of Dynamical Autonomy for whole systems and the criterion of Empirical Adequacy, then one can argue that the transition probabilities for a freely evolving system α are given by $p(C_b^\alpha(t)/C_a^\alpha(s)) = \delta_{ab}$. And if the atomic modal interpretation satisfies Dynamical Autonomy for atomic systems and Empirical Adequacy, one can argue that the transition probabilities for a freely evolving atom α are also given by $p(C_b^\alpha(t)/C_a^\alpha(s)) = \delta_{ab}$.

Using these first results, one can also argue for specific transition probabilities for interacting systems in the special case that there exists a 'snooper' for the interacting system (Section 8.3). However, this second result does not hold for interacting systems in general because snoopers for interacting systems need not always exist.

With these transition probabilities, based on Dynamical Autonomy and Empirical Adequacy in the above cases, one can arrive at some important conclusions about general transition probabilities. The first is that in the spectral and atomic modal interpretations the transition probabilities for subsystems of a freely evolving composite are not, in general, uniquely fixed by the evolution of the state of that composite (failure of the assumption of Dynamical Autonomy for composite systems). Another conclusion is that in the spectral modal interpretation the transition probabilities $p(C_c^\alpha(t), C_d^\beta(t)/C_a^\alpha(s), C_b^\beta(s))$ for two disjoint systems α and β need not exist. A third conclusion is that in the atomic modal interpretation the transition probabilities for molecular systems are not uniquely fixed by the evolution of the state of that molecule (failure of Dynamical Autonomy for whole systems).

The underderminateness of transition probabilities is clearly visible in the work by Bacciagaluppi and Dickson (Sections 8.4 and 8.5). In a number of publications[88] they have constructed by means of the theory of stochastic processes a framework with which one can find transition probabilities for all freely evolving and interacting systems (so the existence of such general transition probabilities is proved). This framework allows the generation of many different expressions for modal transition probabilities, although some of these expressions are ruled out by the already established transition probabilities $p(C_b^\alpha(t)/C_a^\alpha(s)) = \delta_{ab}$ for freely evolving systems.

So, to sum up, there exist in modal interpretations transition probabilities $p(C_b^\alpha(t)/C_a^\alpha(s))$ which describe the dynamics of the actually possessed properties of systems. But, with the exception of a number of cases, there does not yet exist an argument which fixes these transition probabilities uniquely.

8.2 Freely evolving systems: determinism

I start by deriving transition probabilities for freely evolving systems (Vermaas 1996, Sects. 3 and 4). There exist, as I have said, many expressions for transition probabilities which are compatible with modal interpretations. Modal interpretations thus lack the necessary structure to develop them uniquely into fully-fledged theories about possessed properties. In order to proceed I therefore reverse the order of deduction: I require that the correct solution for transition probabilities for freely evolving systems satisfies

---

[88] The original paper is Bacciagaluppi and Dickson (1997) but see also Bacciagaluppi (1996b, 2000) and Dickson (1998a,b). A comprehensive paper is Bacciagaluppi (1998).

the criterion of Empirical Adequacy and then uniquely derive this correct solution from this criterion (see Section 3.3 for this Kantian reversal of deduction).

First a word about notation. In quantum mechanics the state of a freely evolving system α is at time t given by (see Section 2.1)

$$W^\alpha(t) = U^\alpha(t, s) W^\alpha(s) U^\alpha(s, t) \qquad (8.3)$$

with s some initial instant and with $U^\alpha(x, y)$ equal to $\exp([(x - y)/i\hbar] H^\alpha)$. Repeating remarks made in Section 7.2, the eigenprojections of both $W^\alpha(t)$ and $W^\alpha(s)$ lie on continuous trajectories $\{T_q^\alpha(t)\}_q$ given by

$$T_q^\alpha(t) = U^\alpha(t, s) P_q^\alpha(s) U^\alpha(s, t). \qquad (8.4)$$

I now adopt the convention of labelling the eigenprojections of a freely evolving state such that $P_k^\alpha(t)$ of $W^\alpha(t)$ and $P_j^\alpha(s)$ of $W^\alpha(s)$ lie on the same trajectory $T_q^\alpha(t)$ if and only if $k = j$. This implies the labelling

$$P_k^\alpha(t) := U^\alpha(t, s) P_k^\alpha(s) U^\alpha(s, t). \qquad (8.5)$$

In order to execute the announced reverse deduction, I define a model in which the transition probabilities for a freely evolving system uniquely fix the transition probabilities for outcomes possessed by a pointer. By then invoking the criterion of Empirical Adequacy, one can calculate the transition probabilities for the outcomes and hence fix the transition probabilities for the freely evolving system as well. Take, firstly, the bi modal interpretation. The model then consists of a system α and a pointer π. From s_0 to s a measurement is performed on α by means of π and the measured magnitude is represented by the operator $A^\alpha = \sum_j a_j P_j^\alpha(s)$ (A^α is thus an operator which has the same eigenprojections as the state of α at s). Let $\{|p_{jk}^\alpha\rangle\}_{j,k}$ be a set of pair-wise orthogonal vectors in $\mathscr{H}^\alpha$ which satisfy $P_j^\alpha(s) = \sum_k |p_{jk}^\alpha\rangle\langle p_{jk}^\alpha|$. And let the measurement interaction between α and π induce the following evolution

$$U^{\alpha\pi}(s, s_0) |p_{jk}^\alpha\rangle \otimes |R_0^\pi\rangle = |p_{jk}^\alpha\rangle \otimes |R_{jk}^\pi\rangle, \qquad (8.6)$$

where $\{|R_{jk}^\pi\rangle\}_{j,k}$ are pair-wise orthogonal vectors which are related to the reading states $\{R_j^\pi\}_j$ of the pointer as $R_j^\pi = \sum_k |R_{jk}^\pi\rangle\langle R_{jk}^\pi|$, for all j. (I here thus consider the unphysical case of a measurement device consisting of only a pointer, see Section 4.7.) According to the standard formulation of quantum mechanics, this interaction is an acceptable measurement interaction: if α is initially in an eigenstate corresponding to the eigenvalue a_a of A^α, say $|\psi^\alpha(s_0)\rangle = \sum_k c_k |p_{ak}^\alpha\rangle$, the final pointer state is $W^\pi(s) = \sum_k |c_k|^2 |R_{ak}^\pi\rangle\langle R_{ak}^\pi|$;

and given this final state, the probability $\mathrm{Tr}^{\pi}(W^{\pi}(s)\,R_a^{\pi})$ to find the reading R_a^{π} is equal to 1.

Then, from s to t, the systems α and π evolve freely, α evolves by means of some internal dynamics given by $U^{\alpha}(t,s)$ and π evolves by means of the unit operator. The evolution of $\alpha\pi$ is thus given by $U^{\alpha\pi}(t,s)=U^{\alpha}(t,s)\otimes \mathbb{I}^{\pi}$.

Let $|\Psi^{\alpha\pi}(s_0)\rangle = \sum_{j,k}\sqrt{w_j^{\alpha}(s)}\,|p_{jk}^{\alpha}\rangle\otimes|\mathrm{R}_0^{\pi}\rangle$ be the initial state of the model (the values $\{w_j^{\alpha}(s)\}_j$ are the eigenvalues of the state of α at s). At s and t the state of the model is then

$$\left.\begin{aligned}|\Psi^{\alpha\pi}(s)\rangle &= \sum_{j,k}\sqrt{w_j^{\alpha}(s)}\,|p_{jk}^{\alpha}\rangle\otimes|\mathrm{R}_{jk}^{\pi}\rangle,\\[2mm]|\Psi^{\alpha\pi}(t)\rangle &= \sum_{j,k}\sqrt{w_j^{\alpha}(s)}\,U^{\alpha}(t,s)|p_{jk}^{\alpha}\rangle\otimes|\mathrm{R}_{jk}^{\pi}\rangle.\end{aligned}\right\} \tag{8.7}$$

Partial tracing yields that the states of α and π evolve from s to t as

$$\left.\begin{aligned}W^{\alpha}(s) &= \sum_{j}w_j^{\alpha}(s)P_j^{\alpha}(s) \mapsto W^{\alpha}(t)=U^{\alpha}(t,s)W^{\alpha}(s)U^{\alpha}(s,t)=\sum_{j}w_j^{\alpha}(s)P_j^{\alpha}(t),\\[2mm]W^{\pi}(s) &= \sum_{j}w_j^{\alpha}(s)R_j^{\pi} \quad\mapsto W^{\pi}(t)=\sum_{j}w_j^{\alpha}(s)R_j^{\pi},\end{aligned}\right\} \tag{8.8}$$

with $P_j^{\alpha}(t)$ defined as in (8.5). Note that one can embed every possible free evolution (8.3) of α in the model: the model poses no restrictions on the eigenvalues $\{w_j^{\alpha}(s)\}_j$ and the eigenprojections $\{P_j^{\alpha}(s)\}_j$ of the state $W^{\alpha}(s)$, and poses no restrictions on the evolution $U^{\alpha}(t,s)$.

Apply the bi modal interpretation to α and π at s and t. The state of $\alpha\pi$ is pure at all times so one can indeed ascribe properties to α and π as well as correlate them. At s the core projections of α and π are $\{P_j^{\alpha}(s)\}_j$ and $\{R_j^{\pi}\}_j$, respectively, and at t they are $\{P_k^{\alpha}(t)\}_k$ and $\{R_k^{\pi}\}_k$, respectively. Consider now the joint probability $p(P_a^{\alpha}(s),P_b^{\alpha}(t),R_c^{\pi}$ at s, R_d^{π} at $t)$ that the respective actual core properties of α and π are sequentially $P_a^{\alpha}(s)$ and $P_b^{\alpha}(t)$ and R_c^{π} at s and R_d^{π} at t. The joint property ascription (4.10) given on page 51 correlates the properties of α and π one-to-one at both s and t. This implies that $p(P_a^{\alpha}(s),P_b^{\alpha}(t),R_c^{\pi}$ at s, R_d^{π} at $t)$ is zero if either $a\neq c$ or $b\neq d$. One can thus derive that

$$p(P_b^{\alpha}(t)/P_a^{\alpha}(s)) = p(R_b^{\pi}\text{ at }t/R_a^{\pi}\text{ at }s). \tag{8.9}$$

Hence, the desired transition probabilities for α in the model are equal to the conditional probabilities that the pointer possesses the reading R_b^{π} at t, given that it possessed R_a^π at s. These latter conditional probabilities (the right-hand side of (8.9)) yield statistical predictions about measurement outcomes and should according to Empirical Adequacy be equal to the conditional probabilities generated by the standard formulation of quantum mechanics.

Apply therefore the standard formulation to the model. If π assumes the outcome R_a^π at s, then, according to the standard formulation, the state of $\alpha\pi$ collapses to

$$\widehat{W}^{\alpha\pi}(s) = \frac{[\mathbb{I}^\alpha \otimes R_a^\pi] \, W^{\alpha\pi}(s) \, [\mathbb{I}^\alpha \otimes R_a^\pi]}{\mathrm{Tr}^{\alpha\pi}(W^{\alpha\pi}(s) \, [\mathbb{I}^\alpha \otimes R_a^\pi])} = \sum_{q,r} \frac{|p_{aq}^\alpha\rangle\langle p_{ar}^\alpha| \otimes |R_{aq}^\pi\rangle\langle R_{ar}^\pi|}{\mathrm{Tr}^\alpha(P_a^\alpha(s))}. \tag{8.10}$$

This collapsed state evolves to t by $U^{\alpha\pi}(t,s)$ and it follows from the Born rule that

$$p_{\mathrm{Born}}(R_b^\pi \text{ at } t / R_a^\pi \text{ at } s) = \mathrm{Tr}^{\alpha\pi}(\widehat{W}^{\alpha\pi}(t) \, [\mathbb{I}^\alpha \otimes R_b^\pi]) = \delta_{ab}. \tag{8.11}$$

So, by invoking Empirical Adequacy, one can uniquely fix the right-hand side of (8.9) and conclude that

$$p(P_b^\alpha(t)/P_a^\alpha(s)) = \delta_{ab}. \tag{8.12}$$

This result follows uniquely from the criterion of Empirical Adequacy but is still valid only in the special case that α is part of the sketched model. By now using the assumption of Dynamical Autonomy for whole systems, one can turn (8.12) into a generally valid result (here I copy exactly the line of reasoning followed at the end of Section 6.2). Take any system α with a freely evolving state (8.3). If this system is embedded in the above model, the transition probabilities $p(P_b^\alpha(t)/P_a^\alpha(s))$ are equal to δ_{ab}. If this system is not embedded in such a model, the transition probabilities are unknown. Dynamical Autonomy for whole systems demands that in both cases the transition probabilities are the same.[89] Hence, also if a system with a freely evolving state (8.3) is not embedded in the above model, the transition probabilities $p(P_b^\alpha(t)/P_a^\alpha(s))$ are equal to δ_{ab}.

For the spectral modal interpretation one can use a similar argument to derive the same result (8.12) but now with a physically more realistic measurement. Take a model comprising α and a measurement device μ which consists of a pointer π and a mechanism $\bar{\mu}$. Let the pointer reading magnitude be given by $M^\pi = \sum_j m_j |R_j^\pi\rangle\langle R_j^\pi|$ and let $\{|\bar{D}_j^{\bar\mu}\rangle\}_j$ be a set of pairwise orthogonal vectors in $\mathcal{H}^{\bar\mu}$. From s_0 to s a measurement is performed

---

[89] Dynamical Autonomy for whole systems can be used because the transition probabilities $p(P_b^\alpha(t)/P_a^\alpha(s)) = \delta_{ab}$ derived with the model are independent of the state of the pointer π of the model.

on α by means of the evolution

$$U^{\alpha\mu}(s, s_0) |p_{jk}^{\alpha}\rangle \otimes |D_0^{\mu}\rangle = |p_{jk}^{\alpha}\rangle \otimes |\bar{D}_j^{\bar{\mu}}\rangle \otimes |R_j^{\pi}\rangle. \tag{8.13}$$

And from s to t the evolution of the model is governed by $U^{\alpha}(t, s) \otimes \mathbb{I}^{\mu}$ such that all the systems α, $\bar{\mu}$ and π evolve freely.

By starting with an initial state $W^{\alpha\mu}(s_0) = \sum_j w_j^{\alpha}(s) P_j^{\alpha}(s) \otimes |D_0^{\mu}\rangle\langle D_0^{\mu}|$, and by applying the spectral modal interpretation to α and π at s and at t, one again obtains the evolution (8.8) for the states of α and of π such that one can derive the identity (8.9). Repeating the above argument yields the transition probabilities (8.12).

With a few changes one can use this last model also to derive the transition probabilities (8.12) for any freely evolving *atom* in the case of the atomic modal interpretation. These changes concern the pointer vectors $\{|R_j^{\pi}\rangle\}_j$ in (8.13) and are necessary for ascribing readings to the pointer with the atomic modal interpretation (see Section 4.7). Let the atoms in the pointer be given by $\{\beta_q\}_{q=1}^n$. Then, firstly, the vector $|R_j^{\pi}\rangle$ in (8.13) should be equal to the vector $|R_j^{\beta_1}\rangle \otimes \cdots \otimes |R_j^{\beta_n}\rangle$, where the vectors $\{|R_j^{\beta_1}\rangle\}_j, \ldots, \{|R_j^{\beta_n}\rangle\}_j$ are all sets of pair-wise orthogonal vectors. Secondly, the readings should be defined as $R_j^{\pi} = |R_j^{\beta_1}\rangle\langle R_j^{\beta_1}| \otimes \cdots \otimes |R_j^{\beta_n}\rangle\langle R_j^{\beta_n}|$. With these changes one can again derive that the transition probabilities for any atom α embedded in the model, are given by (8.12). Then by using the assumption of Dynamical Autonomy for atomic systems one can turn this result into a generally valid one.

The evolution of the actually possessed core property of a freely evolving *molecule* in the atomic modal interpretation is in general not given by (8.12). Take, for instance, a molecule γ consisting of the atoms α_1 and α_2. The free evolution of γ can then map a state $|e_1^{\alpha_1}\rangle \otimes |f_1^{\alpha_2}\rangle$ at s to a state $c_1 |e_1^{\alpha_1}\rangle \otimes |f_1^{\alpha_2}\rangle + c_2 |e_2^{\alpha_1}\rangle \otimes |f_2^{\alpha_2}\rangle$ at t, where $\langle e_1^{\alpha_1}|e_2^{\alpha_1}\rangle = \langle f_1^{\alpha_2}|f_2^{\alpha_2}\rangle = 0$. The actual core property of γ is $|e_1^{\alpha_1}\rangle\langle e_1^{\alpha_1}| \otimes |f_1^{\alpha_2}\rangle\langle f_1^{\alpha_2}|$ at s with probability 1 and $|e_j^{\alpha_1}\rangle\langle e_j^{\alpha_1}| \otimes |f_j^{\alpha_2}\rangle\langle f_j^{\alpha_2}|$ at t with probability $|c_j|^2$, $j = 1$ or 2. The transition probabilities for γ are thus

$$p(|e_j^{\alpha_1}\rangle\langle e_j^{\alpha_1}| \otimes |f_j^{\alpha_2}\rangle\langle f_j^{\alpha_2}| \text{ at } t \,/\, |e_1^{\alpha_1}\rangle\langle e_1^{\alpha_1}| \otimes |f_1^{\alpha_2}\rangle\langle f_1^{\alpha_2}| \text{ at } s) =$$
$$p(|e_j^{\alpha_1}\rangle\langle e_j^{\alpha_1}| \otimes |f_j^{\alpha_2}\rangle\langle f_j^{\alpha_2}| \text{ at } t) = |c_j|^2 \tag{8.14}$$

which clearly contradicts (8.12).

One can try to determine the transition probabilities for freely evolving molecules in the atomic modal interpretation by means of the evolution of the properties of the atoms. If all the individual atoms $\{\alpha_q\}_q$ in a molecule γ evolve freely (so, $U^{\gamma}(t, s) = U^{\alpha_1}(t, s) \otimes U^{\alpha_2}(t, s) \otimes \cdots$), this indeed works. Loosely, if γ possesses $P_{abc\cdots}^{\gamma}(s)$ at s, the atoms in γ possess with probability 1 simultaneously, respectively, $P_a^{\alpha_1}(s)$, $P_b^{\alpha_2}(s)$, ..., at s. Then, using (8.12), the atoms possess with probability 1 simultaneously, respectively, $U^{\alpha_1}(t,s)P_a^{\alpha_1}(s)U^{\alpha_1}(s,t)$, $U^{\alpha_2}(t,s)P_b^{\alpha_2}(s)U^{\alpha_2}(s,t)$, ..., at t, and it follows with probability 1 that γ possesses $P_{abc\cdots}^{\gamma}(t) := U^{\gamma}(t,s)P_{abc\cdots}^{\gamma}(s)U^{\gamma}(s,t)$ at t. Hence, if the atoms in γ evolve freely, the transition probabilities are

$$p(P_{a'b'c'\cdots}^{\gamma}(t)/P_{abc\cdots}^{\gamma}(s)) = \delta_{aa'}\delta_{bb'}\delta_{cc'}\cdots . \tag{8.15}$$

If, on the other hand, γ evolves freely but its atoms $\{\alpha_q\}_q$ interact with one another, one cannot yet derive transition probabilities for γ from the property evolution of the atoms because transition probabilities for interacting atoms are not yet determined.

Evaluating these results, it can be concluded that the evolution of the actually possessed core property of a freely evolving system α (or of a freely evolving atomic system in the case of the atomic modal interpretation) is with probability 1 confined to only one continuous trajectory $T_q^{\alpha}(t)$ of eigenprojections of the state $W^{\alpha}(t)$. For if at some instant s the actual core property is given by $P_j^{\alpha}(s)$, then the actual core property is according to (8.12) with probability 1 at all times t given by $P_j^{\alpha}(t)$. And all these projections lie on the same trajectory $T_j^{\alpha}(t) = U_j^{\alpha}(t,s)P_j^{\alpha}(s)U_j^{\alpha}(s,t)$.

One can characterise this confined evolution as *deterministic*: if the actual core property lies at one instant on trajectory $T_j^{\alpha}(t)$ it lies with certainty always on that trajectory; the actual core property jumps only with zero probability from one trajectory to another, ruling out a truly stochastic evolution.[90]

Finally, one can observe that the property evolution of freely evolving systems harmonises with the standard formulation of quantum mechanics. If one calculates the transition probabilities $p(P_b^{\alpha}(t)/P_a^{\alpha}(s))$ by means of the Born rule (ignoring the fact that this rule should only be used if measurements are performed), one finds the same result as (8.12).

8.3 Interacting systems: stochasticity

With the help of the deterministic evolution derived for freely evolving systems, one can determine for the spectral and the atomic modal interpretation transition probabilities for special cases of interacting systems (Vermaas 1996, Sect. 5). The trick is to embed the interacting system α in a model in which a freely evolving snooper system σ carries a record of the core property that α actually possesses at instant s from s to a second instant t.

---

[90] In the literature this deterministic evolution is often described by stating that the actual core properties of freely evolving systems exhibit *stability*.

By then determining the correlation between the actual core property of α at t and this record, one indirectly fixes the transition probabilities for α from s to t.

In quantum mechanics the state of a system α that interacts with an environment β is at t given by (see Section 2.1)

$$W^{\alpha}(t) = \text{Tr}^{\beta}(U^{\alpha\beta}(t,s) \, W^{\alpha\beta}(s) \, U^{\alpha\beta}(s,t)), \tag{8.16}$$

with $W^{\alpha\beta}(s)$ the state of the composite $\alpha\beta$ at some initial time s and with $U^{\alpha\beta}(x,y)$ equal to $\exp([(x-y)/i\hbar] \, H^{\alpha\beta})$ (it is thus assumed that $\alpha\beta$ as a whole evolves freely). The results of the last chapter yield that, in general, there are no continuous trajectories which connect the eigenprojections of $W^{\alpha}(t)$ and of $W^{\alpha}(s)$. I therefore do not (or, more accurately, cannot) adopt a convention of labelling the eigenprojections of $W^{\alpha}(t)$ similar to (8.5).

Let's start with the spectral modal interpretation. Take a model with three systems: α, the environment β and a snooper σ. At a first instant s the state of the model is such that there exists a strict correlation between the core properties of α and of σ. The state $W^{\sigma\alpha\beta}(s)$ thus obeys

$$\forall i,j: \qquad i \neq j \;\Rightarrow\; \text{Tr}^{\sigma\alpha\beta}(W^{\sigma\alpha\beta}(s) \, [P^{\sigma}_i(s) \otimes P^{\alpha}_j(s) \otimes \mathbb{I}^{\beta}]) = 0. \tag{8.17}$$

Then, from s to t, the snooper evolves freely by means of the unit operator $\mathbb{I}^{\sigma}$, α interacts with the environment β and the composite $\alpha\beta$ evolves freely with some arbitrary $U^{\alpha\beta}(t,s)$. At t the state of the model is then

$$W^{\sigma\alpha\beta}(t) = [\mathbb{I}^{\sigma} \otimes U^{\alpha\beta}(t,s)] \, W^{\sigma\alpha\beta}(s) \, [\mathbb{I}^{\sigma} \otimes U^{\alpha\beta}(s,t)]. \tag{8.18}$$

By taking partial traces, it follows that the state of α evolves as in (8.16). However, not every possible evolution (8.16) of α (characterised by the pair $W^{\alpha\beta}(s)$ and $U^{\alpha\beta}(t,s)$) can be embedded in this model. One can choose every evolution $U^{\alpha\beta}(t,s)$ but the choice of $W^{\alpha\beta}(s)$ is limited by (8.17). In the MATHEMATICS I prove that if $W^{\sigma\alpha\beta}(s)$ obeys (8.17), then $W^{\alpha\beta}(s)$ obeys

$$W^{\alpha\beta}(s) = \sum_m [P^{\alpha}_m(s) \otimes \mathbb{I}^{\beta}] \, W^{\alpha\beta}(s) \, [P^{\alpha}_m(s) \otimes \mathbb{I}^{\beta}]. \tag{8.19}$$

Conversely, if one has a state $W^{\alpha\beta}(s)$ that obeys (8.19), then there exists a state $W^{\sigma\alpha\beta}(s)$ that obeys (8.17). Hence, one can consider a system α interacting with β if and only if the composite state $W^{\alpha\beta}(s)$ obeys condition (8.19).

Apply the spectral modal interpretation to this new model. At s the core properties of the snooper and α are, respectively, $\{P^{\sigma}_i(s)\}_i$ and $\{P^{\alpha}_j(s)\}_j$. At t these core properties are, respectively, $\{P^{\sigma}_k(t)\}_k$ (with $P^{\sigma}_k(t) = P^{\sigma}_k(s)$ for all k) and $\{P^{\alpha}_l(t)\}_l$. Consider then the joint probability $p(P^{\alpha}_a(s), P^{\alpha}_b(t), P^{\sigma}_c(s), P^{\sigma}_d(t))$

that these systems possess sequentially their respective actual core properties. Due to the one-to-one correlations between the snooper and α at s, this joint probability is zero if $a \neq c$. And by the free evolution of σ and the deterministic transition probabilities (8.12), this joint probability is also zero if $c \neq d$. Hence, one can derive that

$$p(P_a^\alpha(s), P_b^\alpha(t)) = p(P_a^\sigma(t), P_b^\alpha(t)). \tag{8.20}$$

The right-hand side is the probability that α and the snooper possess simultaneously their respective actual core properties at t. These systems are mutually disjoint, so this probability can be calculated with the spectral modal interpretation, yielding

$$p(P_a^\alpha(s), P_b^\alpha(t)) = \text{Tr}^{\sigma\alpha\beta}(W^{\sigma\alpha\beta}(t)\, [P_a^\sigma(t) \otimes P_b^\alpha(t) \otimes \mathbb{I}^\beta]). \tag{8.21}$$

In the MATHEMATICS it is proved from condition (8.17) that one can rewrite this joint probability as

$$p(P_a^\alpha(s), P_b^\alpha(t)) = \text{Tr}^{\alpha\beta}(U^{\alpha\beta}(t,s)\, W^{\alpha\beta}(s)\, [P_a^\alpha(s) \otimes \mathbb{I}^\beta]\, U^{\alpha\beta}(s,t)\, [P_b^\alpha(t) \otimes \mathbb{I}^\beta]). \tag{8.22}$$

With this solution one can derive two sets of transition probabilities. In the model it is assumed that $s < t$. The properties of α thus evolve from s to t and the transition probabilities are equal to

$$p(P_b^\alpha(t)/P_a^\alpha(s)) = \frac{\text{Tr}^{\alpha\beta}(U^{\alpha\beta}(t,s)\, W^{\alpha\beta}(s)\, [P_a^\alpha(s) \otimes \mathbb{I}^\beta]\, U^{\alpha\beta}(s,t)\, [P_b^\alpha(t) \otimes \mathbb{I}^\beta])}{\text{Tr}^\alpha(W^\alpha(s)\, P_a^\alpha(s))}. \tag{8.23}$$

However, one can also assume that $t < s$ (the state of $\sigma\alpha\beta$ then satisfies condition (8.17) at the final instant s) such that the properties of α evolve from t to s. The transition probabilities are then

$$p(P_a^\alpha(s)/P_b^\alpha(t)) = \frac{\text{Tr}^{\alpha\beta}(U^{\alpha\beta}(t,s)\, W^{\alpha\beta}(s)\, [P_a^\alpha(s) \otimes \mathbb{I}^\beta]\, U^{\alpha\beta}(s,t)\, [P_b^\alpha(t) \otimes \mathbb{I}^\beta])}{\text{Tr}^\alpha(W^\alpha(t)\, P_b^\alpha(t))}. \tag{8.24}$$

If one stipulates that the snooper σ is an atomic system, one can also use the above model for deriving transition probabilities for interacting atoms in the atomic modal interpretation. So, take both σ and α atomic (β may be a huge molecule) and apply the atomic modal interpretation to the model. Since the spectral and atomic modal interpretations are equivalent with regard to the property ascription to atoms, one can again use the above argument and conclude that the transition probabilities (8.23) and (8.24) also hold for atoms in the atomic modal interpretation.[91]

These results still hold only for an interacting system α for which there exists a snooper σ at the instant s. The question is now whether these results also hold for interacting systems α if such a snooper is absent.

In the previous section we have seen that the transition probabilities derived for freely evolving systems in the special case that a measurement device is present could be used to fix the transition probabilities for freely evolving systems in the general case that no measurement is present. This generalisation was made possible in the spectral modal interpretation by invoking the assumption of Dynamical Autonomy for whole systems, and was made possible in the atomic modal by invoking the assumption of Dynamical Autonomy for atomic systems. Essentially these two assumptions say that the transition probabilities for freely evolving systems or for freely evolving atoms are uniquely fixed by the evolution of the state of that system or atom, respectively. Hence, if one knows in one special case the transition probabilities for a freely evolving system or atom, one knows them in all cases.

One may now try to do something similar for interacting systems. That is, one may try to argue that the transition probabilities (8.23) and (8.24) also hold for interacting systems in the general case that no snooper for α is present.

A first step towards such a generalisation is to reformulate the above results such that they do not refer to the snooper σ. This first step is indeed possible. The transition probabilities (8.23) and (8.24) are functions of $W^{\alpha\beta}(s)$, $U^{\alpha\beta}(t,s)$, $P_a^{\alpha}(s)$, and $P_b^{\alpha}(t)$ only, so they do not refer to the snooper. And condition (8.17) on the state $W^{\alpha\beta}(s)$ can be replaced by the condition (8.19) (these two conditions were proved to be equivalent) such that this condition does not refer to the snooper either.

The second step is to invoke the assumption of Dynamical Autonomy for composite systems and thus to assume that the transition probabilities for α, the subsystem of a freely evolving composite $\alpha\beta$, are uniquely fixed by the evolution of the state of that composite $\alpha\beta$. However, one can show that

---

[91] One can try to determine with the above model transition probabilities for interacting systems in the bi modal interpretation. Unfortunately this does not work in general. To see this, consider the state of $\alpha\beta$ at s and at t. In order to fix the transition probabilities $p(P_b^{\alpha}(t)/P_a^{\alpha}(s))$, one has to know the correlations between σ and α at s and at t. However, these correlations exist in the bi modal interpretation only if the states $W^{\sigma\alpha}(s)$ and $W^{\sigma\alpha}(t)$ are pure. This implies (see footnote 11) that $\sigma\alpha\beta$ has a state $|\Psi^{\sigma\alpha}\rangle\langle\Psi^{\sigma\alpha}| \otimes W^{\beta}$ at s and at t. Hence, $W^{\alpha\beta} = W^{\alpha} \otimes W^{\beta}$ at s and at t such that one can determine $p(P_b^{\alpha}(t)/P_a^{\alpha}(s))$ only if the state of $\alpha\beta$ is factorised at s and at t. Since interactions between systems in general entangle the states of those systems, one thus cannot, in general, determine transition probabilities in the bi modal interpretation.

if the transition probabilities (8.23) and (8.24) are valid for all interacting systems α with states $W^{\alpha\beta}(s)$ that satisfy the condition (8.19), then the spectral and atomic modal interpretations are contradictory. This is proved by an example, given in Section 9.2 of the next chapter, in which two composites $\alpha\beta$ and $\alpha'\beta'$ of atomic systems evolve freely and with equal states that satisfy the condition (8.19). If one then assumes that the transition probabilities for the systems α and α' are both given by (8.23), one arrives at a contradiction. The transition probabilities for a system α, part of a freely evolving composite $\alpha\beta$, are thus not uniquely fixed by the state of $\alpha\beta$. Hence, the transition probabilities (8.23) and (8.24) are not valid for all interacting systems α that satisfy the condition (8.19); these transition probabilities are valid if there is a snooper σ for α actually present at s.[92]

The example which proves that the spectral and atomic modal interpretations violate Dynamical Autonomy for composite systems[93] can also be used to arrive at two further worrisome conclusions. The first is that if one accepts the deterministic evolution (8.12) of the core properties of freely evolving systems, then the transition probabilities $p(P_c^\alpha(t), P_d^\beta(t)/P_a^\alpha(s), P_b^\beta(s))$, which give the joint evolution of the core properties of two systems, sometimes do not exist in the spectral modal interpretation. And the only way to circumvent this first conclusion is by accepting perspectivalism. The second conclusion is that if one accepts the deterministic evolution (8.12) for freely evolving atoms, then the atomic modal interpretation violates the assumption of Dynamical Autonomy for whole molecular systems. This means that two freely evolving molecules $\alpha\beta$ and $\alpha'\beta'$ can have the same state evolution while the transition probabilities $p(P_c^\alpha(t) \otimes P_d^\beta(t)/P_a^\alpha(s) \otimes P_b^\beta(s))$ and $p(P_c^{\alpha'}(t) \otimes P_d^{\beta'}(t)/P_a^{\alpha'}(s) \otimes P_b^{\beta'}(s))$ for the core properties of these molecules are not the same. (See Section 9.2 for the proofs of these two conclusions.)

From the transition probabilities (8.23) and (8.24) one can reach some general conclusions about the evolution of the core properties of interacting systems. Firstly, this evolution is truly stochastic in the sense that the transition probabilities can have values between 0 and 1. So, if the state of an interacting system α allows a dynamical decomposition in terms of continuous trajectories $\{T_q^\alpha(t)\}_q$ of eigenprojections, then the actual core property

---

[92] In Section 5 of Vermaas (1996) I claimed that the transition probabilities (8.23) and (8.24) could indeed be turned into generally valid ones. I argued for this generalisation by means of an assumption (called **R2″**) which is equivalent to Dynamical Autonomy for composite systems. However, during discussions, Guido Bacciagaluppi (private communication, 1998) made it clear to me that this assumption is untenable. My conclusion in Vermaas (1996) that the transition probabilities (8.23) and (8.24) are generally valid is thus wrong.

[93] One cannot give a proof that Dynamical Autonomy for composite systems is also untenable in the bi modal interpretation because this interpretation is silent about the transition probabilities (8.23) and (8.24) used in this proof.

of α is thus only for finite times confined to one trajectory and then jumps to another trajectory. Secondly, it can be proved that the transition probabilities need not be equal to those derived by the Born rule. The transition probabilities (8.23) are surprisingly exactly equal to those given by the Born rule but the transition probabilities (8.24) can be different. Finally, the evolution of the actual core properties need not be a Markov process. That is,[94] the evolution of the actually possessed core property of α need not satisfy the relation

$$p(P_k^\alpha(t)/P_j^\alpha(s)) = p(P_k^\alpha(t)/P_{i_1}^\alpha(s_1), P_{i_2}^\alpha(s_2), \dots, P_j^\alpha(s)), \tag{8.25}$$

where $\{s_i\}_i$ are ordered instants $s_1 < s_2 < \dots < s < t$. It follows that in order to most precisely determine the probability that α actually possesses $P_k^\alpha(t)$ at t, one should not conditionalise only on the last known actually possessed core property $P_j^\alpha(s)$ of α, but on the whole history of actually possessed core properties $P_{i_1}^\alpha(s_1)$, $P_{i_2}^\alpha(s_2)$, etc. Hence, the last actually possessed core property of α does not carry all information for predicting the future actually possessed properties of α.[95]

These three conclusions (stochasticity, non-Born like and non-Markovian transition probabilities) are proved and illustrated by an example given at the end of the MATHEMATICS.

MATHEMATICS

Firstly, I prove that if $W^{\sigma\alpha\beta}(s)$ satisfies condition (8.17), then its partial trace $W^{\alpha\beta}(s)$ satisfies condition (8.19).

Lemma 8.1
If $\mathrm{Tr}^{\sigma\alpha\beta}(W^{\sigma\alpha\beta}[P_i^\sigma \otimes P_j^\alpha \otimes \mathbb{I}^\beta]) = 0$ *for all* $i \neq j$, *then its partial trace* $W^{\alpha\beta}$ *is equal to* $\sum_m [P_m^\alpha \otimes \mathbb{I}^\beta] W^{\alpha\beta} [P_m^\alpha \otimes \mathbb{I}^\beta].$

Proof: Suppose that $\mathrm{Tr}^{\sigma\alpha\beta}(W^{\sigma\alpha\beta}[P_i^\sigma \otimes P_j^\alpha \otimes \mathbb{I}^\beta]) = 0$ for all $i \neq j$. If one identifies the projections $\{Q_q^\kappa\}_q$ with $\{P_i^\sigma\}_i$ and $\{Q_r^\lambda\}_r$ with $\{P_j^\alpha \otimes \mathbb{I}^\beta\}_j$, Theorem 6.1 on page 91 yields

$$\forall s: \quad \mathrm{Tr}^\sigma(W^{\sigma\alpha\beta}[P_s^\sigma \otimes \mathbb{I}^{\alpha\beta}]) = [P_s^\alpha \otimes \mathbb{I}^\beta] W^{\alpha\beta} = W^{\alpha\beta}[P_s^\alpha \otimes \mathbb{I}^\beta]. \tag{8.26}$$

---

[94] See, for instance, Doob (1953, Sect. II.6) and Feller (1950, Sect. 15.10).

[95] Jeremy Butterfield (private communication, 1995) noted that this absence of the Markov property is in a sense to be expected. Since α is interacting with an environment β, all possible information about the future behaviour of α should be expected to be present in the composite $\alpha\beta$. So, if the evolution of actual core properties should satisfy the Markov property, it should satisfy it with respect to the actually possessed core properties of α, of β and possibly also of $\alpha\beta$.

The operators $\{P_s^\alpha \otimes \mathbb{I}^\beta\}_s$ therefore commute with $W^{\alpha\beta}$ and one can derive that

$$W^{\alpha\beta} = \mathbb{I}^{\alpha\beta} \, W^{\alpha\beta} = \sum_s [P_s^\alpha \otimes \mathbb{I}^\beta] \, W^{\alpha\beta} = \sum_s [P_s^\alpha \otimes \mathbb{I}^\beta]^2 \, W^{\alpha\beta}$$

$$= \sum_s [P_s^\alpha \otimes \mathbb{I}^\beta] \, W^{\alpha\beta} \, [P_s^\alpha \otimes \mathbb{I}^\beta]. \tag{8.27}$$

$\square$

Secondly, I prove that for every state $W^{\alpha\beta}(s)$ satisfying condition (8.19), there exists a state $W^{\sigma\alpha\beta}(s)$ which satisfies condition (8.17) and has $W^{\alpha\beta}(s)$ as a partial trace.

Lemma 8.2
If $W^{\alpha\beta}$ is equal to $\sum_m [P_m^\alpha \otimes \mathbb{I}^\beta] W^{\alpha\beta} [P_m^\alpha \otimes \mathbb{I}^\beta]$, then there exists a state $W^{\sigma\alpha\beta}$ which satisfies that $\mathrm{Tr}^{\sigma\alpha\beta}(W^{\sigma\alpha\beta} [P_i^\sigma \otimes P_j^\alpha \otimes \mathbb{I}^\beta]) = 0$ for all $i \ne j$.

Proof: Take, for instance,

$$W^{\sigma\alpha\beta} = \sum_m P_m^\sigma \otimes \left([P_m^\alpha \otimes \mathbb{I}^\beta] \, W^{\alpha\beta} \, [P_m^\alpha \otimes \mathbb{I}^\beta] \right) \tag{8.28}$$

with $\{P_m^\sigma\}_m$ a set of one-dimensional and pair-wise orthogonal projections. $\square$

Thirdly, I prove from condition (8.17) the equivalence of (8.21) and (8.22):

Theorem 8.1
If $U^{\sigma\alpha\beta}(t,s) = \mathbb{I}^\sigma \otimes U^{\alpha\beta}(t,s)$ and if $W^{\sigma\alpha\beta}(s)$ obeys

$$\forall i, j : \quad i \ne j \;\Rightarrow\; \mathrm{Tr}^{\sigma\alpha\beta}(W^{\sigma\alpha\beta}(s) [P_i^\sigma(s) \otimes P_j^\alpha(s) \otimes \mathbb{I}^\beta]) = 0, \tag{8.29}$$

then

$$\forall a, b : \quad \mathrm{Tr}^{\sigma\alpha\beta}(W^{\sigma\alpha\beta}(t) [P_a^\sigma(t) \otimes P_b^\alpha(t) \otimes \mathbb{I}^\beta]) =$$
$$\mathrm{Tr}^{\alpha\beta}(U^{\alpha\beta}(t,s) \, W^{\alpha\beta}(s) \, [P_a^\alpha(s) \otimes \mathbb{I}^\beta] \, U^{\alpha\beta}(s,t) \, [P_b^\alpha(t) \otimes \mathbb{I}^\beta]). \tag{8.30}$$

Proof: To prove this theorem, rewrite the left-hand side of (8.30) as

$$\mathrm{Tr}^{\sigma\alpha\beta}([\mathbb{I}^\sigma \otimes U^{\alpha\beta}(t,s)] \, W^{\sigma\alpha\beta}(s) \, [\mathbb{I}^\sigma \otimes U^{\alpha\beta}(s,t)] \, [P_a^\sigma(t) \otimes P_b^\alpha(t) \otimes \mathbb{I}^\beta]) =$$
$$\mathrm{Tr}^{\alpha\beta}(U^{\alpha\beta}(t,s) \, \{\mathrm{Tr}^\sigma(W^{\sigma\alpha\beta}(s) \, [P_a^\sigma(t) \otimes \mathbb{I}^{\alpha\beta}])\} \, U^{\alpha\beta}(s,t) \, [P_b^\alpha(t) \otimes \mathbb{I}^\beta]). \tag{8.31}$$

Condition (8.29) is equivalent to (6.14) of Theorem 6.1 if one identifies the projections $\{Q_q^\kappa\}_q$ with $\{P_i^\sigma\}_i$ and $\{Q_r^\lambda\}_r$ with $\{P_j^\alpha \otimes \mathbb{I}^\beta\}_j$. Theorem 6.1

yields that

$$\forall s: \quad \text{Tr}^{\sigma}(W^{\sigma\alpha\beta}(s)\,[P_a^{\sigma}(t)\otimes \mathbb{I}^{\alpha\beta}]) = W^{\alpha\beta}(s)\,[P_a^{\alpha}(t)\otimes \mathbb{I}^{\beta}] \tag{8.32}$$

with which one can derive (8.30). □

Finally, I prove by means of an example that the transition probabilities for interacting systems are stochastic, non-Born like and non-Markovian.

Consider a composite of three atoms σ, α and β. The Hilbert spaces $\mathcal{H}^{\sigma}$, $\mathcal{H}^{\alpha}$ and $\mathcal{H}^{\beta}$ are all two-dimensional an the state of $\sigma\alpha\beta$ at $t = 0$ is given by

$$\begin{aligned}W^{\sigma\alpha\beta}(0) = {} &\tfrac{2}{8}\,|e_1^{\sigma}\rangle\langle e_1^{\sigma}| \otimes |e_1^{\alpha}\rangle\langle e_1^{\alpha}| \otimes |e_1^{\beta}\rangle\langle e_1^{\beta}| + \tfrac{5}{8}\,|e_1^{\sigma}\rangle\langle e_1^{\sigma}| \otimes |e_2^{\alpha}\rangle\langle e_2^{\alpha}| \otimes |e_2^{\beta}\rangle\langle e_2^{\beta}| \\ &+ \tfrac{1}{8}\,|e_2^{\sigma}\rangle\langle e_2^{\sigma}| \otimes |e_2^{\alpha}\rangle\langle e_2^{\alpha}| \otimes |e_2^{\beta}\rangle\langle e_2^{\beta}|\end{aligned} \tag{8.33}$$

with $\{|e_j^{\sigma}\rangle\}_{j=1}^2$, $\{|e_k^{\alpha}\rangle\}_{k=1}^2$ and $\{|e_l^{\beta}\rangle\}_{l=1}^2$ all orthonormal bases. Let the Hamiltonian of $\sigma\alpha\beta$ be given by $H^{\sigma}\otimes\mathbb{I}^{\alpha\beta}+\mathbb{I}^{\sigma}\otimes H^{\alpha\beta}$ (so σ and $\alpha\beta$ do not interact) with

$$H^{\sigma} = \hbar\begin{pmatrix} a & 0 \\ 0 & b \end{pmatrix}, \qquad H^{\alpha\beta} = \hbar\begin{pmatrix} 0 & 0 & 0 & -i \\ 0 & 0 & 0 & 0 \\ 0 & 0 & 0 & 0 \\ i & 0 & 0 & 0 \end{pmatrix}, \tag{8.34}$$

where the matrix of H^{σ} is with respect to the basis $\{|e_1^{\sigma}\rangle,|e_2^{\sigma}\rangle\}$ and where the matrix of $H^{\alpha\beta}$ is with respect to the basis $\{|e_1^{\alpha}\rangle\otimes|e_1^{\beta}\rangle,|e_1^{\alpha}\rangle\otimes|e_2^{\beta}\rangle,|e_2^{\alpha}\rangle\otimes|e_1^{\beta}\rangle,|e_2^{\alpha}\rangle\otimes|e_2^{\beta}\rangle\}$. The evolution of the state of $\sigma\alpha\beta$ is then governed by the operator

$$U^{\sigma\alpha\beta}(t,s) = \begin{pmatrix} e^{-ia(t-s)} & 0 \\ 0 & e^{-ib(t-s)} \end{pmatrix} \otimes \begin{pmatrix} \cos(t-s) & 0 & 0 & -\sin(t-s) \\ 0 & 1 & 0 & 0 \\ 0 & 0 & 1 & 0 \\ \sin(t-s) & 0 & 0 & \cos(t-s) \end{pmatrix}, \tag{8.35}$$

where the matrices are again with respect to the above bases. With this evolution it follows that $W^{\sigma\alpha\beta}(0) = W^{\sigma\alpha\beta}(k\pi)$ with $k\in\mathbb{Z}$.

The state of this composite satisfies condition (8.17) at $s = 0$ and, due to the above remark, also at $s = k\pi$ with $k\in\mathbb{Z}$. Hence, σ acts like a snooper for α at $s = k\pi$ with $k\in\mathbb{Z}$ and one can determine the transition probabilities $p(P_b^{\alpha}(t)/P_a^{\alpha}(s))$ in both the spectral and atomic modal interpretations for all t and all $s = k\pi$ with $k\in\mathbb{Z}$.

The reduced states of $\alpha\beta$ and α evolve like

$$W^{\alpha\beta}(t) = \frac{1}{8}\begin{pmatrix} 1+\cos^2 t & 0 & 0 & \cos t \sin t \\ 0 & 5 & 0 & 0 \\ 0 & 0 & 0 & 0 \\ \cos t \sin t & 0 & 0 & 1+\sin^2 t \end{pmatrix},$$

$$W^{\alpha}(t) = \frac{1}{8}\begin{pmatrix} 6+\cos^2 t & 0 \\ 0 & 1+\sin^2 t \end{pmatrix},$$

$$(8.36)$$

where the last matrix is with respect to the basis $\{|e_1^\alpha\rangle, |e_2^\alpha\rangle\}$.

Since the state of α is at all times diagonal with regard to the this basis, the core properties of α lie at all times on the continuous eigenprojection trajectories

$$T_1^\alpha(t) = \begin{pmatrix} 1 & 0 \\ 0 & 0 \end{pmatrix}, \qquad T_2^\alpha(t) = \begin{pmatrix} 0 & 0 \\ 0 & 1 \end{pmatrix} \tag{8.37}$$

($W^\alpha(t)$ is never degenerate). Let's now fix the transition probabilities from $s = 0$ to any $t > 0$ by means of (8.23) and from any $t < \pi$ to $s = \pi$ by means of (8.24). One obtains

$$
\begin{aligned}
&p(P_1^\alpha(t)/P_1^\alpha(0)) = (5+2\cos^2 t)/7, && p(P_1^\alpha(\pi)/P_1^\alpha(t)) = (5+2\cos^2 t)/(6+\cos^2 t), \\
&p(P_1^\alpha(t)/P_2^\alpha(0)) = \sin^2 t, && p(P_1^\alpha(\pi)/P_2^\alpha(t)) = 2\sin^2 t/(1+\sin^2 t), \\
&p(P_2^\alpha(t)/P_1^\alpha(0)) = 2\sin^2 t/7, && p(P_2^\alpha(\pi)/P_1^\alpha(t)) = \sin^2 t/(6+\cos^2 t), \\
&p(P_2^\alpha(t)/P_2^\alpha(0)) = \cos^2 t, && p(P_2^\alpha(\pi)/P_2^\alpha(t)) = \cos^2 t/(1+\sin^2 t).
\end{aligned}
$$

$$(8.38)$$

It follows that the evolution of the actual core property of α is truly stochastic since the above transition probabilities all can have values between 0 and 1. The evolution of the actual core property thus hops randomly from the one continuous trajectory $T_1^\alpha(t)$ to the other $T_2^\alpha(t)$ and back again.

Secondly, the right-hand transition probabilities differ from the transition probabilities derived with the Born rule. Take, for instance, $p(P_1^\alpha(\pi)/P_1^\alpha(t))$. The Born rule yields

$$
\begin{aligned}
&p_{\text{Born}}(P_1^\alpha(\pi)/P_1^\alpha(t)) \\
&= \frac{\text{Tr}^{\alpha\beta}(U^{\alpha\beta}(\pi,t)\,[P_1^\alpha(t)\otimes\mathbb{I}^\beta]\,W^{\alpha\beta}(t)\,[P_1^\alpha(t)\otimes\mathbb{I}^\beta]\,U^{\alpha\beta}(t,\pi)\,[P_1^\alpha(\pi)\otimes\mathbb{I}^\beta])}{\text{Tr}^{\alpha\beta}(W^{\alpha\beta}(t)\,[P_1^\alpha(t)\otimes\mathbb{I}^\beta])} \\
&= \frac{5+\cos^2 t(1+\cos^2 t)}{6+\cos^2 t}
\end{aligned}
\tag{8.39}
$$

which clearly contradicts (8.38). The evolution of the actual core property of interacting systems is thus non-Born like.

Finally, it can be proved that this evolution is not a Markov process, satisfying (8.25). For if the evolution is Markovian, one can derive for all $t \in (0, \pi)$ the so-called Chapman–Kolmogorov equations:

$$p(P_c^\alpha(\pi)/P_a^\alpha(0)) = \sum_b p(P_c^\alpha(\pi)/P_b^\alpha(t)) \, p(P_b^\alpha(t)/P_a^\alpha(0)). \tag{8.40}$$

Take $a = 1$ and $c = 2$. With the left-hand transition probabilities in (8.38) it follows that $p(P_2^\alpha(\pi)/P_1^\alpha(0))$ is equal to 0, but the above equation yields by (8.38)

$$p(P_c^\alpha(\pi)/P_a^\alpha(0)) = \sin^2(t)\left(\frac{5 + 2\cos^2 t}{7(6 + \cos^2 t)} + \frac{2\cos^2(t)}{1 + \sin^2 t}\right). \tag{8.41}$$

The Chapman–Kolmogorov equations are thus not satisfied and it follows that the evolution of the actual core property of interacting systems is non-Markovian.

Note that the state (8.33) is chosen such that the reduced states of σ, α, $\alpha\beta$ and β never pass a degeneracy in their spectral resolution. Hence, it is not possible to 'blame' the singularities related to degenerate states for the violations of the Born rule and the Chapman–Kolmogorov equations.

8.4 Stochastic processes

A framework for finding transition probabilities for all possible systems has been given by Bacciagaluppi and Dickson (1997). They arrived at this framework by describing the evolution of the core properties by means of the theory of stochastic processes. And they showed that with this framework one can generate a multitude of expressions for general transitions probabilities, even if one requires that these expressions are consistent with the results given in the previous sections. I start by introducing the framework and by briefly discussing stochastic processes.[96] In the next section I present two concrete proposals for general transition probabilities by Bacciagaluppi and Dickson, and make some tentative remarks about how one may devise an argument that singles out one of these proposals.

The framework developed by Bacciagaluppi and Dickson yields general transition probabilities for systems $\{\alpha, \beta, \ldots\}$ for which it holds that: (A) there exist probabilities $p(P_a^\alpha(t), P_b^\beta(t), \ldots)$ for the joint property ascription to $\{\alpha, \beta, \ldots\}$, (B) the core properties $\{P_a^\alpha(t)\}_a$, $\{P_b^\beta(t)\}_b$, etc., lie on trajectories

---

[96] I only discuss the ingredients necessary to present the results by Bacciagaluppi and Dickson. See the references given in footnote 88 for a full treatment of these results. Textbooks on the theory of stochastic processes are Feller (1950) and Doob (1953).

which have well-defined time derivatives, and (C) the joint probabilities $p(P_a^\alpha(t), P_b^\beta(t), \dots)$ have well-defined time derivatives.

Due to the condition (A) the framework is not suitable for the bi and spectral modal interpretations since it was proved in Chapter 6 that these interpretations only yield joint probabilities for sets of disjoint systems. Hence, transition probabilities can only be derived for general sets of systems in the atomic modal interpretation and for sets of disjoint systems in the bi and spectral modal interpretations. And due to the conditions (B) and (C) the framework is suitable only if the evolution of the systems α, β, ..., satifies some constraints since it was proved in Chapter 7 that the core projections $\{P_a^\alpha(t)\}_a$, $\{P_b^\beta(t)\}_b$, etc., need not lie on continuous trajectories. However, if, for instance, the Hamiltonian of the universe is bounded and if one ignores discontinuities related to degenerate states, it follows that the core properties $\{P_a^\alpha(t)\}_a$, $\{P_b^\beta(t)\}_b$, etc., lie on analytic trajectories such that all the derivatives of the trajectories and of the joint probabilities are well defined.[97]

One thus has the following situation: the core properties of the systems $\{\alpha, \beta, \dots\}$ evolve along trajectories $\{P_a^\alpha(t)\}_a$, $\{P_b^\beta(t)\}_b$, ... (I adopt the convention that two eigenprojections $P_a^X(t)$ and $P_{a'}^X(t)$ of a systems X lie on the same trajectory iff $a = a'$) and at each instant t one can give a joint probability $p(P_a^\alpha(t), P_b^\beta(t), \dots)$ that these systems actually possess $P_a^\alpha(t)$, $P_b^\beta(t)$, etc. Let $\mathsf{P}_j(t)$ be shorthand notation of the whole set $\{P_{j_\alpha}^\alpha(t), P_{j_\beta}^\beta(t), \dots\}$ of the core properties of $\{\alpha, \beta, \dots\}$. Then $p(\mathsf{P}_j(t)) = p(P_{j_\alpha}^\alpha(t), P_{j_\beta}^\beta(t), \dots)$ and the evolution of the actually possessed properties is given by the transition probabilities $p(\mathsf{P}_k(t')/\mathsf{P}_j(t))$.

In order to develop the framework, define, firstly, infinitesimal transition probabilities by

$$T_{kj}(t) := \lim_{\varepsilon \to 0} \frac{p(\mathsf{P}_k(t+\varepsilon)/\mathsf{P}_j(t))}{\varepsilon} \tag{8.42}$$

for $j \neq k$ and by

$$T_{jj}(t) := -\sum_{k \neq j} T_{kj}(t) \tag{8.43}$$

such that $\sum_k T_{kj}(t) = 0$. (Loosely speaking, $\varepsilon T_{kj}(t)$ gives the first order contribution to the transition probability from $\mathsf{P}_j(t)$ to $\mathsf{P}_k(t+\varepsilon)$.)

---

[97] Bacciagaluppi and Dickson developed their framework for the atomic modal interpretation. Moreover, they assumed that states evolve analytically such that by Proposition 7.4 on page 116 all trajectories of eigenprojections of states are analytic as well. Finally they removed the discontinuities related to passing degeneracies by defining the atomic core properties by means of these analytic trajectories of eigenprojections, similar to the extended modal interpretation. In the light of the results of Chapter 7 one may question these assumptions.

These infinitesimal transition probabilities satisfy the master equation[98]

$$\dot{p}(\mathsf{P}_k(t)) = \sum_j T_{kj}(t)\, p(\mathsf{P}_j(t)) - T_{jk}(t)\, p(\mathsf{P}_k(t)). \qquad (8.44)$$

Define, secondly, a probability current $J(t)$ as

$$J_{kj}(t) = T_{kj}(t)\, p(\mathsf{P}_j(t)) - T_{jk}(t)\, p(\mathsf{P}_k(t)). \qquad (8.45)$$

This current is antisymmetric ($J_{kj}(t) = -J_{jk}(t)$) and represents the net flow of probability from $\mathsf{P}_j(t)$ to $\mathsf{P}_k(t)$ at t. The master equation can then be rewritten as a continuity equation $\dot{p}(\mathsf{P}_k(t)) = \sum_j J_{kj}(t)$ for the current $J(t)$.

The first step in giving the transition probabilities $p(\mathsf{P}_k(t')/\mathsf{P}_j(t))$ is the determination of a probability current $J(t)$ which satisfies this continuity equation. The second step is the determination of the infinitesimal transition probabilities $\{T_{kj}(t)\}_{j,k}$ by means of this probability current and the relation (8.45). The resulting expressions for $\{T_{kj}(t)\}_{j,k}$ then automatically satisfy the master equation (8.44). The final step is to construct the transition probabilities $p(\mathsf{P}_k(t')/\mathsf{P}_j(t))$ for finite time intervals by means of the infinitesimal transition probabilities $\{T_{kj}(t)\}_{j,k}$.

As is noted by Bacciagaluppi and Dickson, it is not the case that the first and the second steps yield unique solutions for, respectively, the probability current and the infinitesimal transition probabilities. Consider the first step and assume that one has a probability current $J(t)$ which solves the continuity equation $\dot{p}(\mathsf{P}_k(t)) = \sum_j J_{kj}(t)$. Then one can always construct a second solution to this equation by adding to $J(t)$ an antisymmetric term $\tilde{J}(t)$ which satisfies $\sum_j \tilde{J}_{kj}(t) = 0$. That is, $J(t) + \tilde{J}(t)$ is a solution to the continuity equation as well. And one can always find such an extra term $\tilde{J}(t)$. Take, for instance, $\tilde{J}_{12}(t) = \tilde{J}_{31}(t) = -\tilde{J}_{13}(t) = -\tilde{J}_{21}(t) \neq 0$ and take all other values $\tilde{J}_{kj}(t)$ equal to zero.

The second step does not yield a unique solution either. Assume that one has chosen a current $J(t)$. Then a well-known solution to the relation (8.45) is given by

$$T_{kj}(t) = \max\left\{0, \frac{J_{kj}(t)}{p(\mathsf{P}_j(t))}\right\} \qquad (8.46)$$

---

[98] This master equation is derived as follows. By definition $p(\mathsf{P}_k(t')) = \sum_j p(\mathsf{P}_k(t')/\mathsf{P}_j(t))\, p(\mathsf{P}_j(t))$ and $p(\mathsf{P}_k(t)) = \sum_j p(\mathsf{P}_j(t')/\mathsf{P}_k(t))\, p(\mathsf{P}_k(t))$. Hence,

$$p(\mathsf{P}_k(t')) - p(\mathsf{P}_k(t)) = \sum_j p(\mathsf{P}_k(t')/\mathsf{P}_j(t))\, p(\mathsf{P}_j(t)) - p(\mathsf{P}_j(t')/\mathsf{P}_k(t))\, p(\mathsf{P}_k(t)).$$

If one takes $t' = t + \varepsilon$ and divides both sides by ε, the limit $\varepsilon \to 0$ yields the master equation.

for all $k \neq j$. This solution has been adopted by Bell (1987, pages 173-180) and Vink (1993) in the context of Bohmian mechanics and by Bub (1997, 1998b) in the context of his fixed modal interpretation. However, one can again construct a second solution by adding transition probabilities $\{\widetilde{T}_{kj}(t)\}_{j,k}$ which satisfy $\widetilde{T}_{kj}(t) \, p(\mathsf{P}_j(t)) - \widetilde{T}_{jk}(t) \, p(\mathsf{P}_k(t)) = 0$. That is, the transition probabilities $\{T_{kj}(t) + \widetilde{T}_{kj}(t)\}_{j,k}$ form a solution to the relation (8.45) as well. These transition probabilities $\{\widetilde{T}_{kj}(t)\}_{j,k}$ can, for instance, be

$$
\widetilde{T}_{kj}(t) = \begin{cases} p(\mathsf{P}_k(t)) & \text{if } p(\mathsf{P}_j(t)) \neq 0, \\ 0 & \text{if } p(\mathsf{P}_j(t)) = 0. \end{cases} \tag{8.47}
$$

More precisely, the transition probabilities (8.46) only follow uniquely from the probability current $J(t)$ if one assumes that there are no transitions from $\mathsf{P}_j(t)$ to $\mathsf{P}_k(t)$ if the net flow $J_{kj}(t)$ of probability from $\mathsf{P}_j(t)$ to $\mathsf{P}_k(t)$ is negative. For then one takes the net flow $J(t)$ as the real flow of probability and one can only have transitions in the direction of $J(t)$ as is expressed by (8.46). If, on the other hand, one adds transition probabilities like (8.47) to the solution (8.46), then one also has transitions from $\mathsf{P}_j(t)$ to $\mathsf{P}_k(t)$ if $J_{kj}(t)$ is negative.

The final step of constructing the finite time transition probabilities from the infinitesimal transition probabilities is, in the generic case, properly (and uniquely) defined but is, in some special cases, mathematically somewhat problematic.[99] One of the assumptions used by Bacciagaluppi and Dickson in this step is that the evolution of the core properties of $\{\alpha, \beta, \dots\}$ is Markovian. That is, they assume that the transition probabilities $p(\mathsf{P}_k(t')/\mathsf{P}_j(t))$ satisfy the Markov property (8.25). According to Section 8.3, the evolution of the properties of a single interacting system is not Markovian. This does not, however, rule out that the joint evolution of the properties of a set of systems $\{\alpha, \beta, \dots\}$ is Markovian. Moreover, as was noted in footnote 95, if the composite $\alpha\beta\cdots$ evolves freely, it is to be expected that this joint evolution of the properties of $\{\alpha, \beta, \dots\}$ is Markovian.

8.5 Two proposals by Bacciagaluppi and Dickson

With the framework presented in the previous section one can choose transition probabilities for any set of systems $\{\alpha, \beta, \dots\}$ in the atomic modal interpretation, and for sets of disjoint systems in the bi and spectral modal

---

[99] See Bacciagaluppi and Dickson (1997, Sect. 2.2), Bacciagaluppi (1996b, Chap. 7) and Bacciagaluppi (1998, Sect. 4.2) for a full discussion of this step from the infinitesimal to the finite time transition probabilities.

interpretations. The only thing one has to do is to give a probability current $J(t)$ which is compatible with the evolution of the joint probabilities $\{p(\mathsf{P}_j(t))\}_j$, and to specify how this probability current fixes the infinitesimal transition probabilities $\{T_{kj}(t)\}_{j,k}$.

Bacciagaluppi and Dickson themselves considered transition probabilities for sets of disjoint systems $\{\alpha, \beta, \dots\}$ (atoms in the case of the atomic modal interpretation) for which it holds that the composite $\omega = \alpha\beta\cdots$ has a freely evolving pure state $|\Psi^\omega(t)\rangle$. They have given two possible expressions for the probability current $J(t)$ and proposed that this current fixes the infinitesimal transition probabilities by the relation (8.46). Hence, Bacciagaluppi and Dickson take the net probability current $J(t)$ as a real probability current. The first current is the *minimal flow* current, denoted by $J^{\min}(t)$. The second is called the *generalised Schrödinger* current, denoted by $J^{\text{Schr}}(t)$. They are

$$J_{kj}^{\min}(t) = \frac{1}{D}\left[\dot{p}(\mathsf{P}_k(t)) - \dot{p}(\mathsf{P}_j(t))\right], \qquad (8.48)$$

where D the number of different sets of properties in $\mathsf{P}_j(t)$, and

$$J_{kj}^{\text{Schr}}(t) = 2\,\text{Im}\left[\langle\Psi^\omega(t)|(P_{k_\alpha}^\alpha(t) \otimes P_{k_\beta}^\beta(t) \otimes \cdots)H^\omega(P_{j_\alpha}^\alpha(t) \otimes P_{j_\beta}^\beta(t) \otimes \cdots)|\Psi^\omega(t)\rangle\right]$$

$$+ \langle\Psi^\omega(t)|\frac{d(P_{k_\alpha}^\alpha(t) \otimes P_{k_\beta}^\beta(t) \otimes \cdots)}{dt}(P_{j_\alpha}^\alpha(t) \otimes P_{j_\beta}^\beta(t) \otimes \cdots)|\Psi^\omega(t)\rangle$$

$$- \langle\Psi^\omega(t)|\frac{d(P_{j_\alpha}^\alpha(t) \otimes P_{j_\beta}^\beta(t) \otimes \cdots)}{dt}(P_{k_\alpha}^\alpha(t) \otimes P_{k_\beta}^\beta(t) \otimes \cdots)|\Psi^\omega(t)\rangle,$$

$$(8.49)$$

where H^ω is the Hamiltonian of ω.

Bacciagaluppi and Dickson arrived at these two currents as follows. The minimal flow current $J^{\min}(t)$ is the one obtained if one minimises the overall flow of probability between the different sets $\{\mathsf{P}_j(t)\}_j$. Or, more precisely, $J^{\min}(t)$ is obtained if one minimises the value $\sum_{j,k} J_{kj}^2(t)$.[100]

The generalised Schrödinger current $J^{\text{Schr}}(t)$ is a generalisation of a probability current one can define by means of the Schrödinger equation. To illustrate this, note that the probability $p(\mathsf{P}_k(t))$ is equal to

$$p(\mathsf{P}_k(t)) = \langle\Psi^\omega(t)|(P_{k_\alpha}^\alpha \otimes P_{k_\beta}^\beta \otimes \cdots)|\Psi^\omega(t)\rangle. \qquad (8.50)$$

Consider, firstly, the special case in which the properties possessed by the systems $\{\alpha, \beta, \dots\}$ are all time-independent. The time derivative of $p(\mathsf{P}_k)$ is

---

[100] See Appendix 1 of Chapter 7 in Bacciagaluppi (1996b) for the proof.

then by the Schrödinger equation equal to

$$\dot{p}(\mathsf{P}_k) = 2\,\mathrm{Im}\left[\langle\Psi^\omega(t)|(P^\alpha_{k_\alpha}\otimes P^\beta_{k_\beta}\otimes\cdots)H^\omega|\Psi^\omega(t)\rangle\right]. \tag{8.51}$$

If one now uses that $\sum_{j_\alpha,j_\beta,\ldots}P^\alpha_{j_\alpha}\otimes P^\beta_{j_\beta}\otimes\cdots = \mathbb{I}^\omega$, one can in a natural way construct a current which satisfies the continuity equation $\dot{p}(\mathsf{P}_k(t)) = \sum_j J_{kj}(t)$, as

$$J_{kj}(t) = 2\,\mathrm{Im}\left[\langle\Psi^\omega(t)|(P^\alpha_{k_\alpha}\otimes P^\beta_{k_\beta}\otimes\cdots)H^\omega(P^\alpha_{j_\alpha}\otimes P^\beta_{j_\beta}\otimes\cdots)|\Psi^\omega(t)\rangle\right]. \tag{8.52}$$

This current is equal to $J^{\mathrm{Schr}}(t)$ because if the properties of $\{\alpha,\beta,\ldots\}$ are all time-independent, the second and third terms of $J^{\mathrm{Schr}}(t)$ vanish.

Consider, secondly, the general case that the properties possessed by $\{\alpha,\beta,\ldots\}$ are time-dependent. The derivative of $p(\mathsf{P}_k(t))$ contains then an extra term:

$$\dot{p}(\mathsf{P}_k(t)) = 2\,\mathrm{Im}\left[\langle\Psi^\omega(t)|(P^\alpha_{k_\alpha}(t)\otimes P^\beta_{k_\beta}(t)\otimes\cdots)H^\omega|\Psi^\omega(t)\rangle\right]$$
$$+ \langle\Psi^\omega(t)|\frac{\mathrm{d}(P^\alpha_{k_\alpha}(t)\otimes P^\beta_{k_\beta}(t)\otimes\cdots)}{\mathrm{d}t}|\Psi^\omega(t)\rangle. \tag{8.53}$$

Hence, if one again wants to construct a current from this expression which satisfies the continuity equation, one has to add to the current (8.52) an extra term $\tilde{J}_{kj}(t)$ which is real and antisymmetric and which, when summed over the index j, yields $\langle\Psi^\omega(t)|\mathrm{d}(P^\alpha_{k_\alpha}(t)\otimes P^\beta_{k_\beta}(t)\otimes\cdots)/\mathrm{d}t|\Psi^\omega(t)\rangle$. And the sum of the second and third terms of $J^{\mathrm{Schr}}(t)$ is exactly such a term. The generalised Schrödinger current $J^{\mathrm{Schr}}(t)$ is thus a generalisation of the current (8.52) from the case of time-independent properties to the case of time-dependent properties.

From $J^{\mathrm{min}}(t)$ and $J^{\mathrm{Schr}}(t)$ one can derive the infinitesimal transition probabilities by means of relation (8.46). And even though Bacciagaluppi and Dickson do not explicitly calculate these transition probabilities, they do prove that in the case of the generalised Schrödinger current, such transition probabilities are consistent with the deterministic evolution (8.12) for freely evolving systems derived in Section 8.2. This proof proceeds as follows. Take a system α which is part of the set $\{\alpha,\beta,\ldots\}$, and consider all the pairs of sets $\mathsf{P}_j(t) = \{P^\alpha_{j_\alpha}(t),P^\beta_{j_\beta}(t),\ldots\}$ and $\mathsf{P}_k(t) = \{P^\alpha_{k_\alpha}(t),P^\beta_{k_\beta}(t),\ldots\}$ for which it holds that $j_\alpha\neq k_\alpha$. If the net current $J^{\mathrm{Schr}}_{kj}(t)$ is equal to zero for all these pairs of sets, then by the chosen relation (8.46) between the net current and the infinitesimal transition probabilities, there are no transitions from $\mathsf{P}_j(t)$ to $\mathsf{P}_k(t)$ if $j_\alpha\neq k_\alpha$. It then follows that there are also no transitions from the property $P^\alpha_{j_\alpha}(t)$ to the property $P^\alpha_{k_\alpha}(t)$ of α if $j_\alpha \neq k_\alpha$. Bacciagaluppi and Dickson now prove that if α evolves freely, $J^{\text{Schr}}_{kj}(t)$ is indeed equal to zero if $j_\alpha \neq k_\alpha$.

Such a proof cannot be given for the minimal flow current. Take the freely evolving system σ in the composite $\sigma\alpha\beta$ discussed in the MATH-EMATICS of Section 8.3. Consider $P_j(t) = \{|e^\sigma_1\rangle\langle e^\sigma_1|, |e^\alpha_1\rangle\langle e^\alpha_1|, |e^\beta_1\rangle\langle e^\beta_1|\}$ and $P_k(t) = \{|e^\sigma_2\rangle\langle e^\sigma_2|, |e^\alpha_1\rangle\langle e^\alpha_1|, |e^\beta_1\rangle\langle e^\beta_1|\}$. The corresponding joint probabilities are $p(P_j(t)) = \frac{1}{4}\cos^2 t$ and $p(P_k(t)) = \frac{1}{8}\sin^2 t$. Since $D = 8$, it follows that $J^{\min}_{kj}(t) = \frac{3}{32}\cos t \sin t$. Hence, on $t = \pi/6$ the minimal flow current allows transitions from $|e^\sigma_1\rangle\langle e^\sigma_1|$ to $|e^\sigma_2\rangle\langle e^\sigma_2|$.

Reviewing the results of Bacciagaluppi and Dickson, my position is that their proposal to generate transition probabilities with the generalised Schrödinger current shows that one can arrive in a natural way at generally applicable transition probabilities for the bi and atomic modal interpretations (and for the spectral modal interpretation only for sets of disjoint systems). This proposal is formulated while keeping close to the formalism of quantum mechanics and is in that sense attractive. The proposal to generate transition probabilities with the minimal flow current should, on the other hand, be rejected since it contradicts the deterministic evolution of the properties of freely evolving systems. Let's therefore concentrate only on the proposal with the generalised Schrödinger current.

As Bacciagaluppi and Dickson emphasise, the determinations of both the current and the relation (8.46) between the current and the infinitesimal transition probabilities are not unique: one can choose another current $J(t)$ and one can choose another relation between $J_{kj}(t)$ and $T_{kj}(t)$. This non-uniqueness raises the question of whether the choice of $J^{\text{Schr}}(t)$ and the choice for relation (8.46) can be backed up by further arguments or whether they should be challenged.

My present position is that there may exist an argument that singles out the derivation of transitons probabilities by the generalised Schrödinger current as the correct one. The starting point of such an argument could be that in quantum mechanics the net probability current between time-independent properties $\{P^\alpha_{j_\alpha}, P^\beta_{j_\beta}, \ldots\}$ of sets of systems is uniquely given by the 'ungeneralised' Schrödinger current (8.52). The first step of such an argument is then to uniquely fix the generalised Schrödinger current (8.49) from this 'ungeneralised' current. Such a derivation is possible as was proved by James Cushing, who used the fact that one can construct the case of time-dependent properties $\{P^\alpha_{j_\alpha}(t), P^\beta_{j_\beta}(t), \ldots\}$ from the case of time-independent properties $\{P^\alpha_{j_\alpha}, P^\beta_{j_\beta}, \ldots\}$ by means of a unitary transformation $U^\omega(t)$

which satisfies

$$P_{j_\alpha}^{\alpha}(t) \otimes P_{j_\beta}^{\beta}(t) \otimes \cdots = U^{\omega}(t) \left(P_{j_\alpha}^{\alpha} \otimes P_{j_\beta}^{\beta} \otimes \cdots \right) [U^{\omega}(t)]^{\dagger}. \tag{8.54}$$

Under this transformation $U^{\omega}(t)$, the ungeneralised current (8.52) transforms uniquely to the generalised Schrödinger current (8.49).[101]

The second step of this argument is to fix the relation (8.46) between the current and the infinitesimal transition probabilities. As I noted at the end of the previous section, if the probability current $J(t)$ is a real probability current instead of a net current, then the infinitesimal transition probabilities (8.46) follow uniquely from this current. Hence, what is needed is an argument why $J(t)$ is a real current. Such an argument may be possible by using that the properties of freely evolving systems have to evolve deterministically according to the results derived in Section 8.2. The argument may go as follows. If a system α evolves freely, then the transition probabilities are given by the deterministic expression $T_{kj}(t) = 0$ if $j_\alpha \neq k_\alpha$. Bacciagaluppi and Dickson have proved that if α evolves freely, then $J_{kj}^{\text{Schr}}(t) = 0$ if $j_\alpha \neq k_\alpha$. If this current $J_{kj}^{\text{Schr}}(t)$ is a net current, then this zero current $J_{kj}^{\text{Schr}}(t)$ does not imply that there are no transitions between $P_{j_\alpha}^{\alpha}(t)$ and $P_{k_\alpha}^{\alpha}(t)$, which is inconsistent with the determistitic evolution of α. Hence, $J_{kj}^{\text{Schr}}(t)$ cannot be a net current, but must be a real current. (And if $J^{\text{Schr}}(t)$ is a real current, then $J_{kj}^{\text{Schr}}(t) = 0$ does indeed imply that $T_{kj}(t) = 0$.)

This argument is, of course, tentative. My conclusion is therefore that presently we cannot uniquely fix transition probabilities for all interacting systems. Only for freely evolving systems can one argue that the transitions probabilities are uniquely equal to (8.12) and for some interacting systems one can argue that they are uniquely equal to (8.23) or (8.24). However, future work may reveal that generally applicable transition probabilities are uniquely generated by the generalised Schrödinger current.

---

[101] See Section 4.3 in Bacciagaluppi and Dickson (1997) for the derivation of $J^{\text{Schr}}(t)$ by Cushing.

9

Dynamical Autonomy and Locality

In this chapter I prove that the transition probabilities, which describe the dynamics of the actually possessed properties of systems, violate the assumption of Dynamical Autonomy for composite systems. And, related to this violation, it is also proved that the dynamics of the actually possessed properties is non-local in a quite explicit way.

9.1 Introduction

When I introduced the assumptions of Instantaneous and Dynamical Autonomy in Section 3.3, I motivated them by making an appeal to locality. In the case of Instantaneous Autonomy, I argued that a state of a system α should codify all the information about the property ascription to that system. For if a state does not codify all the information, it may happen that a change in the state of some distant system σ (I used the example of a far-away asteroid) means that the properties of α change as well. And this would make modal property ascriptions undesirably non-local. In the case of Dynamical Autonomy one can give a similar motivation. The state of a system α, as it evolves during a time interval I, should codify all the information about the simultaneous and sequential correlations between the properties ascribed to α in that interval I. For if the evolving state does not do so, it may happen that a change in a distant system changes the correlations between the properties of α as well.

Modal interpretations satisfy a number of the Autonomy assumptions. All modal interpretations satisfy Instantaneous Autonomy by construction. The transition probabilities derived in Section 8.2 prove that the bi and spectral modal interpretations satisfy Dynamical Autonomy for whole systems and that the atomic modal interpretation satisfies Dynamical Autonomy for atomic systems. This implies that the properties ascribed by modal inter-

159

pretations are local in a number of cases, as is noted by Dieks (1998a): the properties of a system α (an atom in the case of the atomic modal interpretation) are not affected by changes in the states of distant systems if there is no physical interaction between α and the distant systems, and the dynamics of the properties of a freely evolving system α are also not affected by changes in the states of those distant systems. So, in the context of the Einstein–Podolsky–Rosen experiment, it follows that a measurement performed on one of the particles changes neither the properties ascribed to the other particle, nor the dynamics of these properties.

However, as reported in Section 8.3, the spectral modal interpretation fails to satisfy the assumption of Dynamical Autonomy for composite systems and the atomic modal interpretation fails to satisfy Dynamical Autonomy for composite systems and Dynamical Autonomy for whole molecular systems. This failure opens up the possibility that these two interpretations are non-local in cases different to the ones described in the last paragraph.

In Section 9.2 I prove that the spectral and atomic modal interpretations violate Dynamical Autonomy. Then, in Section 9.3, I give two examples which explicitly demonstrate that these two modal interpretations are non-local in the sense that they violate the following condition

Locality

The dynamics of the properties ascribed to systems, which are part of a freely evolving composite ω that is confined to a space-time region R, is unaffected by any change in the state of another freely evolving composite ω' confined to a second space-time region R', if R' is space-like separated from R.

It may not come as a surprise to the reader that modal interpretations give a description of reality which is non-local. Bohmian mechanics, an interpretation which is identified as a modal interpretation, also violates the above locality condition.[102] Moreover, it is already established that quantum mechanics itself is non-local: if one takes quantum mechanics in the standard way as a theory which ascribed outcomes to measurement devices, then quantum mechanics violates the notion of 'local causality,' as was shown by Bell (1987, pages 52–62).

On the other hand, not all accounts of quantum mechanics violate the above Locality condition. Consider again quantum mechanics in its standard formulation. As I noted in Section 3.3, the standard formulation satisfies Dynamical Autonomy for composite systems and therefore satisfies Locality.

---

[102] In Bohmian mechanics a measurement on one spin $\frac{1}{2}$-particle in the Einstein–Podolsky–Rosen experiment affects the description of the other particle (see, for instance, Albert (1992, pages 145–170) and Holland (1993, Sects. 11.2 and 11.3)).

So, a theory which is non-local in Bell's local causality sense need not necessarily be non-local in the sense of Locality. (In other words, locality comes in grades and 'local causality' is a weaker condition than the Locality condition which I have given above.) It thus may be obvious that modal interpretations are non-local is some sense, but it may still come as a surprise (it was to me, at least) that modal interpretations violate Locality, making them non-local in such an explicit sense.[103]

9.2 The violation of Dynamical Autonomy

In order to prove that the spectral and atomic modal interpretations violate a number of Dynamical Autonomy assumptions, I rephrase an example of four interacting systems by Dickson and Clifton (1998).[104] The example is essentially the Einstein–Podolsky–Rosen–Bohm experiment with two spin $\frac{1}{2}$-particles which interact with two measurement devices. The differences are that here the two spin particles are not in the singlet state but in a more general entangled state, that the measurements do not occur simultaneously but at different times and that the measurement devices are spin $\frac{1}{2}$-particles as well.

Consider four atoms α, β, γ and δ, all defined on two-dimensional Hilbert spaces. Let these systems be spin $\frac{1}{2}$-particles. Initially, at $t = 1$, the particles α and β are in an entangled state $|\psi^{\alpha\beta}(1)\rangle = \sum_{j=1}^{2} c_j |e_j^{\alpha}\rangle \otimes |e_j^{\beta}\rangle$, where the vectors $|e_1^{\alpha}\rangle$ and $|e_2^{\alpha}\rangle$, respectively $|e_1^{\beta}\rangle$ and $|e_2^{\beta}\rangle$, are mutually orthogonal, and γ and δ are initially in the states $|f_1^{\gamma}\rangle$ and $|f_1^{\delta}\rangle$, respectively. The state of the composite $\alpha\beta\gamma\delta$ at $t = 1$ is thus

$$|\Psi^{\alpha\beta\gamma\delta}(1)\rangle = \sum_{j=1}^{2} c_j |e_j^{\alpha}\rangle \otimes |e_j^{\beta}\rangle \otimes |f_1^{\gamma}\rangle \otimes |f_1^{\delta}\rangle. \tag{9.1}$$

From $t = 1$ to $t = 2$, the particles α and γ interact and this interaction satisfies the scheme

$$\left. \begin{array}{l} |g_1^{\alpha}\rangle \otimes |f_1^{\gamma}\rangle \longmapsto |g_1^{\alpha}\rangle \otimes |f_1^{\gamma}\rangle, \\ |g_2^{\alpha}\rangle \otimes |f_1^{\gamma}\rangle \longmapsto |g_2^{\alpha}\rangle \otimes |f_2^{\gamma}\rangle, \end{array} \right\} \tag{9.2}$$

where $|g_1^{\alpha}\rangle$ and $|g_2^{\alpha}\rangle$, respectively $|f_1^{\gamma}\rangle$ and $|f_2^{\gamma}\rangle$, are again mutually orthogonal vectors. The particles β and δ evolve both freely from $t = 1$ to $t = 2$ in

---

[103] For a full discussion of non-locality in modal interpretations, the reader may consult, for instance, Healey (1989), Dieks (1994a, 1998a), Bacciagaluppi (1996b, 2000), Dickson (1998b) and Dickson and Clifton (1998).

[104] Dickson and Clifton used the example to prove that the modal transition probabilities are not Lorentz-invariant.

such a way that their states do not change. Define the coefficients $\{d_{jk}\}_{j,k}$ as $d_{jk} = \langle g_k^\alpha | e_j^\alpha \rangle$. The state of $\alpha\beta\gamma\delta$ at $t = 2$ is then

$$|\Psi^{\alpha\beta\gamma\delta}(2)\rangle = \sum_{j,k=1}^{2} c_j d_{jk} |g_k^\alpha\rangle \otimes |e_j^\beta\rangle \otimes |f_k^\gamma\rangle \otimes |f_1^\delta\rangle. \tag{9.3}$$

From $t = 2$ to $t = 3$ the particles β and δ interact and α and γ evolve freely in such a way that their states do not change. The interaction between β and δ is also given by the scheme (9.2) but now with α replaced by β and with γ replaced by δ. The value $\langle g_l^\beta | e_j^\beta \rangle$ is then equal to the coefficient $d_{jl} = \langle g_l^\alpha | e_j^\alpha \rangle$ and the state of $\alpha\beta\gamma\delta$ at $t = 3$ is

$$|\Psi^{\alpha\beta\gamma\delta}(3)\rangle = \sum_{j,k,l=1}^{2} c_j d_{jk} d_{jl} |g_k^\alpha\rangle \otimes |g_l^\beta\rangle \otimes |f_k^\gamma\rangle \otimes |f_l^\delta\rangle. \tag{9.4}$$

(If γ and δ are taken as pointers of measurement devices one regains the Einstein–Podolsky–Rosen–Bohm experiment because then the interactions (9.2) between α and γ and between β and δ, respectively, count as Von Neumann measurements (2.9) performed on two spin $\frac{1}{2}$-particles α and β which have a joint entangled state.)

Assume that the vectors $|e_1^\alpha\rangle$ and $|e_1^\beta\rangle$, respectively $|e_2^\alpha\rangle$ and $|e_2^\beta\rangle$, represent the same spin magnitudes, say $|e_1^\alpha\rangle$ and $|e_1^\beta\rangle$ both represent spin up in $\check{x}$ direction and $|e_2^\alpha\rangle$ and $|e_2^\beta\rangle$ both represent spin down in $\check{x}$ direction.[105] Assume also that the vectors $|f_1^\gamma\rangle$ and $|f_1^\delta\rangle$, respectively $|f_2^\gamma\rangle$ and $|f_2^\delta\rangle$, represent the same spin magnitudes, say $|f_1^\gamma\rangle$ and $|f_1^\delta\rangle$ represent spin up in $\check{y}$ direction and $|f_2^\gamma\rangle$ and $|f_2^\delta\rangle$ represent spin down. Finally, assume that the vectors $|g_1^\alpha\rangle$ and $|g_1^\beta\rangle$, respectively $|g_2^\alpha\rangle$ and $|g_2^\beta\rangle$, represent the same spin, say spin up and down in direction $\check{v}$ (with $\check{v}$ unspecified). In this example one is then dealing with two interactions in which the states of the composites $\alpha\gamma$ and $\beta\delta$ evolve identically and freely as

$$\sum_{j=1}^{2} |c_j|^2 |e_j^X\rangle\langle e_j^X| \otimes |f_1^Y\rangle\langle f_1^Y| \longmapsto \sum_{j,k,l=1}^{2} |c_j|^2 d_{jk}\bar{d}_{jl} |g_k^X\rangle\langle g_l^X| \otimes |f_k^Y\rangle\langle f_l^Y|, \tag{9.5}$$

where X and Y are, respectively, α and γ in the time interval from $t = 1$ to $t = 2$, and where X and Y are, respectively, β and δ in the interval from $t = 2$ to $t = 3$.

---

[105] The vectors $|e_1^\alpha\rangle$ and $|e_1^\beta\rangle$ (or $|e_2^\alpha\rangle$ and $|e_2^\beta\rangle$, *mutatis mutandis*) cannot strictly speaking represent the same spin magnitude because the first vector represents a magnitude pertaining to α and the second represents one pertaining to β. However, they can represent the same magnitude in the sense of that they both represent magnitudes which are measured by the same procedure. The spins represented by $|e_1^\alpha\rangle$ and $|e_1^\beta\rangle$ are, for instance, both measured by a Stern–Gerlach device oriented in the $\check{x}$ direction.

The property ascription to α and β in the spectral and atomic modal interpretations is straightforward. The states of α and β are

$$\left.\begin{array}{l} W^{\alpha}(1) = \displaystyle\sum_{j=1}^{2} |c_j|^2 |e_j^{\alpha}\rangle\langle e_j^{\alpha}|, \quad W^{\alpha}(2) = W^{\alpha}(3) = \displaystyle\sum_{j,k=1}^{2} |c_j|^2 |d_{jk}|^2 |g_k^{\alpha}\rangle\langle g_k^{\alpha}|, \\[4mm] W^{\beta}(1) = W^{\beta}(2) = \displaystyle\sum_{j'=1}^{2} |c_{j'}|^2 |e_{j'}^{\beta}\rangle\langle e_{j'}^{\beta}|, \quad W^{\beta}(3) = \displaystyle\sum_{j',l=1}^{2} |c_{j'}|^2 |d_{j'l}|^2 |g_l^{\beta}\rangle\langle g_l^{\beta}|. \end{array}\right\}$$

$$(9.6)$$

(I used here that $\sum_{k=1}^{2} d_{jk}\bar{d}_{j'k} = \delta_{jj'}$.)

The given decompositions are spectral resolutions if the two values $\{|c_1|^2, |c_2|^2\}$, respectively $\{\sum_{j=1}^{2} |c_j|^2 |d_{j1}|^2, \sum_{j=1}^{2} |c_j|^2 |d_{j2}|^2\}$, are not degenerate.[106] So, the core properties of α are $\{|e_1^{\alpha}\rangle\langle e_1^{\alpha}|, |e_2^{\alpha}\rangle\langle e_2^{\alpha}|\}$ at $t = 1$ and $\{|g_1^{\alpha}\rangle\langle g_1^{\alpha}|, |g_2^{\alpha}\rangle\langle g_2^{\alpha}|\}$ at $t = 2$ and at $t = 3$, and the core properties of β are $\{|e_1^{\beta}\rangle\langle e_1^{\beta}|, |e_2^{\beta}\rangle\langle e_2^{\beta}|\}$ at $t = 1$ and at $t = 2$ and $\{|g_1^{\beta}\rangle\langle g_1^{\beta}|, |g_2^{\beta}\rangle\langle g_2^{\beta}|\}$ at $t = 3$.

One can furthermore calculate the simultaneous correlations between the core properties of α and β at $t = 1$ and at $t = 3$. They are

$$\left.\begin{array}{l} p(|e_j^{\alpha}\rangle\langle e_j^{\alpha}| \text{ at } 1, |e_{j'}^{\beta}\rangle\langle e_{j'}^{\beta}| \text{ at } 1) = \delta_{jj'} |c_j|^2, \\[3mm] p(|g_k^{\alpha}\rangle\langle g_k^{\alpha}| \text{ at } 3, |g_l^{\beta}\rangle\langle g_l^{\beta}| \text{ at } 3) = |c_1 d_{1k} d_{1l} + c_2 d_{2k} d_{2l}|^2. \end{array}\right\}$$

$$(9.7)$$

Consider then the joint transition probabilities

$$p(|g_k^{\alpha}\rangle\langle g_k^{\alpha}| \text{ at } 3, |g_l^{\beta}\rangle\langle g_l^{\beta}| \text{ at } 3 \, / \, |e_j^{\alpha}\rangle\langle e_j^{\alpha}| \text{ at } 1, |e_{j'}^{\beta}\rangle\langle e_{j'}^{\beta}| \text{ at } 1) \qquad (9.8)$$

for α and β from $t = 1$ to $t = 3$. If these transition probabilities exist, then the joint probabilities

$$p(|e_j^{\alpha}\rangle\langle e_j^{\alpha}| \text{ at } 1, |e_{j'}^{\beta}\rangle\langle e_{j'}^{\beta}| \text{ at } 1, |g_k^{\alpha}\rangle\langle g_k^{\alpha}| \text{ at } 3, |g_l^{\beta}\rangle\langle g_l^{\beta}| \text{ at } 3) \qquad (9.9)$$

exist as well: just multiply the above transition probabilities with the joint probability $p(|e_j^{\alpha}\rangle\langle e_j^{\alpha}| \text{ at } 1, |e_{j'}^{\beta}\rangle\langle e_{j'}^{\beta}| \text{ at } 1)$.

Assume now that the core properties of freely evolving systems evolve deterministically, that is, by means of the transition probabilities (8.12). Then, if one also assumes that Dynamical Autonomy for composite systems holds, one can prove that the joint probabilities (9.9), and consequently the transition probabilities (9.8), sometimes do *not* exist. This proof proceeds in three steps. Firstly, the transition probabilities for α from $t = 1$ to $t = 2$ are determined with the results of Section 8.3. Secondly, application of Dynamical Autonomy for composite systems yields that the transition

---

[106] Below I give values $\{c_j\}_j$ and $\{d_{jk}\}_{jk}$ such that degeneracies do not occur.

probabilities for β from $t = 2$ to $t = 3$ are equal to those of α from $t = 1$ to $t = 2$. Finally, it is shown that given these transition probabilities for α and β, the joint probabilities (9.9) cannot always exist. More specifically it is proved that these joint probabilities are sometimes not classical probabilities in the sense that they violate the Bell–Wigner inequalities (6.23) given on page 94.

The first step of the proof goes as follows. At $t = 1$ the system β is a snooper for α. Hence, at $t = 1$ the state of $\alpha\beta\gamma$ satisfies condition (8.17) with β in the rôle of the snooper and $\alpha\gamma$ in the rôle of the freely evolving composite containing α. One can thus calculate the joint probabilities and the transition probabilities for α from $t = 1$ to $t = 2$ with (8.22) and (8.23). One obtains

$$\left.\begin{aligned}
p(|e_j^\alpha\rangle\langle e_j^\alpha| \text{ at } 1, |g_k^\alpha\rangle\langle g_k^\alpha| \text{ at } 2) &= |c_j|^2|d_{jk}|^2, \\
p(|g_k^\alpha\rangle\langle g_k^\alpha| \text{ at } 2 \,/\, |e_j^\alpha\rangle\langle e_j^\alpha| \text{ at } 1) &= |d_{jk}|^2.
\end{aligned}\right\} \qquad (9.10)$$

As has been said, the evolution of the state $W^{\alpha\gamma}(t)$ from $t = 1$ to $t = 2$ is equal to the evolution of $W^{\beta\delta}(t)$ from $t = 2$ to $t = 3$. Hence, by invoking Dynamical Autonomy for composite systems, one can conclude (and this is the second step of the proof) that the joint probabilities and the transition probabilities for β from $t = 2$ to $t = 3$ are equal to, respectively, the joint probabilities and the transition probabilities for α from $t = 1$ to $t = 2$. So,

$$\left.\begin{aligned}
p(|e_{j'}^\beta\rangle\langle e_{j'}^\beta| \text{ at } 2, |g_l^\beta\rangle\langle g_l^\beta| \text{ at } 3) &= |c_{j'}|^2|d_{j'l}|^2, \\
p(|g_l^\beta\rangle\langle g_l^\beta| \text{ at } 3 \,/\, |e_{j'}^\beta\rangle\langle e_{j'}^\beta| \text{ at } 2) &= |d_{j'l}|^2.
\end{aligned}\right\} \qquad (9.11)$$

For the third step, return to the joint probabilities (9.9). Due to the one-to-one correlations between the core properties possessed by α and β at $t = 1$, these joint probabilities are zero for $j \neq j'$ and equal to

$$p(|e_j^\alpha\rangle\langle e_j^\alpha| \text{ at } 1, |g_k^\alpha\rangle\langle g_k^\alpha| \text{ at } 3, |g_l^\beta\rangle\langle g_l^\beta| \text{ at } 3) \qquad (9.12)$$

if $j = j'$. These latter joint probabilities should be classical probabilities satisfying the Bell–Wigner inequalities (6.23). Specifically they should satisfy the fourth inequality:

$$\begin{aligned}
&p(|e_j^\alpha\rangle\langle e_j^\alpha| \text{ at } 1) - p(|e_j^\alpha\rangle\langle e_j^\alpha| \text{ at } 1, |g_k^\alpha\rangle\langle g_k^\alpha| \text{ at } 3) \\
&\quad - p(|e_j^\alpha\rangle\langle e_j^\alpha| \text{ at } 1, |g_l^\beta\rangle\langle g_l^\beta| \text{ at } 3) + p(|g_k^\alpha\rangle\langle g_k^\alpha| \text{ at } 3, |g_l^\beta\rangle\langle g_l^\beta| \text{ at } 3) \geq 0. \quad (9.13)
\end{aligned}$$

The terms in this inequality can be calculated. Firstly, the probability $p(|e_j^\alpha\rangle\langle e_j^\alpha| \text{ at } 1)$ is equal to $|c_j|^2$. Secondly, α evolves freely from $t = 2$ to $t = 3$, so its core properties evolve deterministically. Hence, α possesses $|g_k^\alpha\rangle\langle g_k^\alpha|$ at

$t = 3$ with probability 1 if and only if α possesses $|g_k^\alpha\rangle\langle g_k^\alpha|$ at $t = 2$ (assuming that $U^\alpha(3,2) = \mathbb{I}^\alpha$). The joint probability $p(|e_j^\alpha\rangle\langle e_j^\alpha|$ at $1, |g_k^\alpha\rangle\langle g_k^\alpha|$ at 3) is thus equal to the joint probability given in (9.10), that is, it is equal to $|c_j|^2|d_{jk}|^2$. Thirdly, the core properties possessed by α and β are one-to-one correlated at $t = 1$. The core properties of β evolve deterministically from $t = 1$ to $t = 2$. Hence, α possesses $|e_j^\alpha\rangle\langle e_j^\alpha|$ at $t = 1$ with probability 1 if and only if β possesses $|g_j^\beta\rangle\langle g_j^\beta|$ at $t = 2$. It follows that $p(|e_j^\alpha\rangle\langle e_j^\alpha|$ at $1, |g_l^\beta\rangle\langle g_l^\beta|$ at 3) is equal to the joint probability given in (9.11), that is, it is equal to $|c_j|^2|d_{jl}|^2$. Finally, the joint probability $p(|g_k^\alpha\rangle\langle g_k^\alpha|$ at $3, |g_l^\beta\rangle\langle g_l^\beta|$ at 3) is given in (9.7) and is equal to $|c_1 d_{1k}d_{1l} + c_2 d_{2k}d_{2l}|^2$. The fourth Bell–Wigner inequality thus becomes

$$|c_j|^2 - |c_j|^2|d_{jk}|^2 - |c_j|^2|d_{jl}|^2 + |c_1 d_{1k}d_{1l} + c_2 d_{2k}d_{2l}|^2 \geq 0. \qquad (9.14)$$

By choosing the appropriate coefficients $\{c_j\}_j$ and $\{d_{jk}\}_{j,k}$ one can now violate this inequality. Take, for instance, $c_1 = -\frac{\sqrt{3}}{2}$, $c_2 = \frac{1}{2}$, $d_{11} = \frac{\sqrt{3}}{2}$, $d_{12} = \frac{1}{2}$, $d_{21} = \frac{1}{2}$ and $d_{22} = -\frac{\sqrt{3}}{2}$.[107] Then, for $j = k = l = 1$, one obtains that the left-hand side is equal to $\frac{2-3\sqrt{3}}{32} \approx -0,1$. So, the joint probabilities (9.9) and, consequently, the transition probabilities (9.8) do not exist.

This result has a number of consequences for the spectral and atomic modal interpretations. I start with the atomic modal interpretation. The assumptions leading to the contradiction are: (A) the existence of the transition probabilities (9.8), (B) the deterministic evolution for freely evolving atoms and (C) Dynamical Autonomy for composite systems. Giving up assumption (A) does not seem to be attractive because transition probabilities like (9.8) do exist: if one adopts the proposals by Bacciagaluppi and Dickson (1997) and defines transition probabilities by means of the generalised Schrödinger current $J^{\text{Schr}}(t)$ (see Section 8.5), then these transition probabilities (9.8) can be derived in the atomic modal interpretation. Giving up (B) is not attractive either: the deterministic evolution for freely evolving systems is a straightforward consequence of Instantaneous Autonomy and Empirical Adequacy (see Section 8.2). Hence, the assumption which should be dropped is (C) such that it is proved that the atomic modal interpretation violates Dynamical Autonomy for composite systems. Without this assumption, the atomic modal interpretation is again free of contradiction: in the above example one can then still derive the transition probabilities (9.10) for α from $t = 1$ to $t = 2$ because β serves at $t = 1$ as a snooper for α. However, the transition probabilities for β from $t = 2$ to $t = 3$ can

---

[107] With these coefficients the states of α and β are not degenerate at $t = 1, 2, 3$. The values $\{d_{jk}\}_{j,k}$ are obtained by taking $|g_1^X\rangle = \frac{\sqrt{3}}{2}|e_1^X\rangle + \frac{1}{2}|e_2^X\rangle$ and $|g_2^X\rangle = \frac{1}{2}|e_1^X\rangle - \frac{\sqrt{3}}{2}|e_2^X\rangle$ for X both α and β.

no longer be determined because at $t = 2$ or at $t = 3$ none of α or γ or $\alpha\gamma$ as a whole serves as a snooper for β. The only thing one knows is that the transition probabilities for β are not given by (9.11), because if they are given by (9.11), one would again violate the fourth Bell–Wigner inequality.

A second consequence is that the atomic modal interpretation also violates Dynamical Autonomy for whole molecular systems. The core properties of the molecule $\alpha\gamma$ are given by $\{|e_j^\alpha\rangle\langle e_j^\alpha| \otimes |f_0^\gamma\rangle\langle f_0^\gamma|\}_j$ at $t = 1$ and by $\{|g_k^\alpha\rangle\langle g_k^\alpha| \otimes |f_k^\gamma\rangle\langle f_k^\gamma|\}_k$ at $t = 2$. Now, since $\alpha\gamma$ possesses $|e_j^\alpha\rangle\langle e_j^\alpha| \otimes |f_0^\gamma\rangle\langle f_0^\gamma|$ at $t = 1$ if and only if α possesses $|e_j^\alpha\rangle\langle e_j^\alpha|$ at $t = 1$, and since $\alpha\gamma$ possesses $|g_k^\alpha\rangle\langle g_k^\alpha| \otimes |f_k^\pi\rangle\langle f_k^\pi|$ at $t = 2$ if and only if α possesses $|g_k^\alpha\rangle\langle g_k^\alpha|$ at $t = 2$, it follows that the evolution of the core properties of $\alpha\gamma$ from $t = 1$ to $t = 2$ is given by

$$p(|g_k^\alpha\rangle\langle g_k^\alpha| \otimes |f_k^\gamma\rangle\langle f_k^\gamma| \text{ at } 2 \,/\, |e_j^\alpha\rangle\langle e_j^\alpha| \otimes |f_0^\gamma\rangle\langle f_0^\gamma| \text{ at } 1) =$$
$$p(|g_k^\alpha\rangle\langle g_k^\alpha| \text{ at } 2 \,/\, |e_j^\alpha\rangle\langle e_j^\alpha| \text{ at } 1). \qquad (9.15)$$

Analogously one can derive that the evolution of the core properties of $\beta\delta$ from $t = 2$ to $t = 3$ is given by

$$p(|g_l^\beta\rangle\langle g_l^\beta| \otimes |f_l^\delta\rangle\langle f_l^\delta| \text{ at } 3 \,/\, |e_{j'}^\beta\rangle\langle e_{j'}^\beta| \otimes |f_0^\delta\rangle\langle f_0^\delta| \text{ at } 2) =$$
$$p(|g_l^\beta\rangle\langle g_l^\beta| \text{ at } 3 \,/\, |e_{j'}^\beta\rangle\langle e_{j'}^\beta| \text{ at } 2). \qquad (9.16)$$

We have established that the transition probabilities for α from $t = 1$ to $t = 2$ have to be different to the transition probabilities for β from $t = 2$ to $t = 3$ (the first are given by (9.10) whereas the second are not given by (9.11)). So, due to the equalities (9.15) and (9.16), it follows that the transition probabilities for the molecule $\alpha\gamma$ from $t = 1$ to $t = 2$ are also different to the transition probabilities for the molecule $\beta\delta$ from $t = 2$ to $t = 3$. And since these molecules both evolve freely and since the evolution of their states is identical (they are both given by (9.5)), one can conclude that there is no unique relationship between the state of a freely evolving molecule and the transition probabilities for the properties possessed by that molecule. Hence, the atomic modal interpretation violates Dynamical Autonomy for whole molecular systems.

Consider now the spectral modal interpretation. If one again doesn't want to give up assumptions (A) and (B), one is again forced to the conclusion that the assumption of Dynamical Autonomy for composite systems is violated. Unfortunately, if one drops this assumption, the spectral modal interpretation is still not free of contradiction because one can also derive the transition probabilities (9.10) and (9.11) without Dynamical Autonomy for composite systems. The transition probabilities (9.10) for α follow with the results of Section 8.3 if one identifies β as a snooper for α at $t = 1$. The transition probabilities (9.11), however, also follow from these results because the composite $\alpha\gamma$ serves in the spectral modal interpretation as a snooper for β at $t = 2$ (the state of $\alpha\beta\gamma\delta$ at $t = 2$ satisfies condition (8.17) if one takes $\alpha\gamma$ as the snooper for β and if one takes $\beta\delta$ as the freely evolving composite comprising β). If one then applies (8.22) and (8.23) to β, one again obtains (9.11). The proof that the transition probabilities (9.8) do not exist can thus be given for the spectral modal interpretation even if one denies Dynamical Autonomy for composite systems.

In order to escape the above contradiction in the spectral modal interpretation without giving up the deterministic property evolution for freely evolving systems, one can return to perspectivalism. If one accepts perspectivalism, one can only simultaneously consider the properties of the systems α and β from the perspectives '$\alpha\beta\gamma\delta$ divided by α, β, γ and δ' or '$\alpha\beta\gamma\delta$ divided by α, β and $\gamma\delta$.' Hence, if one wants to consider the transition probabilities (9.8) or the joint probabilities (9.9), one has to adopt one of these perspectives. One can then argue that from both these two perspectives it is impossible to fix the transition probabilities (9.11) for β during the second measurement, but the costs of this argument are high.

The argument goes as follows. The transition probabilities (9.11) can only be derived because the composite $\alpha\gamma$ serves as a snooper for β at $t = 2$. Because the properties of $\alpha\gamma$ are one-to-one correlated at $t = 2$ with the properties of β at $t = 2$, and because the properties of $\alpha\gamma$ evolve deterministically from $t = 2$ to $t = 3$, one can fix the transition probabilities for β from $t = 2$ to $t = 3$. Hence, these transition probabilities for β are derived by simultaneously considering the properties of β and of $\alpha\gamma$. However, these properties can only be considered simultaneously from the perspective '$\alpha\beta\gamma\delta$ divided by β, $\alpha\gamma$ and δ' and thus not in the perspectives necessary to consider the transition probabilities (9.8) and the joint probabilities (9.9). Hence, if one adopts the perspectives needed to consider these latter probabilities, one cannot derive the transition probabilities for β during the second measurement.

The consequence of this argument is that transition probabilities become perspective-dependent because now the transition probabilities for β seen from the perspective '$\alpha\beta\gamma\delta$ divided by β, $\alpha\gamma$ and δ' are equal to (9.11) whereas these same transition probabilities seen from the perspectives '$\alpha\beta\gamma\delta$ divided by α, β, γ and δ' or '$\alpha\beta\gamma\delta$ divided by α, β and $\gamma\delta$' are not equal to (9.11) (for if the transition probabilities for β seen from the perspectives '$\alpha\beta\gamma\delta$ divided by α, β, γ and δ' or '$\alpha\beta\gamma\delta$ divided by α, β and $\gamma\delta$' are equal to (9.11), one would again violate the fourth Bell–Wigner inequality within these two perspectives). Hence, if one accepts the argument, one also has to accept that there don't exist unique and perspective-independent transition probabilities for systems but that transition probabilities are only defined relative to a perspective.

A final remark concerns the assumption of Dynamical Autonomy for measurements. A slightly stronger version of this assumption is given by

Assumption of Dynamical Autonomy for measurements, #2

If two composite systems evolve freely and have, during an interval, equal states and equal Hamiltonians and if these composite systems consist of an object system and a measurement device and one is dealing with a measurement of *any* property of the object system, then the correlations between the initial properties of the object system and the finally possessed outcome of the measurement device are also equal.

(In the formulation given in Section 3.3, Dynamical Autonomy is only assumed for measurements of the *initially possessed properties* of the object system.)

Using the above example it is easy to prove that both the spectral and atomic modal interpretations violate this stronger assumption. Firstly, one has to take the systems γ and δ as pointers of measurement devices. Secondly, one has to assume that the vectors $\{|f_1^\gamma\rangle, |f_2^\gamma\rangle\}$ are eigenvectors of the pointer reading magnitude $M^\gamma = \sum_{k=1}^{2} m_k |R_k^\gamma\rangle\langle R_k^\gamma|$, that is, $|f_1^\gamma\rangle = |R_1^\gamma\rangle$ and $|f_2^\gamma\rangle = |R_2^\gamma\rangle$ and similarly one has to assume for δ that $|f_1^\delta\rangle = |R_1^\delta\rangle$ and $|f_2^\delta\rangle = |R_2^\delta\rangle$. Thirdly, one has to assume in the case of the atomic modal interpretation that these pointer reading states are given by (4.33) such that the atomic modal interpretation ascribes the readings to the pointers γ and δ (the spectral modal interpretation ascribes these readings without further assumptions). Fourthly, one has to assume that the transition probabilities

$$p(|R_k^\gamma\rangle\langle R_k^\gamma| \text{ at } 3, |R_l^\delta\rangle\langle R_l^\delta| \text{ at } 3 \,/\, |e_j^\alpha\rangle\langle e_j^\alpha| \text{ at } 1, |e_{j'}^\beta\rangle\langle e_{j'}^\beta| \text{ at } 1) \tag{9.17}$$

exist. One can then rerun the above argument and prove that these transition probabilities sometimes do not exist (one has to recalculate all the above expressions (9.10)–(9.14) with $|g_k^\alpha\rangle\langle g_k^\alpha|$ replaced by $|R_k^\gamma\rangle\langle R_k^\gamma|$ and with $|g_l^\beta\rangle\langle g_l^\beta|$ replaced by $|R_l^\delta\rangle\langle R_l^\delta|$).

So, if the transition probabilities (9.17) do exist, the spectral and atomic modal interpretations violate the stronger assumption of Dynamical Autonomy for measurements. This is the reason why I restrict Dynamical

Autonomy for measurements to only measurements of the initially possessed properties of an object system.[108]

9.3 The violation of Locality

By means of the example given in the previous section, one can demonstrate that the spectral and atomic modal interpretations are non-local in the sense that they violate the Locality condition (Vermaas 1999).

Consider, firstly, the example described in the previous section. Let ω be the composite $\alpha\gamma$ and let ω' be the composite $\beta\delta$ and assume that there is a huge distance between these two composites $\omega = \alpha\gamma$ and $\omega' = \beta\delta$. Say, this distance is such that the space-time region R which is occupied by $\alpha\gamma$ in the time interval $I = [-1, 4]$ is space-like separated from the space-time region R' which is occupied by $\beta\delta$ in the time interval $I = [-1, 4]$.[109] Hence, no signal emitted by ω' after $t = -1$ and travelling with the speed of light, or more slowly, can reach ω before $t = 4$, and *vice versa*.

The derivation of the transition probabilities for α from $t = 1$ to $t = 2$ is still valid under these extra assumptions. So, by (9.10) one has that both the spectral and atomic modal interpretations predict that

$$p(|g_k^\alpha\rangle\langle g_k^\alpha| \text{ at } 2 \,/\, |e_j^\alpha\rangle\langle e_j^\alpha| \text{ at } 1) = |d_{jk}|^2. \tag{9.18}$$

The derivation of the transition probabilities for β from $t = 2$ to $t = 3$ is no longer valid because Dynamical Autonomy for composite systems is untenable. However, the result (9.11) of this derivation cannot be right, for if this result were right, one would again violate the fourth Bell–Wigner inequality (9.13) and such a violation contradicts the fact that transition probabilities like (9.8) exist. Hence, one may conclude that the transition probabilities for β from $t = 2$ to $t = 3$ satisfy the constraint

$$p(|g_l^\beta\rangle\langle g_l^\beta| \text{ at } 3 \,/\, |e_{j'}^\beta\rangle\langle e_{j'}^\beta| \text{ at } 2) \neq |d_{j'l}|^2. \tag{9.19}$$

Consider, secondly, a modification of the example of the previous section. Namely, let β and δ not interact with one another in the time interval from $t = 2$ to $t = 3$ but in the time interval from $t = 0$ to $t = 1$. Let this interaction

---

[108] If one measures the initially possessed properties of α and of β, that is, if one takes $|g_j^X\rangle = |e_j^X\rangle$ for X both α and β, one can derive that

$$p(|\mathbf{R}_k^\gamma\rangle\langle \mathbf{R}_k^\gamma| \text{ at } 2 \,/\, |e_j^\alpha\rangle\langle e_j^\alpha| \text{ at } 1) = \delta_{jk}, \qquad p(|\mathbf{R}_l^\delta\rangle\langle \mathbf{R}_l^\delta| \text{ at } 3 \,/\, |e_{j'}^\beta\rangle\langle e_{j'}^\beta| \text{ at } 2) = \delta_{j'l},$$

and that the transition probabilities (9.17) are equal to $\delta_{jk}\delta_{jl}$ if $j = j'$. In this case the above transition probabilities thus satisfy Dynamical Autonomy for measurements without the consequence that the transition probabilities (9.17) do not exist.

[109] I here consider only one inertial coordinate system with t the time coordinate.

again be given by the scheme (9.2). So, in the modified example the order
of the interactions between α and γ and between β and δ, respectively,
is reversed as compared to the original example. That is, in the modified
example β and δ interact first from $t = 0$ to $t = 1$, and α and γ interact
second from $t = 1$ to $t = 2$.

Due to the symmetry between the original example and the modified
example (the modified example is obtained from the original one by inter-
changing $\alpha\gamma$ and $\beta\delta$ and by subtracting 1 from the time coordinate) the
results (9.18) and (9.19) for the transition probabilities are, *mutatis mutandis*,
still valid for the modified example. So,

$$\left.\begin{array}{l} p(|g_l^\beta\rangle\langle g_l^\beta| \text{ at } 1 \, / \, |e_{j'}^\beta\rangle\langle e_{j'}^\beta| \text{ at } 0) = |d_{j'l}|^2, \\[2mm] p(|g_k^\alpha\rangle\langle g_k^\alpha| \text{ at } 2 \, / \, |e_j^\alpha\rangle\langle e_j^\alpha| \text{ at } 1) \; \neq |d_{jk}|^2. \end{array}\right\} \qquad (9.20)$$

If one now compares the original and the modified examples, one can
prove the stated violation of Locality. The evolution of the state of the
composite $\omega = \alpha\gamma$ from $t = -1$ to $t = 4$ is equal in both examples. The
only difference between the two examples is the evolution of the state of the
composite $\omega' = \beta\delta$ in the interval from $t = 0$ to $t = 3$. In the first example the
interaction (9.2) between β and δ changes the state $W^{\omega'}(t)$ in the time interval
$I_1 = (2,3)$ and in the second modified example this interaction changes the
state $W^{\omega'}(t)$ in the interval $I_2 = (0,1)$. By assumption this difference in the
evolution of $W^{\omega'}(t)$ takes place in a region space-like separated from the
region in which $\omega = \alpha\gamma$ is confined from $t = -1$ to $t = 4$. If the spectral and
atomic modal interpretations satisfy Locality, this difference in the evolution
of $W^{\omega'}(t)$ may thus not affect the transition probabilities for α from $t = 1$
to $t = 2$. But these transition probabilities are affected: in the first example
they are by (9.18) equal to $|d_{jk}|^2$, whereas in the second example they are
by (9.20) equal to values different to $|d_{jk}|^2$. Hence, the spectral and atomic
modal interpretations violate the Locality condition.

Put more straightforwardly, these two modal interpretation are non-local
because one can manipulate the evolution of the properties ascribed to a
system α confined to a space-time region R, by changing the state evolution
of a freely evolving composite ω' confined to a second space-time region R'
which is space-like separated from R.

Part two

Physics

In Part one the description of reality by modal interpretations has been developed as far as possible. In this part I determine whether modal interpretations are empirically adequate by considering their descriptions of measurements.

In Chapter 10 I focus on the measurement problem. After a measurement, according to our observations, the measurement device displays a definite outcome. Such an outcome is traditionally called a pointer reading and the question is whether modal interpretations manage to ascribe such pointer readings.

In the standard formulation of quantum mechanics one predicts by means the Born rule the probabilities and the correlations with which measurements have outcomes. In Chapter 11 I discuss whether modal interpretations can reproduce these empirically correct predictions.

10

The measurement problem

In Part one of this book I have considered on occasions special measurement interactions for which modal interpretations solve the measurement problem by ascribing readings to the pointers of the measurement devices. In this chapter I address the question of whether modal interpretation are empirically adequate by considering whether they solve the measurement problem for all possible measurement interactions.

10.1 Introduction

One of the facts of physics is that measurements produce outcomes in the form of, say, pointers that assume readings, counters which indicate numbers, or plotters that write something on paper. These outcomes correspond to properties possessed by the pointer, the counter or the ink, etc., and are observed by us. So, if modal interpretations are to be empirically adequate, they should in some way confirm that pointers, counters, etc., have these outcome properties after measurements.[110]

For the measurement interactions that I have considered up to now, modal interpretations indeed yield that measurements have outcomes. Unfortunately, however, there also exist measurement interactions for which this is not the case. Consider, for instance, a measurement of the

---

[110] One may claim that modal interpretations are already empirically adequate if they yield that after measurements an observer of the measurement device acquires a brain-state that corresponds to the observer's belief that the pointer, counter, etc., displays an outcome. However fruitful this latter approach may be, I ignore it here since it diverges too much from the idea maintained in this book that interpretations of quantum mechanics aim at describing reality in terms of (empirically adequate) properties which are possessed by systems independently of the presence of observers.

second kind:[111]

$$|\Psi^{\sigma\pi}\rangle = \sum_j c_j |a_j^\sigma\rangle \otimes |R_0^\pi\rangle \longmapsto |\tilde{\Psi}^{\sigma\pi}\rangle = \sum_j c_j |\tilde{a}_j^\sigma\rangle \otimes |R_j^\pi\rangle. \qquad (10.1)$$

Here, the vectors $\{|a_j^\sigma\rangle\}_j$ are the pair-wise orthogonal eigenvectors of the measured magnitude A^σ of the object system σ and the vectors $\{|\tilde{a}_j^\sigma\rangle\}_j$ are the not necessarily pair-wise orthogonal perturbed eigenvectors of A^σ. That is, if before the measurement σ is in the eigenstate $|a_k^\sigma\rangle$, then the measurement interaction changes this eigenstate to the state $|\tilde{a}_k^\sigma\rangle$. The state of the pointer π after the measurement is $W^\pi = \sum_{j,j'} c_j \bar{c}_{j'} \langle \tilde{a}_{j'}^\sigma | \tilde{a}_j^\sigma \rangle |R_j^\pi\rangle\langle R_{j'}^\pi|$. And because this state is, in general, not diagonal with respect to the reading states $\{|R_j^\pi\rangle\langle R_j^\pi|\}_j$, the bi and spectral modal interpretations do not, in general, ascribe these readings.

For the atomic modal interpretation one can consider a Von Neumann measurement (2.9), where the pointer π consists of two atoms α and β and where the reading states are given by $|R_j^\pi\rangle = |e_j^\alpha\rangle \otimes |e_j^\beta\rangle$ for all j. The interaction then becomes

$$|\Psi^{\sigma\pi}\rangle = \sum_j c_j |a_j^\sigma\rangle \otimes |e_0^\alpha\rangle \otimes |e_0^\beta\rangle \longmapsto |\tilde{\Psi}^{\sigma\pi}\rangle = \sum_j c_j |a_j^\sigma\rangle \otimes |e_j^\alpha\rangle \otimes |e_j^\beta\rangle. \qquad (10.2)$$

The vectors $\{|R_j^\pi\rangle\}_j$ are eigenvectors of the reading magnitude $M^\pi = \sum_j m_j |R_j^\pi\rangle\langle R_j^\pi|$ and are therefore pair-wise orthogonal. Hence, assume that the vectors $\{|e_j^\alpha\rangle\}_j$ are pair-wise orthogonal and that the vectors $\{|e_j^\beta\rangle\}_j$ are pair-wise *non-orthogonal* (the vectors $\{|R_j^\pi\rangle\}_j$ are under these assumptions indeed pair-wise orthogonal). One can then prove that the pointer does not, in general, assume its readings after the interaction (10.2).[112]

So, according to modal interpretations not all possible measurements have outcomes. And, to make things worse, Albert and Loewer (1990, 1993) argued that measurements which are performed in the real world are typically those measurements for which modal interpretations fail to ascribe outcomes. Consider an ideal measurement which establishes strict correlations between the pointer readings $\{|R_j^\pi\rangle\}_j$ and the eigenvectors $\{|a_j^\sigma\rangle\}_j$ of the measured

---

[111] Bacciagaluppi and Hemmo (1996, Sect. 3.3) attribute this example to Harvey Brown.

[112] Proof: Assume that there exists a state of π for which the atomic modal interpretation ascribes with non-trivial probabilities $\{p_j\}_j$ the readings $\{|R_j^\pi\rangle\langle R_j^\pi|\}_j$ to π. If π is ascribed the reading $|R_k^\pi\rangle\langle R_k^\pi|$, then, according to the Weakening condition (see page 72), π is ascribed a core property C^π that satisfies $|R_k^\pi\rangle\langle R_k^\pi| C^\pi = C^\pi$. Since $|R_k^\pi\rangle\langle R_k^\pi|$ is a one-dimensional projection, the only projection C^π that satisfies this relation is $|R_k^\pi\rangle\langle R_k^\pi|$ itself. So, π is ascribed the core property $|R_k^\pi\rangle\langle R_k^\pi| = |e_k^\alpha\rangle\langle e_k^\alpha| \otimes |e_k^\beta\rangle\langle e_k^\beta|$. This implies that the atom β is ascribed the core property $|e_k^\beta\rangle\langle e_k^\beta|$ and that $|e_k^\beta\rangle\langle e_k^\beta|$ is an eigenprojection of the state W^β of this atom. Repeating the argument for all the readings $\{|R_j^\pi\rangle\langle R_j^\pi|\}_j$ yields that all the projections $\{|e_j^\beta\rangle\langle e_j^\beta|\}_j$ are eigenprojections of W^β. This is, however, impossible since the vectors $\{|e_j^\beta\rangle\}_j$ are assumed to be non-orthogonal. $\square$

magnitude, and which does not perturb these eigenvectors $\{|a_j^\sigma\rangle\}_j$ to vectors $\{|\tilde{a}_j^\sigma\rangle\}_j \neq \{|a_j^\sigma\rangle\}_j$. Such a measurement can be modelled by a Von Neumann measurement interaction given by

$$|\Psi^{\sigma\pi}\rangle = \sum_j c_j\,|a_j^\sigma\rangle \otimes |R_0^\pi\rangle \longmapsto |\tilde{\Psi}^{\sigma\pi}\rangle = \sum_j c_j\,|a_j^\sigma\rangle \otimes |R_j^\pi\rangle. \qquad (10.3)$$

After this interaction the bi and spectral modal interpretations generally do ascribe the readings to the pointer, but Albert and Loewer now argue that real-life measurements almost never satisfy the idealisations with which we describe them. So, an ideal measurement, which is meant to strictly correlate the pointer readings and the eigenvectors of the measured magnitude, in reality makes small errors by correlating $|R_k^\pi\rangle$ with the wrong eigenvector $|a_j^\sigma\rangle$. A more realistic model for the interaction is therefore

$$|\Psi^{\sigma\pi}\rangle = \sum_j c_j\,|a_j^\sigma\rangle \otimes |R_0^\pi\rangle \longmapsto |\tilde{\Psi}^{\sigma\pi}\rangle = \sum_{j,k} c_{jk}\,|a_j^\sigma\rangle \otimes |R_k^\pi\rangle \qquad (10.4)$$

with $c_{jj} \approx c_j$ and $c_{jk} \approx 0$ for all $j \neq k$ (the coefficients $\{c_{jk}\}_{j\neq k}$ model the errors). The state of π after this more realistic interaction is $W^\pi = \sum_{j,k,k'} c_{jk}\,\bar{c}_{jk'}|R_k^\pi\rangle\langle R_{k'}^\pi|$. And this state is generally not diagonal with respect to the pointer readings. So, for realistic measurements, the bi and spectral modal interpretations do not ascribe readings, in general, to the pointers. Moreover, Albert and Loewer pointed out that even if the errors $\{c_{jk}\}_{j\neq k}$ are arbitrarily small, the pointer properties can still be substantially different to the readings.[113] Hence, even if modal interpretations ascribe readings to pointers after measurements, this ascription is unstable in the sense discussed in Section 7.4. That is, small changes in the final pointer state may

---

[113] Take, for instance, a pointer π with two possible readings and choose

$$c_{11} = (16\sqrt{1+\varepsilon} + 9\sqrt{1-\varepsilon})/25\sqrt{2},$$
$$c_{22} = (9\sqrt{1+\varepsilon} + 16\sqrt{1-\varepsilon})/25\sqrt{2},$$
$$c_{12} = c_{21} = 12(\sqrt{1+\varepsilon} - \sqrt{1-\varepsilon})/25\sqrt{2}.$$

The coefficients c_{12} and c_{21} modelling the errors of the measurement are in a first order expansion equal to $\frac{12}{25\sqrt{2}}\varepsilon + O(\varepsilon^2)$ so can be taken arbitrarily small by taking ε arbitrarily small.

The state of the pointer after the measurement (10.4) is

$$W^\pi = \frac{1}{50}\begin{pmatrix} 25+7\varepsilon & 24\varepsilon \\ 24\varepsilon & 25-7\varepsilon \end{pmatrix}$$

with respect to the pointer basis $\{|R_1^\pi\rangle, |R_2^\pi\rangle\}$. The eigenvalues of W^π are $\frac{1}{2}(1+\varepsilon)$ and $\frac{1}{2}(1-\varepsilon)$, the corresponding eigenvectors are $(\frac{4}{5}, \frac{3}{5})$ and $(\frac{3}{5}, -\frac{4}{5})$, respectively. So, the core properties of the pointer are in the bi and spectral modal interpretations for every value ε substantially different to the pointer readings represented by the vectors $(1, 0)$ and $(0, 1)$.

radically change the pointer properties from readings to properties different to readings, and back again.[114]

Similar remarks can be found in Elby (1993) who argued that a real-life measurement always makes small errors due to the infinitely spread 'tails' of the state of the object system σ. Ruetsche (1995) showed that modal interpretations in general do not ascribe readings after normal unitary measurements and Bacciagaluppi and Hemmo (1996) extended this conclusion to unsharp measurements.

Modal interpretations thus do not confirm that every measurement imaginable in quantum mechanics has an outcome. Some authors, like Albert (1992, page 191), therefore conclude that modal interpretations do not solve the measurement problem and should thus be discarded as empirically inadequate interpretations. However, if this conclusion is based on measurement interactions like (10.4), as Albert does, it seems that this conclusion is unfounded exactly for the same reasons with which Albert and Loewer criticise the ability of modal interpretations to solve the measurement problem. Albert and Loewer argued that in real life ideal measurements are properly modelled not by interaction (10.3) but by interaction (10.4). And for this last interaction, modal interpretetations do not ascribe outcomes. However, this last interaction is also quite unrealistic. Measurement devices μ consist in real life of a pointer π (which may also be a counter, a dot of ink, etc.) plus some supporting mechanism $\bar{\mu}$ (see Section 4.7). A realistic model of a measurement therefore includes interactions not only between the object system σ and the pointer π, as in the case of interaction (10.4), but between σ, π, the mechanism $\bar{\mu}$ and most probably also between those systems and the surrounding environment. Hence, if one takes Albert and Loewer's criticism seriously one should also reject interaction (10.4) as a realistic model of a measurement.

Thus, to make the argument precise, one can say that it is too strong to assume that an interpretation is empirically adequate only if it ascribes outcomes after every *imaginable* measurement interaction. Instead, empirical adequacy of an interpretation is already obtained if one can prove that it yields that every *physically realistic* measurement has an outcome. After all it is only a fact of physics that measurements which are realistic in the sense

---

[114] Consider, for instance, a pointer state which changes as

$$W^{\pi}(t_1) = \tfrac{1}{50} \begin{pmatrix} 25 + 7\varepsilon & 0 \\ 0 & 25 - 7\varepsilon \end{pmatrix} \longmapsto W^{\pi}(t_2) = \tfrac{1}{50} \begin{pmatrix} 25 + 7\varepsilon & 24\varepsilon \\ 24\varepsilon & 25 - 7\varepsilon \end{pmatrix}.$$

This change can occur in a very small interval because the difference between $W^{\pi}(t_1)$ and $W^{\pi}(t_2)$ is arbitrarily small if one takes ε arbitrarily small. But the eigenprojections of π change substantially: at t_1 they are $(1, 0)$ and $(0, 1)$ and at t_2 they are according to footnote 113 equal to $(\tfrac{4}{5}, \tfrac{3}{5})$ and $(\tfrac{3}{5}, -\tfrac{4}{5})$.

that they can actually be performed in the real world have outcomes. Hence, an interpretation which fails by not ascribing outcomes after inoperable measurements like (10.1) or (10.4) is not immediately empirically inadequate.

In Section 10.2 I present a well-elaborated solution to the measurement problem by Bacciagaluppi and Hemmo (1994, 1996). Their solution can be understood along the lines of the ideas sketched above because Bacciagaluppi and Hemmo prove that if the pointer of the measurement device interacts with the environment (and if the dimension of the Hilbert space associated with the pointer may be taken as effectively finite), then after any measurement the pointer state has eigenprojections $\{P_j^{\pi}(t)\}_j$ which become very fast very close to the readings $\{|R_j^{\pi}\rangle\langle R_j^{\pi}|\}_j$. Hence, for physically more realistic models of measurements in which the pointer interacts with the environment, the pointer possesses in the bi and spectral modal interpretation properties which are close to readings.

In Sections 10.3 and 10.4 I derive necessary and sufficient conditions for solving the measurement problem with the bi, spectral and atomic modal interpretations (Vermaas 1998c). These conditions apply to the internal structure of a measurement device. A measurement device is taken as a composite $\bar{\mu}\pi$ of a mechanism and a pointer and if this mechanism and pointer satisfy the sufficient conditions, modal interpretations ascribe after any measurement exactly the readings to the pointer. These results also draw upon the idea that not every imaginable measurement is physically realistic in the sense that not any system counts as a physically realistic measurement device. (And it is, of course, my hope that all realistic devices satisfy the sufficient conditions.)

I believe it is not yet decided whether or not modal interpretations solve the measurement problem for physically realistic measurements. There are some strong indications that modal interpretations do not solve this problem (see Section 14.5) so the possibility that modal interpretations may be proved to be empirically inadequate seems present. However, this proof should be based on realistic measurement schemes. The discarding of modal interpretations by Albert, based on a model in which a measurement device consists only of a pointer, is thus far too rash for me.

10.2 Bacciagaluppi and Hemmo: decoherence

Healey (1989) and Dieks (1994a,b) argued that a solution to the measurement problem can be found if one includes the environment in the description of measurements. Interactions between the environment and the pointer would then make by decoherence that the state of the pointer becomes (almost) diagonal with respect to the reading states very fast. And this would imply that the bi and spectral modal interpretations ascribe properties to the pointer which become (almost) the readings very fast.

Bacciagaluppi and Hemmo took up this idea and in two excellent papers (Bacciagaluppi and Hemmo 1994, 1996)[115] gave a quantitative analysis of the effects of decoherence on the pointer. They considered in particular a model of a measurement in which the pointer is (effectively) described by a few-dimensional Hilbert space. Then, by applying the results of decoherence theory as developed by Zeh (1970, 1973) and Zurek (1981, 1982),[116] they proved that the eigenprojections of the pointer state indeed converge to the reading states.

The model by Bacciagaluppi and Hemmo is roughly as follows. Before the measurement, the object system σ has an arbitrary state $\sum_j c_j |a_j^\sigma\rangle$ and the pointer is in its 'ready to measure' state $|R_0^\pi\rangle$. The environment, denoted by ω, initially has an arbitrary pure state $|\mathscr{E}^\omega\rangle$. Then, a measurement takes place by a measurement interaction similar to the interaction (10.4) considered by Albert and Loewer. During this measurement interaction the pointer also interacts with the environment and this second interaction is governed by a Hamiltonian $H^{\pi\omega}$ which commutes with the pointer reading projections $\{|R_j^\pi\rangle\langle R_j^\pi| \otimes \mathbb{I}^\omega\}_j$ (which expresses that the pointer readings are conserved quantities during the evolution induced by the interaction between the pointer and the environment). Finally, the measurement interaction is assumed to be 'slow' compared to the interaction between the pointer and the environment.[117] The evolution of the state of $\sigma\pi\omega$ then becomes

$$
\begin{aligned}
|\Psi^{\sigma\pi\omega}\rangle &= \sum_j c_j |a_j^\sigma\rangle \otimes |R_0^\pi\rangle \otimes |\mathscr{E}^\omega\rangle \longmapsto |\widetilde{\Psi}^{\sigma\pi\omega}(t)\rangle \\
&= \sum_{j,k} c_{jk} |a_j^\sigma\rangle \otimes |R_k^\pi\rangle \otimes |\mathscr{E}_k^\omega(t)\rangle
\end{aligned}
\tag{10.5}
$$

and the state of the pointer after the measurement is therefore

$$
W^\pi(t) = \sum_{j,k,k'} c_{jk} \bar{c}_{jk'} \langle \mathscr{E}_{k'}^\omega(t)|\mathscr{E}_k^\omega(t)\rangle |R_k^\pi\rangle\langle R_{k'}^\pi|.
\tag{10.6}
$$

It is a result of decoherence theory that the environment states $\{|\mathscr{E}_k^\omega(t)\rangle\}_k$ which are coupled to the reading states $\{|R_k^\pi\rangle\}_k$, respectively, become almost orthogonal very fast. More precisely, for $k \neq k'$ the inner product $|\langle \mathscr{E}_{k'}^\omega(t)|\mathscr{E}_k^\omega(t)\rangle|$ approaches zero exponentially, that is, proportional to

---

[115] See also Bacciagaluppi (1996b, Chap. 5) and Hemmo (1996, Chap. I.1).
[116] See footnote 21 in Bacciagaluppi and Hemmo (1996) for more references.
[117] See Bacciagaluppi and Hemmo (1996) for a more precise exposition of the various assumptions.

$\exp(-t/\tau_D)$, with a decoherence time τ_D that can be of the order of 10^{-25} seconds. By this exponential decay, the pointer state $W^\pi(t)$ rapidly converges to the state $\widehat{W}^\pi = \sum_{j,k} |c_{jk}|^2 |R_k^\pi\rangle\langle R_k^\pi|$ which is diagonal with respect to the reading states.

This convergence of $W^\pi(t)$ to $\widehat{W}^\pi$ need not imply that the eigenprojections of $W^\pi(t)$ converge to the eigenprojections $\{|R_j^\pi\rangle\langle R_j^\pi|\}_j$ of $\widehat{W}^\pi$ because states which are close to $\widehat{W}^\pi$ need not have the same eigenprojections as $\widehat{W}^\pi$ (see footnote 114). The remarkable result by Baciagaluppi and Hemmo is now that they were able to prove that, for their model, the eigenprojections of $W^\pi(t)$ do converge to the eigenprojections of $\widehat{W}^\pi$. Hence, they were able to prove that after the measurement interaction (10.5) the eigenprojections of $W^\pi(t)$ converge to the readings $\{|R_j^\pi\rangle\langle R_j^\pi|\}_j$. The proof consists of the following steps.

Consider the operator $W^\pi(t) - \widehat{W}^\pi$. The operator norm of this difference has an upper bound given by[118]

$$\|W^\pi(t) - \widehat{W}^\pi\| < \dim(\mathscr{H}^\pi) \max_{a,b\,(a\neq b)} \left| \sum_j c_{ja}\bar{c}_{jb}\langle \mathscr{E}_b^\omega(t)|\mathscr{E}_a^\omega(t)\rangle \right|. \tag{10.7}$$

Assume that the eigenvalues $\{\sum_j |c_{jk}|^2\}_k$ of $\widehat{W}^\pi$ are not degenerate and define the nearest neighbour eigenvalue distance δ_a as

$$\delta_a := \min_{b(\neq a)} \left\| \sum_j |c_{jb}|^2 - \sum_j |c_{ja}|^2 \right\|. \tag{10.8}$$

Finally, let $\{|w_k^\pi(t)\rangle\}_k$ be the set of eigenvectors of $W^\pi(t)$. Baciagaluppi and Hemmo then prove the following relation between the eigenprojections of $W^\pi(t)$ and $\widehat{W}^\pi$:[119]

Proposition 10.1

Let $0 < \varepsilon < 1$. *Then,*

$$\|W^\pi(t) - \widehat{W}^\pi\|^2 < \frac{\delta_k^2 \varepsilon^2}{4} \implies |\langle w_k^\pi(t)|R_k^\pi\rangle|^2 > 1 - \varepsilon^2. \tag{10.9}$$

---

[118] Baciagaluppi and Hemmo posit this upper bound of the norm of $W^\pi(t) - \widehat{W}^\pi$ as a crude estimate without further argument. A derivation which nearly does the trick goes as follows. The operator $W^\pi(t)$ can be represented by a matrix A with the elements $A_{ab} = \sum_j c_{ja}\bar{c}_{jb}\langle \mathscr{E}_b^\omega(t)|\mathscr{E}_a^\omega(t)\rangle$ and $\widehat{W}^\pi$ can be represented by a matrix B with the elements $B_{ab} = \delta_{ab}\sum_j c_{ja}\bar{c}_{jb}\langle \mathscr{E}_b^\omega(t)|\mathscr{E}_a^\omega(t)\rangle$. The operator $W^\pi(t) - \widehat{W}^\pi$ is thus represented by a matrix M with the elements $M_{aa} = 0$ and $M_{ab} = \sum_j c_{ja}\bar{c}_{jb}\langle \mathscr{E}_b^\omega(t)|\mathscr{E}_a^\omega(t)\rangle$ if $a \neq b$. (All matrices are with respect to the pointer basis $\{|R_k^\pi\rangle\}_k$.) For any operator M^π represented by a matrix M it can be proved that $\|M^\pi\| = \|M\| \leq \|\tilde{M}\|$, where $\tilde{M}$ is the matrix with the elements $\tilde{M}_{cd} = \max_{a,b} |M_{ab}|$ for all c and d. It holds that $\|\tilde{M}\| = \dim(\mathscr{H}^\alpha)\max_{a,b} |M_{ab}|$ because $\tilde{M}$ has only two eigenvalues, namely 0 and $\dim(\mathscr{H}^\alpha)\max_{a,b} |M_{ab}|$. It thus follows that $\|W^\pi(t) - \widehat{W}^\pi\| \leq \dim(\mathscr{H}^\pi)\max_{a,b\,(a\neq b)} |\sum_j c_{ja}\bar{c}_{jb}\langle \mathscr{E}_b^\omega(t)|\mathscr{E}_a^\omega(t)\rangle|$.

[119] This proposition is Proposition 5.2 in Baciagaluppi (1996b, Sect. 5.4.2).

By (10.7) it follows that one can choose ε^2 to be

$$\varepsilon^2 = \frac{4(\dim(\mathscr{H}^\pi))^2}{\delta_k^2} \max_{a,b\,(a\neq b)} \left(\left| \sum_j c_{ja}\bar{c}_{jb}\langle \mathscr{E}_b^\omega(t)|\mathscr{E}_a^\omega(t)\rangle \right| \right)^2. \qquad (10.10)$$

The factor $4\,(\dim(\mathscr{H}^\pi))^2/\delta_k^2$ is constant and can be enormously large. However, because the inner product $\langle \mathscr{E}_b^\omega(t)|\mathscr{E}_a^\omega(t)\rangle$ approaches zero so fast for $a \neq b$, ε^2 will rapidly become smaller than 1 (such that the above proposition holds) and will one instant later be so small that $|\langle w_k^\pi(t)|R_k^\pi\rangle|$ is virtually equal to 1. Hence, by decoherence, the eigenprojections of $W^\pi(t)$ converge to the projections $\{|R_k^\pi\rangle\langle R_k^\pi|\}_k$. It follows that the bi and spectral modal interpretations ascribe after the measurement interaction (10.5) properties which in Hilbert space are very fast very close to the pointer readings.

This nice result has, however, a number of drawbacks. Firstly, Bacciagaluppi and Hemmo note themselves that this result holds only if the pointer π is indeed (effectively) described by a few-dimensional Hilbert space. For measurements in which the pointer is described by an infinite-dimensional Hilbert space, there are indications that decoherence will not make that the eigenprojections of the pointer state converge to reading states (Bacciagaluppi 1996a).

Secondly, as pointed out by Bacciagaluppi and Hemmo as well, it is assumed in the above model of a measurement that the environmental influences do not affect the diagonal matrix elements $\{\langle R_a^\pi|W^\pi(t)|R_a^\pi\rangle\}_a$ of the pointer state. The eigenvalues of $\widehat{W^\pi}$ are therefore constant in time and the distances $\{\delta_k\}_k$ are therefore constant as well. If, however, these diagonal matrix elements are also perturbed, the distances $\{\delta_k\}_k$ fluctuate such that ε as given in (10.10) is no longer fixed. It might, for instance, happen that during (small) time intervals, δ_k approaches zero faster than $|\langle \mathscr{E}_b^\omega(t)|\mathscr{E}_a^\omega(t)\rangle|$ does. Then ε becomes larger in time and the eigenvectors of $W^\pi(t)$ can again differ substantially from the reading states. Hence, if the values $\{\langle R_a^\pi|W^\pi(t)|R_a^\pi\rangle\}_a$ change in time, the properties ascribed to the pointer after measurement interaction (10.5) can become unstable.

Thirdly, it is assumed in the model that before a measurement the state of the pointer and the environment factorises, that is, $W^{\pi\omega} = |R_0^\pi\rangle\langle R_0^\pi| \otimes |\mathscr{E}^\omega\rangle\langle \mathscr{E}^\omega|$. This factorisability now seems possible only if the interaction between the pointer and the environment in the period before the measurement satisfies strong constraints. See Arntzenius (1998) for an extensive discussion of this assumption of factorisability in decoherence theory.

Fourthly, there is not yet consensus about whether the ascription to a pointer of properties, which are represented by projections that are in

Hilbert space close to projections $\{|R_j^\pi\rangle\langle R_j^\pi|\}_j$, counts as a proper solution to the measurement problem. Bacciagaluppi and Hemmo (1994, 1996) and also Dieks (1994a,b) argue that the ascription of properties which in Hilbert space are approximately pointer readings is sufficient but Ruetsche (1998), for instance, has questioned this position.

Finally, the above results yield only that the bi and spectral modal interpretations ascribe properties to the pointer that are close to readings. For the atomic modal interpretation one is still empty-handed. Consider again the measurement interaction (10.2). The proof, given in footnote 112, that the pointer cannot assume its readings in the atomic modal interpretation is independent of the specific state of the pointer. Hence, decoherence can change the final state of this pointer to whatever one likes, but the atomic modal interpretation remains incapable of ascribing the readings.

10.3 Exact solutions for the atomic modal interpretation

According to the last remark of the previous section one cannot solve the measurement problem for the atomic modal interpretation by using that pointers of realistic measurement devices interact with the environment. Hence, in order to obtain empirical adequacy for this interpretation, one has to use something else. In this section I show that if the pointer of a measurement device satisfies certain conditions, then the atomic modal interpretation ascribes exactly the readings to this pointer (Vermaas 1998c, Sects. 4.1 and 4.2).

The idea that the measurement problem can be solved by focusing on the internal structure of measurement devices is not new. Healey (1989) and Dieks (1994a,b) have already argued in this direction. In a sense it is even preferable if a pointer assumes its readings, not because of the interaction between the pointer and the environment, but because of the internal make-up of the measurement device. For if the interaction between the pointer and the environment must induce that the pointer possesses its readings after a measurement, one explicitly has to assume that this interaction can change the properties of the pointer. And it is this assumption which leads to the problem that the properties of the pointer are sometimes only close to the pointer readings and that these properties are unstable: if the interaction between the pointer and the environment can change the properties of the pointer, then, if the pointer possesses its readings at one instant after a measurement, this interaction can induce that the pointer possesses at the next instant properties which are close to readings or which are substantially different to readings; if, on the other hand, the pointer always exactly possesses its readings because of the way the measurement device is structured, then the interaction between the pointer and the environment cannot affect the properties of the pointer — this interaction can then at most change the probabilities with which the pointer possesses its readings. Hence, if the internal structure of a measurement device makes the pointer exactly assumes its readings, the measurement problem is solved exactly. One then is saved from instabilities and one need not argue that the ascription of properties which are close to the readings also counts as a solution to the measurement problem.

I start by deriving necessary conditions for ascribing pointer readings by means of the atomic modal interpretation. These conditions apply to the eigenprojections $\{R_j^\pi\}_j$ of the pointer magnitude M^π. Then it is argued that for physically realistic pointers these necessary conditions can be slightly strengthened. These strengthened conditions are proved to be (with a few exceptions) sufficient for ascribing readings to pointers by means of the atomic modal interpretation.

So, let's see under which conditions the atomic modal interpretation ascribes readings to a pointer π. Let the pointer magnitude be $M^\pi = \sum_j m_j R_j^\pi$, where the projections $\{R_j^\pi\}_j$ which represent the readings are pair-wise orthogonal but not necessarily one-dimensional. Let the atoms in the pointer be α, β, γ, etc.

Suppose that there exists at least one pointer state for which the atomic modal interpretation ascribes non-trivially all readings to π. That is, if the pointer is in this state, then there is for every possible reading a non-zero probability that π possesses it and with probability 1 the pointer actually possesses exactly one of these readings. Denote this state by $\widehat{W}^\pi$. The states of the atoms in the pointer are then the partial traces $\widehat{W}^\alpha$, $\widehat{W}^\beta$, etc. Let $\{P_a^\alpha\}_a$ be the set of the eigenprojections of $\widehat{W}^\alpha$, let $\{P_b^\beta\}_b$ be the set of the eigenprojections of $\widehat{W}^\beta$, and so on. Finally let $P_{ab\cdots}^{\alpha\beta\cdots}$ be shorthand for $P_a^\alpha \otimes P_b^\beta \otimes \cdots$. The atomic core property ascription to the pointer is then

$$[P_{ab\cdots}^{\alpha\beta\cdots}] = 1 \text{ with probability } \text{Tr}^\pi(\widehat{W}^\pi P_{ab\cdots}^{\alpha\beta\cdots}). \tag{10.11}$$

This core property ascription should yield that π assumes with probability 1 exactly one of its readings as well as that all the readings are with non-zero probability possibly possessed. Hence, each assignment $[P_{ab\cdots}^{\alpha\beta\cdots}] = 1$ occurring in (10.11) with non-zero probability should induce that exactly one reading R_j^π is assigned the value 1. (If $[P_{ab\cdots}^{\alpha\beta\cdots}] = 1$ does not induce that exactly one R_j^π has the value 1, then either there exists a non-zero probability that π assumes no reading, or there exists a non-zero probability that π assumes two or more readings simultaneously; both cases block that with probability 1 all the readings are properly ascribed.) So,

$$\forall P_{ab\cdots}^{\alpha\beta\cdots} \text{ with } \text{Tr}^{\pi}(\widehat{W}^{\pi} P_{ab\cdots}^{\alpha\beta\cdots}) \neq 0 \; \exists! R_j^{\pi} \; : \; R_j^{\pi} P_{ab\cdots}^{\alpha\beta\cdots} = P_{ab\cdots}^{\alpha\beta\cdots}. \tag{10.12}$$

And because there should exist for every reading a non-zero probability that π assumes it, it also holds that

$$\forall R_j^{\pi} \; \exists P_{ab\cdots}^{\alpha\beta\cdots} \text{ with } \text{Tr}^{\pi}(\widehat{W}^{\pi} P_{ab\cdots}^{\alpha\beta\cdots}) \neq 0 \; : \; R_j^{\pi} P_{ab\cdots}^{\alpha\beta\cdots} = P_{ab\cdots}^{\alpha\beta\cdots}. \tag{10.13}$$

Define the index-sets $\{I_j\}_j$ of tuples $\langle a, b, \ldots \rangle$ belonging to the strings $\{P_{ab\cdots}^{\alpha\beta\cdots}\}_{ab\cdots}$ as

$$\langle a, b, \ldots \rangle \in I_j \text{ iff } \left[\text{Tr}^{\pi}(\widehat{W}^{\pi} P_{ab\cdots}^{\alpha\beta\cdots}) \neq 0 \text{ and } R_j^{\pi} P_{ab\cdots}^{\alpha\beta\cdots} = P_{ab\cdots}^{\alpha\beta\cdots} \right]. \tag{10.14}$$

From (10.12) it follows that these index-sets are mutually disjoint, and from (10.13) it follows that they are non-empty.

If two projections R^{π} and P^{π} obey the relation $R^{\pi} P^{\pi} = P^{\pi}$, then R^{π} can be written as $P^{\pi} + \widetilde{R}^{\pi}$ with $\widetilde{R}^{\pi}$ the projection equal to $R^{\pi} - P^{\pi}$. Hence, using relations (10.12) and (10.13) as well as the index-sets $\{I_j\}_j$, one can expand all R_j^{π}s as

$$R_j^{\pi} = \sum_{\langle a, b, \ldots \rangle \in I_j} P_{ab\cdots}^{\alpha\beta\cdots} + \widetilde{R}_j^{\pi}. \tag{10.15}$$

This is the first necessary condition that the projections $\{R_j^{\pi}\}_j$ have to obey. Furthermore, these projections $\{R_j^{\pi}\}_j$ are pair-wise orthogonal, which implies that $P_{ab\cdots}^{\alpha\beta\cdots} P_{a'b'\cdots}^{\alpha\beta\cdots} = 0$ if $\langle a, b, \ldots \rangle \in I_j$ and $\langle a', b', \ldots \rangle \in I_k$ with $j \neq k$. This product $P_{ab\cdots}^{\alpha\beta\cdots} P_{a'b'\cdots}^{\alpha\beta\cdots}$ is equal to $P_a^{\alpha} P_{a'}^{\alpha} \otimes P_b^{\beta} P_{b'}^{\beta} \otimes \cdots$ and is thus zero only if there exists at least one atom X for which the product $P_x^X P_{x'}^X$ is equal to zero. A second necessary condition is thus

$$\forall \langle a, b, \ldots \rangle \in I_j, \langle a', b', \ldots \rangle \in I_k \text{ with } j \neq k :$$

$$P_x^X P_{x'}^X = 0 \text{ for at least one atom } X \in \{\alpha, \beta, \ldots\}. \tag{10.16}$$

To sum up: if there exists at least one pointer state for which the atomic modal interpretation solves the measurement problem, the projections $\{R_j^{\pi}\}_j$ necessarily obey relations (10.15) and (10.16). The index-sets $\{I_j\}_j$ occurring in these conditions are non-empty and mutually disjoint. The projections $P_{ab\cdots}^{\alpha\beta\cdots}$ are strings $P_a^{\alpha} \otimes P_b^{\beta} \otimes \cdots$ generated by sets of pair-wise orthogonal projections $\{P_a^{\alpha}\}_a$, $\{P_b^{\beta}\}_b$, etc.

These two necessary conditions are not yet sufficient conditions for ascribing readings with the atomic modal interpretation. Consider, for instance, a measurement (10.1) of the second kind and assume that all the readings are given by

$$R_j^\pi = |e_j^\alpha\rangle\langle e_j^\alpha| \otimes |e^\beta\rangle\langle e^\beta| \otimes |e^\gamma\rangle\langle e^\gamma| \otimes \cdots \tag{10.17}$$

with $\{|e_j^\alpha\rangle\}_j$ a set of pair-wise orthogonal vectors. Define P_j^α as $|e_j^\alpha\rangle\langle e_j^\alpha|$ and define P_1^β as $|e^\beta\rangle\langle e^\beta|$, P_1^γ as $|e^\gamma\rangle\langle e^\gamma|$, etc. These readings then satisfy the conditions (10.15) and (10.16) if one takes $I_j = \langle j, 1, 1, \ldots \rangle$. The pointer state after the interaction (10.1) is

$$W^\pi = \sum_{j,j'} c_j \bar{c}_{j'} \langle \tilde{a}_{j'}^\sigma | \tilde{a}_j^\sigma \rangle \, |e_j^\alpha\rangle\langle e_{j'}^\alpha| \otimes |e^\beta\rangle\langle e^\beta| \otimes |e^\gamma\rangle\langle e^\gamma| \otimes \cdots . \tag{10.18}$$

Since the eigenprojections of the state of α are generally not given by $\{|e_j^\alpha\rangle\langle e_j^\alpha|\}_j$, the atomic modal interpretation generally does not ascribe the readings to the pointer. Hence, sufficient conditions for ascribing readings should be stronger than (10.15) and (10.16). I propose now to strengthen condition (10.16) to

$$\forall \langle a, b, \ldots \rangle \in I_j, \langle a', b', \ldots \rangle \in I_k \text{ with } j \neq k :$$
$$P_x^X P_{x'}^X = 0 \text{ for at least } n \geq 2 \text{ atoms } X \in \{\alpha, \beta, \ldots\} \tag{10.19}$$

and my argumentation for this strengthening goes as follows.

Condition (10.16) permits two different readings to be represented by projections R_j^π and R_k^π, which are mutually orthogonal only with respect to the state of one atom. Take a second look at the reading states given by (10.17). If, in that example, the state of the pointer changes from one reading to another, only the state of atom α changes; the states of the other atoms remain constant. The readings R_j^π and R_k^π are thus orthogonal only because the states $|e_j^\alpha\rangle\langle e_j^\alpha|$ and $|e_k^\alpha\rangle\langle e_k^\alpha|$ are orthogonal. This implies that the (macroscopic) difference between two readings R_j^π and R_k^π supervenes only a difference in the state of one atom. And if one removes this one atom from the pointer, the 'pointer readings' of the pointer-minus-one-atom become identical (in the case of (10.17) the 'readings' $\{R_j^{\pi/\alpha}\}_j$ of the stripped pointer π/α are all represented by $|e^\beta\rangle\langle e^\beta| \otimes |e^\gamma\rangle\langle e^\gamma| \otimes \cdots$). For physically realistic measurements it is, however, very implausible that if the pointer loses one atom, the readings of that pointer will become identical. Hence, for a realistic pointer, two readings R_j^π and R_k^π should be pair-wise orthogonal with respect to the states of at least two (but most probably many more) atoms, as is expressed by the strengthened condition (10.19); the pointer can then indeed lose an atom while its readings remain orthogonal.

One can now prove that the strengthened conditions (10.15) and (10.19)

are sufficient conditions for ascribing readings with the atomic modal interpretation. Consider any measurement performed on an object system σ. The
measurement device can be divided into a pointer π (or some other indicator)
to display the outcome, and some supporting mechanism $\bar{\mu}$ which consists
of wires, electronics, magnets, etc. The device is thus the composite $\mu = \bar{\mu}\pi$.
Assume that the measured magnitude is represented by an operator A^σ with
a discrete and non-degenerate spectral resolution $A^\sigma = \sum_{p=1} a_p |a_p^\sigma\rangle\langle a_p^\sigma|$.[120]
Before the measurement, π displays that the device is ready to measure. Let
this 'ready' reading be represented by the projection R_0^π. After the measurement π is supposed to display outcomes that correspond to the eigenspaces
$\{|a_p^\sigma\rangle\langle a_p^\sigma|\}_p$ of A^σ. Let these 'outcome' readings be represented by the projections $\{R_j^\pi\}_{j=1}$ and assume that all readings $\{R_j^\pi\}_{j=0}$ obey the conditions
(10.15) and (10.19).

Before the measurement, σ is in some arbitrary state W_0^σ and the device μ
is in an initial state $|\psi_0^\mu\rangle\langle\psi_0^\mu|$. Define the coefficients W_{pq}^σ as $\langle a_p^\sigma|W_0^\sigma|a_q^\sigma\rangle$. The
initial state of the composite is then[121]

$$W_0^{\sigma\mu} = \sum_{p,q} W_{pq}^\sigma \,|a_p^\sigma\rangle\langle a_q^\sigma| \otimes |\psi_0^\mu\rangle\langle\psi_0^\mu|. \tag{10.20}$$

To make sure that before the measurement the pointer possesses the reading
R_0^π with probability 1, one has to demand that the initial device state obeys

$W_0^\pi = \mathrm{Tr}^{\bar{\mu}}(|\psi_0^\mu\rangle\langle\psi_0^\mu|)$ yields the core property ascription

$\quad [P_{ab\cdots}^{\alpha\beta\cdots}] = 1$ with probability $\mathrm{Tr}^\pi(W_0^\pi P_{ab\cdots}^{\alpha\beta\cdots})$ if $\langle a, b, \dots\rangle \in I_0$

$\quad$ and satisfies $\sum_{\langle a,b,\dots\rangle \in I_0} \mathrm{Tr}^\pi(W_0^\pi P_{ab\cdots}^{\alpha\beta\cdots}) = 1$. $\tag{10.21}$

From condition (10.15) it then follows that initially $[R_0^\pi] = 1$ with probability
1.

Take firstly the special case of a perfect measurement. So, if the ini-
tial state W_0^σ obeys the relation $W_{jj}^\sigma = 1$ (such that the Born probability
$\mathrm{Tr}^\sigma(W_0^\sigma |a_j^\sigma\rangle\langle a_j^\sigma|)$ of finding an outcome corresponding to the eigenvalue
a_j is 1), then after the measurement the pointer possesses the reading R_j^π
with probability 1. This perfectness can be obtained by assuming that the
measurement interaction is given by

$$|a_p^\sigma\rangle \otimes |\psi_0^\mu\rangle \longmapsto |\Psi_p^{\sigma\mu}\rangle, \qquad\qquad p = 1, 2, \dots, \tag{10.22}$$

---

[120] I consider for simplicity only non-degenerate magnitudes A^σ but all the results generalise to degenerate
magnitudes (Vermaas 1998b, Sect. 9.3), although the formulae become a bit lengthy.
[121] The state of μ is pure. Hence, $W_0^{\sigma\mu}$ is uniquely given by $W_0^\sigma \otimes W_0^\mu$ (see footnote 11).

where the pair-wise orthogonal vectors $\{|\Psi_p^{\sigma\mu}\rangle\}_p$ meet the condition

$W_j^\pi = \text{Tr}^{\sigma\bar{\mu}}(|\Psi_j^{\sigma\mu}\rangle\langle\Psi_j^{\sigma\mu}|)$ yields the core property ascription

$\qquad [P_{ab\cdots}^{\alpha\beta\cdots}] = 1$ with probability $\text{Tr}^\pi(W_j^\pi P_{ab\cdots}^{\alpha\beta\cdots})$ if $\langle a, b, \ldots \rangle \in I_j$

$\qquad$ and satisfies $\sum_{\langle a,b,\ldots\rangle \in I_j} \text{Tr}^\pi(W_j^\pi P_{ab\cdots}^{\alpha\beta\cdots}) = 1.$ $\hfill$ (10.23)

Again from condition (10.15) it then follows that finally $[R_j^\pi] = 1$ with probability 1.

It can be proved that, barring degeneracies, the atomic modal interpretation solves the measurement problem for this perfect measurement: for every initial state W_0^σ the atomic modal interpretation yields that the pointer finally possesses the reading R_j^π with probability $\text{Tr}^\sigma(W_0^\sigma|a_j^\sigma\rangle\langle a_j^\sigma|) = W_{jj}^\sigma$ (see the MATHEMATICS).

Take secondly the general case of a measurement in which errors can also occur in the sense that an initial state W_0^σ with $W_{jj}^\sigma = 1$ can sometimes yield that the pointer possesses after the measurement a reading R_k^π with $k \neq j$. An interaction for such a general measurement is given by

$$|a_{p'}^\sigma\rangle \otimes |\psi_0^\mu\rangle \longmapsto |\widetilde{\Psi}_{p'}^{\sigma\mu}\rangle = \sum_p \lambda_{p'p} |\Psi_p^{\sigma\mu}\rangle, \qquad p' = 1, 2, \ldots, \qquad (10.24)$$

where the vectors $\{|\Psi_p^{\sigma\mu}\rangle\}_p$ still obey condition (10.23). The values $\{\lambda_{p'p}\}_{p',p}$ satisfy $\sum_p |\lambda_{p'p}|^2 = 1$ and model the erroneous correlations between the eigenvector $|a_{p'}^\sigma\rangle$ and the final state $|\Psi_p^{\sigma\mu}\rangle$. In the MATHEMATICS it is proved that the atomic modal interpretation also solves the measurement problem for this interaction (barring degeneracies): for every initial state W_0^σ the pointer possesses the reading R_j^π with probability $\sum_{p',q'} W_{p'q'}^\sigma \lambda_{p'j} \bar{\lambda}_{q'j}$. Furthermore, environmental influences may distort interaction (10.24) by freely changing the values $\{\lambda_{p'p}\}_{p',p}$ and by changing the vectors $\{|\Psi_p^{\sigma\mu}\rangle\}_p$ provided they still satisfy condition (10.23).

MATHEMATICS

Let a pointer π consist of the atoms α, β, $\ldots$ and let its readings $\{R_j^\pi\}_j$ be represented by projections obeying the strengthened conditions (10.15) and (10.19). So

$$R_j^\pi = \sum_{\langle a,b,\ldots\rangle \in I_j} P_{ab\cdots}^{\alpha\beta\cdots} + \widetilde{R}_j^\pi, \qquad j = 1, 2, \ldots, \qquad (10.25)$$

and

$$\forall \langle a, b, \ldots \rangle \in I_j, \langle a', b', \ldots \rangle \in I_k \text{ with } j \neq k :$$
$$P_x^X P_{x'}^X = 0 \text{ for at least } n \geq 2 \text{ atoms } X \in \{\alpha, \beta, \ldots\}. \qquad (10.26)$$

Here, $\{I_j\}_j$ are non-empty and mutually disjoint sets of ordered indices $\langle a, b, \ldots \rangle$ and $P_{ab\cdots}^{\alpha\beta\cdots}$ is equal to $P_a^\alpha \otimes P_b^\beta \otimes \cdots$, where $\{P_a^\alpha\}_a$, $\{P_b^\beta\}_b$, … are sets of pair-wise orthogonal projections.

Given the initial state (10.20) and the interaction (10.24), the final pointer state is

$$W^\pi = \mathrm{Tr}^{\sigma\bar\mu}(\sum_{p,q} Y_{pq} |\Psi_p^{\sigma\mu}\rangle\langle\Psi_q^{\sigma\mu}|) \tag{10.27}$$

with Y_{pq} equal to $\sum_{p',q'} W_{p'q'}^\sigma \lambda_{p'p} \bar\lambda_{q'q}$ and with $\{|\Psi_p^{\pi\mu}\rangle\}_p$ pair-wise orthogonal vectors that obey the condition (10.23). So,

$W_j^\pi = \mathrm{Tr}^{\sigma\bar\mu}(|\Psi_j^{\sigma\mu}\rangle\langle\Psi_j^{\sigma\mu}|)$ yields the core property ascription

$[P_{ab\cdots}^{\alpha\beta\cdots}] = 1$ with probability $\mathrm{Tr}^\pi(W_j^\pi P_{ab\cdots}^{\alpha\beta\cdots})$ if $\langle a, b, \ldots \rangle \in I_j$

and satisfies $\sum_{\langle a,b,\ldots\rangle \in I_j} \mathrm{Tr}^\pi(W_j^\pi P_{ab\cdots}^{\alpha\beta\cdots}) = 1$. $\tag{10.28}$

This condition has two consequences which I use later. The first is that the state $|\Psi_p^{\sigma\mu}\rangle\langle\Psi_p^{\sigma\mu}|$ must yield 'pointer-atom' states W^α, W^β, … with eigenprojections in the sets $\{P_a^\alpha\}_a$, $\{P_b^\beta\}_b$, …, respectively. If not, the core property ascription to the pointer cannot be in terms of the projections $P_{ab\cdots}^{\alpha\beta\cdots}$. For the second consequence I need

Lemma 10.1
Let $|\Psi^{\alpha\beta}\rangle$ be a normalised vector with a partial trace $W^\beta = \mathrm{Tr}^\alpha(|\Psi^{\alpha\beta}\rangle\langle\Psi^{\alpha\beta}|)$. Then $\mathrm{Tr}^\beta(W^\beta Q^\beta) = 1$ and $[Q^\beta]^2 = Q^\beta$ implies that $|\Psi^{\alpha\beta}\rangle = (\mathbb{I}^\alpha \otimes Q^\beta)|\Psi^{\alpha\beta}\rangle$.

Proof: One has $\langle\Psi^{\alpha\beta}|(\mathbb{I}^\alpha \otimes Q^\beta)|\Psi^{\alpha\beta}\rangle = \mathrm{Tr}^{\alpha\beta}(|\Psi^{\alpha\beta}\rangle\langle\Psi^{\alpha\beta}|(\mathbb{I}^\alpha \otimes Q^\beta)) = \mathrm{Tr}^\beta(W^\beta Q^\beta)$ and by assumption this is equal to 1. Consider the identity $|\Phi^{\alpha\beta}\rangle = (\mathbb{I}^\alpha \otimes Q^\beta)|\Phi^{\alpha\beta}\rangle + (\mathbb{I}^{\alpha\beta} - \mathbb{I}^\alpha \otimes Q^\beta)|\Phi^{\alpha\beta}\rangle$. The squared norm of the last vector is

$$\langle\Phi^{\alpha\beta}|(\mathbb{I}^{\alpha\beta} - \mathbb{I}^\alpha \otimes Q^\beta)^2|\Phi^{\alpha\beta}\rangle = \langle\Phi^{\alpha\beta}|\Phi^{\alpha\beta}\rangle - \langle\Phi^{\alpha\beta}|(\mathbb{I}^\alpha \otimes Q^\beta)|\Phi^{\alpha\beta}\rangle = 0. \tag{10.29}$$

This last vector is thus the null vector, yielding $|\Phi^{\alpha\beta}\rangle = (\mathbb{I}^\alpha \otimes Q^\beta)|\Phi^{\alpha\beta}\rangle$. $\square$

By this lemma and condition (10.28) it follows that the vector $|\Psi_j^{\sigma\mu}\rangle$ is equal to $(\mathbb{I}^{\sigma\bar\mu} \otimes \sum_{\langle a,b,\ldots\rangle \in I_j} P_{ab\cdots}^{\alpha\beta\cdots})|\Psi_j^{\sigma\mu}\rangle$. And because $\sum_{\langle a,b,\ldots\rangle \in I_j} P_{ab\cdots}^{\alpha\beta\cdots}$ and $\sum_{\langle a',b',\ldots\rangle \in I_k} P_{a'b'\cdots}^{\alpha\beta\cdots}$ are orthogonal if $j \neq k$, one can conclude that

$$\delta_{jp} |\Psi_p^{\sigma\mu}\rangle = (\mathbb{I}^{\sigma\bar\mu} \otimes \sum_{\langle a,b,\ldots\rangle \in I_j} P_{ab\cdots}^{\alpha\beta\cdots})|\Psi_p^{\sigma\mu}\rangle, \qquad \forall j, p. \tag{10.30}$$

Theorem 10.1
*If the pointer state is equal to (10.27), the atomic modal interpretation yields
(barring degeneracies) that $[R_j^\pi] = 1$ with probability Y_{jj} for all j.*

Proof: In order to apply the atomic modal interpretation to state (10.27)
one firstly has to determine the states of the atoms in the pointer. The state
of α is

$$W^\alpha = \text{Tr}^{\sigma\mu/\alpha}(\sum_{p,q} Y_{pq} |\Psi_p^{\sigma\mu}\rangle\langle\Psi_q^{\sigma\mu}|). \tag{10.31}$$

By using relation (10.30) and cyclic permutation of the operator $\mathbb{I}^{\sigma\bar\mu\alpha} \otimes P_b^\beta \otimes P_c^\gamma \otimes \cdots,$[122] one can deduce that

$$\text{Tr}^{\sigma\mu/\alpha}(|\Psi_p^{\sigma\mu}\rangle\langle\Psi_q^{\sigma\mu}|) = \sum_{\langle a,b,\dots\rangle\in I_p} \sum_{\langle a',b',\dots\rangle\in I_q}$$

$$\text{Tr}^{\sigma\mu/\alpha}((\mathbb{I}^{\sigma\mu/\alpha} \otimes P_a^\alpha) |\Psi_p^{\sigma\mu}\rangle\langle\Psi_q^{\sigma\mu}| (\mathbb{I}^{\sigma\bar\mu} \otimes P_{a'}^\alpha \otimes P_{b'}^\beta P_b^\beta \otimes P_{c'}^\gamma P_c^\gamma \otimes \cdots)). \tag{10.32}$$

Since there exist at least two atoms $X \in \{\alpha, \beta, \dots\}$ with $P_x^X P_{x'}^X = 0$ (condition
(10.28)), there exists at least one atom $X \in \{\beta, \gamma, \dots\}$ with $P_x^X P_{x'}^X = 0$.
Therefore (10.32) is equal to zero if $p \neq q$ and W^α becomes

$$W^\alpha = \sum_p Y_{pp} \text{Tr}^{\sigma\mu/\alpha}(|\Psi_p^{\sigma\mu}\rangle\langle\Psi_p^{\sigma\mu}|). \tag{10.33}$$

The state $|\Psi_p^{\sigma\mu}\rangle\langle\Psi_p^{\sigma\mu}|$ is subject to condition (10.28). The eigenprojections
of every term $\text{Tr}^{\sigma\mu/\alpha}(|\Psi_p^{\sigma\mu}\rangle\langle\Psi_p^{\sigma\mu}|)$ are therefore in the set $\{P_a^\alpha\}_a$ and it fol-
lows that the eigenprojections of W^α itself are in the set $\{P_a^\alpha\}_a$ (barring
degeneracies).
    Repeating the same argument for the other atoms $\beta, \gamma, \dots$ yields, barring
degeneracies, that the eigenprojections of W^β are in the set $\{P_b^\beta\}_b$, that the
eigenprojections of W^γ are in the set $\{P_c^\gamma\}_c$, etc. The core property ascription
to π is thus

$$[P_{ab\cdots}^{\alpha\beta\cdots}] = 1 \text{ with probability } \text{Tr}^\pi(W^\pi P_{ab\cdots}^{\alpha\beta\cdots}). \tag{10.34}$$

The probability $\text{Tr}^\pi(W^\pi P_{ab\cdots}^{\alpha\beta\cdots})$ is equal to

$$\text{Tr}^{\sigma\mu}(\sum_{p,q} Y_{pq} |\Psi_p^{\sigma\mu}\rangle\langle\Psi_q^{\sigma\mu}| (\mathbb{I}^{\sigma\bar\mu} \otimes P_{ab\cdots}^{\alpha\beta\cdots})). \tag{10.35}$$

Using that $\mathbb{I}^{\sigma\bar\mu} \otimes P_{ab\cdots}^{\alpha\beta\cdots} = (\mathbb{I}^{\sigma\bar\mu} \otimes P_{ab\cdots}^{\alpha\beta\cdots})^2$ and cyclic permutation, this probability

---

[122] For a partial trace one has that $\text{Tr}^\tau([\mathbb{I}^\alpha \otimes P^\tau] Q^{\alpha\tau})$ and $\text{Tr}^\tau(Q^{\alpha\tau} [\mathbb{I}^\alpha \otimes P^\tau])$ are equal.

can be rewritten as

$$\mathrm{Tr}^{\pi}(W^{\pi}\,P_{ab\cdots}^{\alpha\beta\cdots}) = \mathrm{Tr}^{\sigma\mu}\Big(\sum_{p,q} Y_{pq}\,(\mathbb{I}^{\sigma\bar{\mu}}\otimes P_{ab\cdots}^{\alpha\beta\cdots})\,|\Psi_p^{\sigma\mu}\rangle\langle\Psi_q^{\sigma\mu}|\,(\mathbb{I}^{\sigma\bar{\mu}}\otimes P_{ab\cdots}^{\alpha\beta\cdots})\Big).$$

$$(10.36)$$

Let's assume that $\langle a,b,\ldots\rangle \in I_j$. By (10.30) it then follows that $(\mathbb{I}^{\sigma\bar{\mu}}\otimes P_{ab\cdots}^{\alpha\beta\cdots})\,|\Psi_p^{\sigma\mu}\rangle = 0$ if $p\neq j$, and that $\langle\Psi_q^{\sigma\mu}|\,(\mathbb{I}^{\sigma\bar{\mu}}\otimes P_{ab\cdots}^{\alpha\beta\cdots}) = 0$ if $q\neq j$. So, again using cyclic permutation,

$$\mathrm{Tr}^{\pi}(W^{\pi}\,P_{ab\cdots}^{\alpha\beta\cdots}) = Y_{jj}\,\mathrm{Tr}^{\sigma\mu}(|\Psi_j^{\sigma\mu}\rangle\langle\Psi_j^{\sigma\mu}|\,(\mathbb{I}^{\sigma\bar{\mu}}\otimes P_{ab\cdots}^{\alpha\beta\cdots})). \qquad (10.37)$$

The state $|\Psi_j^{\sigma\mu}\rangle\langle\Psi_j^{\sigma\mu}|$ is still subject to condition (10.28), so one can derive that the core property ascription to π satisfies

$$\sum_{\langle a,b,\ldots\rangle\in I_j}\mathrm{Tr}^{\pi}(W^{\pi}P_{ab\cdots}^{\alpha\beta\cdots}) = Y_{jj}\sum_{\langle a,b,\ldots\rangle\in I_j}\mathrm{Tr}^{\pi}(W_j^{\sigma}P_{ab\cdots}^{\alpha\beta\cdots}) = Y_{jj}. \qquad (10.38)$$

The reading R_j^{π} is assigned the value 1 if and only if a string $P_{ab\cdots}^{\alpha\beta\cdots}$ with $\langle a,b,\ldots\rangle \in I_j$, is assigned value 1. So, with this last result it follows that $[R_j^{\pi}] = 1$ with probability Y_{jj}. □

This theorem yields that the atomic modal interpretation ascribes after measurement (10.24) reading R_j^{π} with probability $Y_{jj} = \sum_{p',q'} W_{p'q'}^{\sigma}\lambda_{p'j}\bar{\lambda}_{q'j}$ to the pointer. For perfect measurements this probability simplifies to W_{jj}^{σ}.

10.4 Exact solutions for the bi and spectral modal interpretations

The results by Bacciagaluppi and Hemmo (1994, 1996), presented in Section 10.2, proved that the bi and spectral modal interpretations are able to (almost) ascribe readings to pointers after measurements. However, in their approach it is the interactions between the pointers and the environment which induce the pointers to possess their readings approximately. And, as argued in the beginning of the previous section, it is preferable if it is the internal structure of the measurement devices that makes pointers to assume their readings. In this section I show that, analogously to the case of the atomic modal interpretation, one can formulate conditions pertaining to this internal structure which are sufficient for exactly solving the measurement problem for the bi and spectral modal interpretations (Vermaas 1998c, Sect. 4.3).

The derivation of necessary conditions for ascribing readings to a pointer is trivial within the bi and spectral modal interpretations. Any choice of the reading projections $\{R_j^{\pi}\}_j$ will do, for there exists for every possible set $\{R_j^{\pi}\}_j$

a state such that π assumes with probability 1 exactly one of its readings (take $W^\pi = \sum_j w_j R_j^\pi$).

Sufficient conditions for solving the measurement problem can be derived as follows: Consider again the measurement sketched in the previous section.[123] Thus, let the initial state $W_0^{\sigma\mu}$ again be given by (10.20). In the bi and spectral modal interpretations π possesses with probability 1 reading R_0^π before the measurement if and only if[124]

$$W_0^\pi = \mathrm{Tr}^{\bar\mu}(|\psi_0^\mu\rangle\langle\psi_0^\mu|) \text{ satisfies } R_0^\pi W_0^\pi = W_0^\pi. \tag{10.39}$$

Consider firstly the special case of a perfect measurement. Let the measurement interaction be given by (10.22). Perfectness is then obtained if and only if[124]

$$W_j^\pi = \mathrm{Tr}^{\sigma\bar\mu}(|\Psi_j^{\sigma\mu}\rangle\langle\Psi_j^{\sigma\mu}|) \text{ satisfies } R_j^\pi W_j^\pi = W_j^\pi. \tag{10.40}$$

Now take the property ascription to the mechanism $\bar\mu$ also into account. Let $\{D_j^{\bar\mu}\}_j$ (projections on $\mathscr{H}^{\bar\mu}$) represent properties of the mechanism for which it holds that if the final state of $\sigma\mu$ is $|\Psi_j^{\sigma\mu}\rangle\langle\Psi_j^{\sigma\mu}|$ then the mechanism possesses $D_j^{\bar\mu}$. This implies that

$$W_j^{\bar\mu} = \mathrm{Tr}^{\sigma\pi}(|\Psi_j^{\sigma\mu}\rangle\langle\Psi_j^{\sigma\mu}|) \text{ satisfies } D_j^{\bar\mu} W_j^{\bar\mu} = W_j^{\bar\mu}. \tag{10.41}$$

Thus, after a measurement with an initial state $W_0^\sigma = |a_j^\alpha\rangle\langle a_j^\alpha|$, the pointer possesses R_j^π and $\bar\mu$ possesses $D_j^{\bar\mu}$, both with probability 1. It is natural to assume that the projections $\{D_j^{\bar\mu}\}_j$ are pair-wise orthogonal, so

$$D_j^{\bar\mu} D_k^{\bar\mu} = \delta_{jk} D_j^{\bar\mu}, \qquad\qquad j,k = 1,2,3,\dots \tag{10.42}$$

and my motivation for this is as follows. Pointers of measurement devices acquire their readings because they are in a certain way, mechanically or electronically, driven by the mechanism of the device. If one then assumes that the properties of the mechanism which drive the pointer to reading R_j^π, are macroscopically distinguishable from the properties of the mechanism which drive the pointer to R_k^π, $j \neq k$, those mechanism properties are represented by pair-wise orthogonal projections as expressed by condition (10.42).

It can be proved that, barring degeneracies, the bi and spectral modal

---

[123] All the results generalise to degenerate measured magnitudes (Vermaas 1998b, Sect. 9.4).

[124] Proof: 'If': If $R_j^\pi W^\pi = W^\pi$, then for every eigenprojection P_a^π of W^π it holds that $R_j^\pi P_a^\pi = P_a^\pi$. The core property ascription yields with probability 1 that one of the eigenprojections $\{P_a^\pi\}_a$ has the value 1. Hence, the full property ascription yields with probability 1 that $[R_j^\pi] = 1$. 'Only if': If $[R_j^\pi] = 1$ with probability 1, then all core properties $\{P_a^\pi\}_a$ ascribed to π mean that $[R_j^\pi] = 1$. So, for every eigenprojection P_a^π of W^π it follows that $R_j^\pi P_a^\pi = P_a^\pi$ and thus that $R_j^\pi W^\pi = W^\pi$. □

interpretations solve the measurement problem for the measurement inter-action (10.22): given the conditions (10.39)–(10.42), these two interpretations yield for any initial state W_0^σ that π possesses the reading R_j^π with probability $\mathrm{Tr}^\sigma(W_0^\sigma \, |a_j^\sigma\rangle\langle a_j^\sigma|) = W_{jj}^\sigma$ (see the MATHEMATICS).

These results also hold for the erroneous measurement interaction (10.24). As long as the vectors $\{|\Psi_p^{\sigma\mu}\rangle\}_p$ meet conditions (10.40)–(10.42), the bi and spectral modal interpretations ascribe the reading R_j^π with probability $\sum_{p',q'} W_{p'q'}^\sigma \lambda_{p'j} \bar\lambda_{q'j}$. Environmental influences may again distort the interac-tion (10.24) by freely changing the values $\{\lambda_{p',p}\}_{p',p}$ and by changing the vectors $\{|\Psi_p^{\sigma\mu}\rangle\}_p$ provided they still satisfy the mentioned conditions.

MATHEMATICS

Given the initial state (10.20) and the interaction (10.24), the final pointer state is

$$W^\pi = \mathrm{Tr}^{\sigma\bar\mu}(\sum_{p,q} Y_{pq} \, |\Psi_p^{\sigma\mu}\rangle\langle\Psi_q^{\sigma\mu}|) \tag{10.43}$$

with Y_{pq} equal to $\sum_{p',q'} W_{p'q'}^\sigma \lambda_{p'p} \bar\lambda_{q'q}$ and with $\{|\Psi_p^{\sigma\mu}\rangle\}_p$ pair-wise orthogonal vectors that obey the conditions (10.40), (10.41) and (10.42). So,

$$\left.\begin{aligned} W_j^\pi &= \mathrm{Tr}^{\sigma\bar\mu}(|\Psi_j^{\sigma\mu}\rangle\langle\Psi_j^{\sigma\mu}|) \text{ satisfies } R_j^\pi \, W_j^\pi = W_j^\pi, \\ W_j^{\bar\mu} &= \mathrm{Tr}^{\sigma\pi}(|\Psi_j^{\sigma\mu}\rangle\langle\Psi_j^{\sigma\mu}|) \text{ satisfies } D_j^{\bar\mu} \, W_j^{\bar\mu} = W_j^{\bar\mu}, \end{aligned}\right\} \tag{10.44}$$

with $\{R_j^\pi\}_j$ and $\{D_j^{\bar\mu}\}_j$ sets of pair-wise orthogonal projections.

From the second condition one can derive that

$$\delta_{jp} \, |\Psi_p^{\sigma\mu}\rangle = (\mathbb{I}^{\sigma\pi} \otimes D_j^{\bar\mu}) \, |\Psi_p^{\sigma\mu}\rangle, \qquad\qquad \forall j, p. \tag{10.45}$$

Proof: The second condition of (10.44) yields that $W_j^{\bar\mu} = \mathrm{Tr}^{\sigma\pi}(|\Psi_j^{\sigma\mu}\rangle\langle\Psi_j^{\sigma\mu}|)$ satisfies $D_j^{\bar\mu} \, W_j^{\bar\mu} = W_j^{\bar\mu}$. The trace $\mathrm{Tr}^{\bar\mu}$ of this last equality yields $\mathrm{Tr}^{\bar\mu}(D_j^{\bar\mu} \, W_j^{\bar\mu}) = 1$. Lemma 10.1 on page 187 then gives (10.45). $\qquad\square$

Theorem 10.2
If the pointer state is equal to (10.43), the bi and spectral modal interpretations yield (barring degeneracies) that $[R_j^\pi] = 1$ with probability Y_{jj} for all j.

Proof: By relation (10.45) the final pointer state (10.43) is equal to

$$W^\pi = \mathrm{Tr}^{\sigma\bar\mu}(\sum_{p,q} Y_{pq} \, (\mathbb{I}^{\sigma\pi} \otimes D_p^{\bar\mu}) \, |\Psi_p^{\sigma\mu}\rangle\langle\Psi_q^{\sigma\mu}|). \tag{10.46}$$

Cyclic permutation of $\mathbb{I}^\sigma \otimes D_p^{\bar\mu}$ and using relation (10.45) again yields that W^π can be written as a sum $\sum_j Y_{jj} \mathrm{Tr}^{\sigma\bar\mu}(|\Psi_j^{\sigma\mu}\rangle\langle\Psi_j^{\sigma\mu}|)$. Let the spectral resolutions of the terms in this sum be $\mathrm{Tr}^{\sigma\bar{\mu}}(|\Psi_j^{\sigma\mu}\rangle\langle\Psi_j^{\sigma\mu}|) = \sum_k w_{j,k} P_{j,k}^\pi$. The state $|\Psi_j^{\sigma\mu}\rangle\langle\Psi_j^{\sigma\mu}|$ satisfies condition (10.44). Consequently, the eigenprojections $\{P_{j,k}^\pi\}_{j,k}$ obey

$$R_j^\pi P_{j,k}^\pi = P_{j,k}^\pi, \qquad j = 1,2,\dots\ ; k = 1,2,\dots . \tag{10.47}$$

From this, one can conclude that the projections $\{P_{j,k}^\pi\}_{j,k}$ are pair-wise orthogonal (by using $R_j^\pi R_{j'}^\pi = 0$ one can derive that $P_{j,k}^\pi P_{j',k'}^\pi = \delta_{jj'}\delta_{kk'} P_{j,k}^\pi$ for all j,j',k,k'). The expansion $W^\pi = \sum_{j,k} Y_{jj} w_{j,k} P_{j,k}^\pi$ is therefore, barring degeneracies, a spectral resolution and it follows that, barring degeneracies, $\{P_{j,k}^\pi\}_{j,k}$ are the eigenprojections of W^π. The core property ascription to π in the bi and spectral modal core property ascriptions is thus, barring degeneracies, $[P_{j,k}^\pi] = 1$ with probability $\mathrm{Tr}^\pi(W^\pi P_{j,k}^\pi)$. Using relation (10.47) it follows that a reading R_j^π is assigned the value 1 if and only if $P_{j,k}^\pi$ (with k free) is assigned the value 1. So, $[R_j^\pi] = 1$ with probability $\sum_k \mathrm{Tr}^\pi(W^\pi P_{j,k}^\pi)$. Using the expansion $W^\pi = \sum_{j,k} Y_{jj} w_{j,k} P_{j,k}^\pi$ one can calculate that this probability is equal to Y_{jj}. $\square$

This theorem yields that after the measurement interaction (10.24) the bi and spectral modal interpretations ascribe the reading R_j^π with probability $\sum_{p',q'} W_{p'q'}^\sigma \lambda_{p'j} \bar{\lambda}_{q'j}$. For the perfect measurement interaction (10.22) this probability simplifies to W_{jj}^σ.

10.5 Degeneracies and a continuous solution

The exact solutions to the measurement problem given in the last two sections, still have two blanks. Firstly, pointers assume their readings only before and after the measurement interaction, and not during the interaction. Secondly, the ascription of readings is always 'barring degeneracies.' I start by making some tentative remarks about the first blank.

It is natural to assume that the pointer of a measurement device always possesses readings, that is, not only at the beginning and at the end of, but also during the measurement interaction. One may expect, for instance, that initially the actual pointer reading will be the ready reading R_0^π and that during the measurement this actual reading will jump from R_0^π to R_1^π to R_2^π, etc. (a 'rotation') until it finally reaches the outcome R_k^π.

It is now possible to give measurement interactions such that the pointer indeed always possesses, barring degeneracies, its readings. (I consider for simplicity only the perfect measurements (10.22) but the erroneous ones (10.24) can be treated similarly.) Suppose that the measurement interaction (10.22) starts at t_0 and ends at t_1 and suppose that the composite $\sigma\mu$ evolves freely. Then, the state $W^{\sigma\mu}(t)$ evolves with a unitary operator $U^{\sigma\mu}(t,t_0)$ for which it holds that $U^{\sigma\mu}(t_1,t_0)$ maps the initial state $|a_p^\sigma\rangle \otimes |\psi_0^\mu\rangle$ to the final state $|\Psi_p^{\sigma\mu}\rangle$. Let this unitary operator yield

$$U^{\sigma\mu}(t,t_0)|a_p^\sigma\rangle \otimes |\psi_0^\mu\rangle = \sqrt{1-g_p(t)}\,|\tilde{a}_p^\sigma(t)\rangle \otimes |\psi_0^\mu\rangle + \sqrt{g_p(t)}\,|\Psi_p^{\sigma\mu}\rangle \qquad (10.48)$$

for all p, where $\{|\tilde{a}_p^\sigma(t)\rangle\}_p$ are sufficiently smooth vector-valued functions with $|\tilde{a}_p^\sigma(t_0)\rangle = |a_p^\sigma\rangle$, and where $\{g_p(t)\}_p$ are functions which obey $g_p(t_0) = 0$, $g_p(t_1) = 1$ and $0 \le g_p(t) \le 1$. For instance, if one takes $g_p(t) = \sin((\pi/2)\,[(t-t_0)/(t_1-t_0)])$ and $|\tilde{a}_p^\sigma(t)\rangle = |a_p^\sigma\rangle$, the evolution consists of simple rotations in Hilbert space.

With this interaction one can derive the evolution of state of $\sigma\mu$ and determine for the various modal interpretations which properties are possessed in time by the pointer. By again imposing the sufficient conditions, one can prove that the pointer always possesses, barring degeneracies, its readings. To be precise, if $W^\sigma(t_0)$ is the initial state of the object system σ, then, barring degeneracies, the pointer possesses at all times $t \in [t_0,t_1]$ reading R_j^π with probability $g_j(t)W_{jj}^\sigma(t_0)$ for all $j = 1,2,\ldots$ and reading R_0^π with probability $1 - \sum_p g_p(t)W_{pp}^\sigma(t_0)$. Thus, given interaction (10.48), the pointer nearly always possesses readings: initially it assumes the ready reading R_0^π and finally an outcome reading R_j^π with probability $W_{jj}^\sigma(t_0)$. (The specific dynamics of the actual readings during the interaction is given by the transition probabilities discussed in Chapter 8.)

Consider, secondly, the problem of degeneracies.[125] In the case of the atomic modal interpretation, degeneracies can frustrate the ascription of readings as follows. At the end of the measurement the state of an atom in the pointer, say α, can according to (10.33) be written as the sum $W^\alpha = \sum_j Y_{jj}\,\mathrm{Tr}^{\sigma\mu/\alpha}(|\Psi_j^{\sigma\mu}\rangle\langle\Psi_j^{\sigma\mu}|)$, where $Y_{jj} = \langle a_j^\sigma|W_0^\sigma|a_j^\sigma\rangle$. The eigenprojections of all the terms $\{\mathrm{Tr}^{\sigma\mu/\alpha}(|\Psi_j^{\sigma\mu}\rangle\langle\Psi_j^{\sigma\mu}|)\}_j$ are in the set of projections $\{P_a^\alpha\}_a$. So, the spectral resolution of each term is given by $\mathrm{Tr}^{\sigma\mu/\alpha}(|\Psi_j^{\sigma\mu}\rangle\langle\Psi_j^{\sigma\mu}|) = \sum_k w_{j,k}\,P_k^\alpha$ and one obtains

$$W^\alpha = \sum_{j,k} Y_{jj}w_{j,k}\,P_k^\alpha. \qquad (10.49)$$

In general the values $\{\sum_j Y_{jj}w_{j,k}\}_k$ are all different, yielding that the eigenprojections of W^α are also in the set $\{P_a^\alpha\}_a$. The core properties of the atom are then also in the set $\{P_a^\alpha\}_a$. Hence, the core properties of the pointer as a whole can then be given by $\{P_{ab\cdots}^{\alpha\beta\cdots}\}_{a,b,\ldots}$ such that the atomic modal interpretation can ascribe the readings $\{R_j^\pi\}_j$. If, however, two values $\sum_j Y_{jj}w_{j,k}$ and

---

$\sum_j Y_{jj}w_{j,l}$ (with $k \neq l$) are equal, then P_k^α and P_l^α are not eigenprojections of the state of the atom α at the end of the measurement ($P_k^\alpha + P_l^\alpha$ is the eigenprojection). The core properties of the pointer are then not given by the projections $\{P_{ab\cdots}^{\alpha\beta\cdots}\}_{a,b,\ldots}$ and the atomic modal interpretation does not ascribe the pointer readings.

In the case of the bi and spectral modal interpretations there is an analogous problem. According to the MATHEMATICS of the previous section, W^π can be written as

$$W^\pi = \sum_{j,k} Y_{jj}w_{j,k} P_{j,k}^\pi \tag{10.50}$$

with $P_{j,k}^\pi P_{j',k'}^\pi = \delta_{jj'}\delta_{kk'} P_{j,k}^\pi$ and $R_j^\pi P_{j,k}^\pi = P_{j,k}^\pi$. If the set of values $\{Y_{jj}w_{j,k}\}_{j,k}$ is non-degenerate, the projections $\{P_{j,k}^\pi\}_{j,k}$ are the eigenprojections of W^π and the bi and spectral modal interpretations ascribe readings to the pointer. If, however, this set of values is degenerate at the end of the measurement, say $Y_{jj}w_{j,k} = Y_{ll}w_{l,m}$, then $P_{j,k}^\pi + P_{l,m}^\pi$ is an eigenprojection of the state of π. In this case the bi and spectral modal interpretations ascribe the property $R_j^\pi + R_l^\pi$ to the pointer and not the individual readings.

To my knowledge this problem cannot be resolved. Two attempts to improve on the interpretation of degenerate states (the tri modal interpretation and the extended modal interpretation discussed in the beginning of Chapter 7) both failed. Hence, there is presently no way to modify the atomic modal interpretation such that it ascribes the readings if, after a measurement (10.22) or (10.24), the state of one of the atoms in the pointer is degenerate in the way described above. There is presently also no way to modify the bi and specral modal interpretations such that they ascribe the readings if the state of the pointer itself is degenerate in the way described above.

The question is now whether this problem of degeneracies proves that the bi, spectral and atomic modal interpretations are empirically inadequate. From a principal point of view it indeed does prove this inadequacy because it is in principle possible that the mentioned degeneracies occur after measurement (10.22) or (10.24). However, from a practical point of view one may try to maintain that it doesn't. As I noted in Section 7.4, it holds for states defined on finite-dimensional Hilbert spaces that if the dimensionality of the set of all states is N, then the dimensionality of the set of degenerate states is $N - 3$. One may thus argue that with regard to any sensible measure on the set of all states, states are not degenerate. Hence, after measurement (10.22) or (10.24), the pointer still in practice possesses with probability 1 a reading.

11

The Born rule

Measurements not only have outcomes, their outcomes also occur with certain frequencies and, in the case of a number of measurements, with certain correlations. The standard formulation of quantum mechanics generates by means of the Born rule predictions about these frequencies and correlations and these predictions are empirically correct. In the previous chapter I have discussed whether modal interpretations confirm that measurements have outcomes. In this chapter I consider the question of whether modal interpretations reproduce the empirically adequate frequencies and correlations.

11.1 Probabilities for single outcomes

Consider a single measurement which establishes an outcome at instant t. Let this outcome be a property of a system π which may be a pointer, a counter, a piece of magnetic tape, a dot of ink, etc. and let the different possible outcomes be represented by the pair-wise orthogonal projections $\{R_j^\pi\}_j$. The standard formulation then predicts by means of the Born rule that we observe with probability

$$p_{\text{Born}}(R_j^\pi \text{ at } t) = \text{Tr}^\pi(W^\pi(t) R_j^\pi) \tag{11.1}$$

that π possesses at t the outcome represented by R_j^π.

Assume now that the bi, spectral and atomic modal interpretations solve the measurement problem by ascribing at t exactly one of the possible outcomes to π. It can then be proved that these modal interpretations ascribe the outcomes with probabilities which are equal to the Born probabilities (11.1). This can be proved from three features which the bi, spectral and atomic modal interpretations share, namely: (A) the core projections $\{C_a^\pi(t)\}_a$ ascribed to π are pair-wise orthogonal, (B) the core projection $C_a^\pi(t)$ is ascribed to π with probability $\text{Tr}^\pi(W^\pi(t) C_a^\pi(t))$, and (C) all the core projections

195

$\{C_a^\pi(t)\}_a$ which are ascribed with a non-zero probability mean that exactly one of the outcomes is actually possessed by π. Feature (C) follows from the assumption that the modal interpretations solve the measurement problem because if there exists a core projection $C_b^\pi(t)$ which means that either no or more than one outcome is possessed by π, then there exists a non-zero probability that, respectively, no outcome is possessed or two or more outcomes are simultaneously possessed. Both cases contradict a solution to the measurement problem so (C) holds.

A consequence of (C) is that for every core projection $C_a^\pi(t)$ there exists exactly one outcome projection R_j^π such that $R_j^\pi C_a^\pi(t) = C_a^\pi(t)$. Using assumption (A) one can then expand the outcome projections as

$$R_j^\pi = \sum_{a \in I_j} C_a^\pi(t) + \tilde{R}_j^\pi. \tag{11.2}$$

The index-sets $\{I_j\}_j$ are defined by $a \in I_j$ iff $R_j^\pi C_a^\pi(t) = C_a^\pi(t)$. The projections $\{\tilde{R}_j^\pi\}_j$ are defined by $\tilde{R}_j^\pi = R_j^\pi - \sum_{a \in I_j} C_a^\pi(t)$ and satisfy the relation $\tilde{R}_j^\pi C_b^\pi(t) = 0$ for all j and b. From (A) and (B) one can furthermore prove that $W^\pi(t) \sum_a C_a^\pi(t) = W^\pi(t)$, where the sum $\sum_a$ runs over all the core projections $\{C_a^\pi(t)\}_a$ which are ascribed to π with a non-zero probability (see the MATHEMATICS for the proofs of these relations).

Let's now calculate the probability with which modal interpretations ascribe the outcomes to π at t. It is clear that π possesses R_j^π only if π possesses a core projection $C_a^\pi(t)$ that makes π possess R_j^π. So,

$$p(R_j^\pi \text{ at } t) = \sum_{a \in I_j} p(C_a^\pi(t)) = \sum_{a \in I_j} \text{Tr}^\pi(W^\pi(t) C_a^\pi(t)). \tag{11.3}$$

The Born probability (11.1) can in its turn be rewritten as

$$p_{\text{Born}}(R_j^\pi \text{ at } t) = \sum_a \text{Tr}^\pi(W^\pi(t) C_a^\pi(t) R_j^\pi) = \sum_{a \in I_j} \text{Tr}^\pi(W^\pi(t) C_a^\pi(t)). \tag{11.4}$$

(In the first step I used that $W^\pi(t) = W^\pi(t) \sum_a C_a^\pi(t)$, in the second I substituted (11.2) and used that $\tilde{R}_j^\pi C_a^\pi(t) = 0$ for all j and a.)

Hence, if the bi, spectral or atomic modal interpretations solve the measurement problem, they ascribe outcomes to π with the same probabilities as predicted by the Born rule. This result is obtained without making any assumption about how the measurement outcomes come about; the result thus hold for every type of measurement.

MATHEMATICS

Firstly, I prove the relation $\tilde{R}_j^\pi C_b^\pi(t) = 0$ for all j and b.

Proof: If $b \in I_j$, then $R_j^\pi C_b^\pi(t) = C_b^\pi(t)$ and $\sum_{a \in I_j} C_a^\pi(t) C_b^\pi(t) = C_b^\pi(t)$. From the definition of $\tilde{R}_j^\pi$ it then follows that $\tilde{R}_j^\pi C_b^\pi(t) = 0$. If, on the other hand, $b \notin I_j$, then $b \in I_k$ with $k \neq j$ because for every core projection there exists exactly one outcome such that $R_k^\pi C_b^\pi(t) = C_b^\pi(t)$. If one multiplies this last relation from the left-hand side with R_j^π, one obtains $R_j^\pi C_b^\pi(t) = 0$ because the projections R_j^π and R_k^π are orthogonal. Furthermore, one has $\sum_{a \in I_j} C_a^\pi(t) C_b^\pi(t) = 0$ because $\sum_{a \in I_j} C_a^\pi(t)$ and $C_b^\pi(t)$ are orthogonal if $b \neq I_j$. Hence, it again follows that $\tilde{R}_j^\pi C_b^\pi(t) = 0$. $\square$

Secondly, I prove from assumptions (A) and (B) the relation $W^\pi(t) \sum_a C_a^\pi(t) = W^\pi(t)$, where the sum runs over all the core projections ascribed to π with the non-zero probability $p(C_a^\pi(t)) = \mathrm{Tr}^\pi(W^\pi(t) C_a^\pi(t))$. The sum of all the non-zero probabilities $p(C_a^\pi(t))$ should be equal to 1, so $\sum_a \mathrm{Tr}^\pi(W^\pi(t) C_a^\pi(t)) = \mathrm{Tr}^\pi(W^\pi(t) \sum_a C_a^\pi(t)) = 1$. From the following lemma it then follows that $W^\pi(t) \sum_a C_a^\pi(t) = W^\pi(t)$.

Lemma 11.1
If W is a density operator and if Q is a projection, then

$$\mathrm{Tr}(W Q) = 1 \quad \text{if and only if} \quad W Q = W. \tag{11.5}$$

Proof: The 'only if' part is trivial, so take the 'if' part. Expand Q as $\sum_k |q_{1k}\rangle\langle q_{1k}|$ and expand $\mathbb{I} - Q$ as $\sum_k |q_{2k}\rangle\langle q_{2k}|$. Then $\{|q_{jk}\rangle\}_{j,k}$ is an orthonormal basis for the Hilbert space on which W and Q are defined and one can expand W as

$$W = \sum_{j,j'=1}^{2} \sum_{k,k'} |q_{jk}\rangle\langle q_{jk}|W|q_{j'k'}\rangle\langle q_{j'k'}|. \tag{11.6}$$

From the left-hand side of (11.5), one can deduce that $\mathrm{Tr}(W [\mathbb{I} - Q]) = 0$ and thus that $\langle q_{2k}|W|q_{2k}\rangle = 0$ for all k. Lemma 6.1 on page 91 then yields that $\langle q_{2k}|W|q_{jk'}\rangle = 0$ and $\langle q_{jk'}|W|q_{2k}\rangle = 0$ for all j, k, k'. The expansion of W thus simplifies to

$$W = \sum_{k,k'} |q_{1k}\rangle\langle q_{1k}|W|q_{1k'}\rangle\langle q_{1k'}| \tag{11.7}$$

and it is easy to check that $W Q = W$. $\square$

11.2 Correlations between multiple outcomes

In the standard formulation one can also correlate the outcomes of a number of measurements. Consider, firstly, a series of measurements which simultaneously yield outcomes. Let the pointers or the counters, etc., be given by π_1, π_2, π_3, ..., and assume that all these systems are pair-wise disjoint.[126] Let the different possible outcomes be represented by the pair-wise orthogonal projections $\{R_j^{\pi_1}\}_j$, $\{R_k^{\pi_2}\}_k$, ..., respectively. The Born rule predicts that the probability that π_1, π_2, ... simultaneously possess their outcomes at instant t is given by

$$p_{\mathrm{Born}}(R_j^{\pi_1}, R_k^{\pi_2}, \cdots \text{ at } t) = \mathrm{Tr}^{\pi_1 \pi_2 \cdots}(W^{\pi_1 \pi_2 \cdots}(t)\, [R_j^{\pi_1} \otimes R_k^{\pi_2} \otimes \cdots]) \qquad (11.8)$$

with $W^{\pi_1 \pi_2 \cdots}(t)$ the state of the composite $\pi_1 \pi_2 \cdots$ at t.

The bi modal interpretation cannot reproduce these joint probabilities because it does not correlate the properties of three or more systems. So, consider the spectral and atomic modal interpretations.

Assume that these two interpretations solve the measurement problem for π_1, π_2, π_3, ... at t. Following the line of reasoning given in the previous section, one can decompose the outcome projections as

$$R_j^{\pi_1} = \sum_{a \in I_j^1} C_a^{\pi_1}(t) + \widetilde{R}_j^{\pi_1}, \quad R_k^{\pi_2} = \sum_{b \in I_k^2} C_b^{\pi_2}(t) + \widetilde{R}_k^{\pi_2}, \qquad \text{etc.} \qquad (11.9)$$

and $\widetilde{R}_j^{\pi_1}\, C_a^{\pi_1}(t) = 0$ for all j and a, and $\widetilde{R}_k^{\pi_2}\, C_b^{\pi_2}(t) = 0$ for all k and b, etc. Furthermore, it holds that[127]

$$\left. \begin{aligned} W^{\pi_1 \pi_2 \cdots}(t) \sum_a [C_a^{\pi_1}(t) \otimes \mathbb{I}^{\pi_2 \pi_3 \cdots}] &= W^{\pi_1 \pi_2 \cdots}(t), \\[2mm] W^{\pi_1 \pi_2 \cdots}(t) \sum_b [C_b^{\pi_2}(t) \otimes \mathbb{I}^{\pi_1 \pi_3 \cdots}] &= W^{\pi_1 \pi_2 \cdots}(t), \qquad \text{etc.} \end{aligned} \right\} \qquad (11.10)$$

The Born probability (11.8) can thus be rewritten as

$$p_{\mathrm{Born}}(R_j^{\pi_1}, R_k^{\pi_2}, \cdots \text{ at } t) =$$
$$\mathrm{Tr}^{\pi_1 \pi_2 \cdots}(W^{\pi_1 \pi_2 \cdots}(t)\, [\sum_a C_a^{\pi_1}(t)\, R_j^{\pi_1} \otimes \sum_b C_b^{\pi_2}(t)\, R_k^{\pi_2} \otimes \cdots]), \qquad (11.11)$$

---

[126] In the case that one considers outcomes displayed by a set of pointers, it is clear that these pointers are pair-wise disjoint. It is, on the other hand, also clear that in quantum mechanics there exist situations in which non-disjoint systems simultaneously record outcomes of measurements. A famous example is Einstein's photon box (Bohr 1949): the hands of the clock in the box register the time of the opening of the shutter and the oscillation of the centre of mass of the box, which includes the mass of the hands of the clock, indicates the frequency of the emitted photon. As discussed in Chapter 6, only the atomic modal interpretation can in general give joint property ascriptions to non-disjoint systems. It would thus be of interest to see how this interpretation describes this experiment.

[127] To prove the first relation of (11.10), note that $\mathrm{Tr}^{\pi_1 \pi_2 \cdots}(W^{\pi_1 \pi_2 \cdots}(t) \sum_a [C_a^{\pi_1}(t) \otimes \mathbb{I}^{\pi_2 \pi_3 \cdots}])$ is equal to $\mathrm{Tr}^{\pi_1}(W^{\pi_1}(t) \sum_a C_a^{\pi_1}(t))$ and is thus equal to 1. The first relation then follows by Lemma 11.1.

which again simplifies to

$$p_{\text{Born}}(R_j^{\pi_1}, R_k^{\pi_2}, \cdots \text{ at } t)$$
$$= \sum_{a \in I_j^1} \sum_{b \in I_k^2} \cdots \text{Tr}^{\pi_1 \pi_2 \cdots}(W^{\pi_1 \pi_2 \cdots}(t) \, [C_a^{\pi_1}(t) \otimes C_b^{\pi_2}(t) \otimes \cdots]). \qquad (11.12)$$

This latter probability is exactly the joint probability with which the spectral and atomic modal interpretations ascribe at t the outcomes $R_j^{\pi_1}$, $R_k^{\pi_2}$, ... to π_1, π_2, ..., respectively. So, these two interpretations yield the same predictions for the correlations between the outcomes of simultaneously performed measurements as the Born rule.

Consider, secondly, a series of measurements which sequentially yield outcomes. The outcomes are again recorded by the mutually disjoint systems π_1, π_2, π_3, etc. Assume that π_1 acquires its outcome at a first instant t_1, that π_2 acquires its outcome at a second instant $t_2 \geq t_1$, and so on. One may think, for instance, of a series of measurements performed by different measurement devices (such that π_1, π_2, ... refer to the different pointers) or of a series of measurements performed by one device (such that π_1, π_2, ... refer to the different bits of paper or magnetic tape on which the outcomes are registered). The outcomes are again represented by the pair-wise orthogonal projections $\{R_j^{\pi_1}\}_j$, $\{R_k^{\pi_2}\}_k$, ..., respectively.

The standard formulation fixes by means of the Born rule and the projection postulate the correlations between these sequentially established outcomes. Let the systems π_1, π_2, ... be part of a larger composite ω which, as a whole, evolves freely (ω could ultimately be the whole universe). The evolution of the state of this composite from one instant s to a second instant t is then given by

$$W^\omega(t) = U^\omega(t,s) \, W^\omega(s) \, U^\omega(s,t) \qquad (11.13)$$

with $U^\omega(x,y)$ equal to $\exp([(x-y)/i\hbar] \, H^\omega)$.

At t_1 a measurement outcome $R_j^{\pi_1}$ is recorded by π_1 and the standard formulation predicts via the Born rule that

$$p_{\text{Born}}(R_j^{\pi_1} \text{ at } t_1) = \text{Tr}^\omega(W^\omega(t_1) \, [R_j^{\pi_1} \otimes \mathbb{I}^{\omega/\pi_1}]). \qquad (11.14)$$

If $R_j^{\pi_1}$ is indeed recorded, the state of ω collapses to

$$\widehat{W}^\omega(t_1) = \frac{[R_j^{\pi_1} \otimes \mathbb{I}^{\omega/\pi_1}] \, W^\omega(t_1) \, [R_j^{\pi_1} \otimes \mathbb{I}^{\omega/\pi_1}]}{\text{Tr}^\omega(W^\omega(t_1) \, [R_j^{\pi_1} \otimes \mathbb{I}^{\omega/\pi_1}])}. \qquad (11.15)$$

At t_2 a second outcome $R_k^{\pi_2}$ is recorded by π_2 and the standard formulation gives via the Born rule a probability for the occurrence of this outcome conditional on the occurrence of the first outcome $R_j^{\pi_1}$. Let $R_j^{\pi_1;\omega}$ be shorthand for $R_j^{\pi_1} \otimes \mathbb{I}^{\omega/\pi_1}$ and let $R_k^{\pi_2;\omega}$ be shorthand for $R_k^{\pi_2} \otimes \mathbb{I}^{\omega/\pi_2}$. Then

$$p_{\text{Born}}(R_k^{\pi_2} \text{ at } t_2/R_j^{\pi_1} \text{ at } t_1) = \text{Tr}^\omega(U^\omega(t_2,t_1)\,\widehat{W}^\omega(t_1)\,U^\omega(t_1,t_2)\,R_k^{\pi_2;\omega}), \quad (11.16)$$

where $U^\omega(t_2,t_1)\,\widehat{W}^\omega(t_1)\,U^\omega(t_1,t_2)$ is the state of ω which evolved unitarily from the collapsed state (11.15). From this conditional probability one can calculate a joint probability for the outcomes (substitute (11.15) into (11.16) and multiply by (11.14)):

$$p_{\text{Born}}(R_j^{\pi_1} \text{ at } t_1, R_k^{\pi_2} \text{ at } t_2) = \text{Tr}^\omega(U^\omega(t_2,t_1)\,R_j^{\pi_1;\omega}\,W^\omega(t_1)\,R_j^{\pi_1;\omega}\,U^\omega(t_1,t_2)\,R_k^{\pi_2;\omega}).$$
$$(11.17)$$

If $R_j^{\pi_1}$ and $R_k^{\pi_2}$ are indeed recorded at t_1 and at t_2, respectively, the state of ω collapses at t_2 to

$$\widehat{W}^\omega(t_2) = \frac{R_k^{\pi_2;\omega}\,U^\omega(t_2,t_1)\,R_j^{\pi_1;\omega}\,W^\omega(t_1)\,R_j^{\pi_1;\omega}\,U^\omega(t_1,t_2)\,R_k^{\pi_2;\omega}}{\text{Tr}^\omega(U^\omega(t_2,t_1)\,R_j^{\pi_1;\omega}\,W^\omega(t_1)\,R_j^{\pi_1;\omega}\,U^\omega(t_1,t_2)\,R_k^{\pi_2;\omega})}. \quad (11.18)$$

From this new collapsed state one can again determine the probability for a third outcome $R_l^{\pi_3}$, registered by π_3, conditional on the occurrences of the two previous outcomes. And with this conditional probability one can determine the joint probability for the three outcomes $R_j^{\pi_1}$, $R_k^{\pi_2}$ and $R_l^{\pi_3}$, write down the collapsed state at t_3 and consider a fourth outcome, etc., etc. For N outcomes the resulting joint probability is

$$p_{\text{Born}}(R_j^{\pi_1} \text{ at } t_1, R_k^{\pi_2} \text{ at } t_2, \dots, R_m^{\pi_N} \text{ at } t_N) =$$
$$\text{Tr}^\omega(U_{t_N,t_{N-1}}^\omega \cdots R_k^{\pi_2;\omega}\,U_{t_2,t_1}^\omega\,R_j^{\pi_1;\omega}\,W_{t_1}^\omega\,R_j^{\pi_1;\omega}\,U_{t_1,t_2}^\omega\,R_k^{\pi_2;\omega} \cdots U_{t_{N-1},t_N}^\omega\,R_m^{\pi_N;\omega}).$$
$$(11.19)$$

I am not able to generally determine the modal counterpart of this joint Born probability. Results about correlations between properties possessed by systems at different times were given in Chapter 8 when discussing the evolution of the actually possessed properties, and these results were rather poor. However, in special cases one can calculate the joint probability (11.19) within the spectral and atomic modal interpretations (Vermaas 1996, Sect. 7) and if one does, one obtains the same results as obtained by the Born rule.

Consider the case in which the systems π_1, π_2, ..., π_N evolve freely from the time that they register their outcomes. So, π_1 may interact with the rest of ω before t_1 but after t_1 it evolves freely; and π_2 evolves freely after t_2, etc.

This implies that the evolution operator $U^\omega(t, s)$ factorises after $t_1, t_2, \ldots,$ so

$$\left.\begin{aligned}
U^\omega(t, t_1) &= U^{\pi_1}(t, t_1) \otimes U^{\omega/\pi_1}(t, t_1), \\
U^\omega(t, t_2) &= U^{\pi_1}(t, t_2) \otimes U^{\pi_2}(t, t_2) \otimes U^{\omega/\pi_1\pi_2}(t, t_2), \qquad \text{etc.}
\end{aligned}\right\} \qquad (11.20)$$

Let's assume for simplicity that during their free evolution the states of the systems $\pi_1, \pi_2, \ldots, \pi_N$ remain fixed, so $U^{\pi_1}(t, t_1) = \mathbb{I}^{\pi_1}$, $U^{\pi_2}(t, t_2) = \mathbb{I}^{\pi_2}$, etc. Physically this assumption means that one freezes the states of the systems $\pi_1, \pi_2, \ldots, \pi_N$, once they have registered their respective outcomes. (What follows can also be derived without this simplifying assumption.)

A first consequence of this free and constant evolution of $\pi_1, \pi_2, \ldots, \pi_N$ is that the projection $R_j^{\pi_1} \otimes \mathbb{I}^{\omega/\pi_1}$ commutes with the evolution operator $U^\omega(t, t_1)$, that $R_k^{\pi_2} \otimes \mathbb{I}^{\omega/\pi_2}$ commutes with $U^\omega(t, t_2)$, etc. Hence, the Born probability (11.19) can be rewritten as

$$p_{\text{Born}}(R_j^{\pi_1} \text{ at } t_1, R_k^{\pi_2} \text{ at } t_2, \ldots, R_m^{\pi_N} \text{ at } t_N) =$$
$$\text{Tr}^\omega(W^\omega(t_N) [R_j^{\pi_1} \otimes R_k^{\pi_2} \otimes \cdots \otimes R_m^{\pi_N} \otimes \mathbb{I}^{\omega/\pi_1\pi_2\cdots\pi_N}]) \qquad (11.21)$$

(using that $U^\omega(t'', t') U^\omega(t', t)$ equals $U^\omega(t'', t)$, using (11.13) and using cyclic permutation).

Another consequence of the free and constant evolution of $\pi_1, \pi_2, \ldots, \pi_N$ is that the properties modal interpretations ascribe to $\pi_1, \pi_2, \ldots, \pi_N$ evolve deterministically. So, if π_1 possesses the outcome $R_j^{\pi_1}$ at t_1, then, with probability 1, π_1 continues to possess that outcome $R_j^{\pi_1}$ at any instant t later than t_1. The property that π_1 possesses at $t > t_1$ is thus a faithful record of the outcome registered at t_1. The same result holds *mutatis mutandis* for the other pointers. So, the joint probability that modal interpretations yield for the sequence of measurement outcomes satisfies

$$p(R_j^{\pi_1} \text{ at } t_1, R_k^{\pi_2} \text{ at } t_2, \ldots, R_m^{\pi_N} \text{ at } t_N) =$$
$$p(R_j^{\pi_1} \text{ at } t_N, R_k^{\pi_2} \text{ at } t_N, \ldots, R_m^{\pi_N} \text{ at } t_N). \qquad (11.22)$$

This latter probability is the joint probability that the disjoint systems $\pi_1, \pi_2, \ldots, \pi_N$ possess simultaneously their outcomes and can be calculated with the spectral and atomic modal interpretations. One obtains

$$p(R_j^{\pi_1} \text{ at } t_1, R_k^{\pi_2} \text{ at } t_2, \ldots, R_m^{\pi_N} \text{ at } t_N) =$$
$$\sum_{a \in I_j^1} \sum_{b \in I_k^2} \cdots \sum_{d \in I_m^N} \text{Tr}^\omega(W^\omega(t_N) [C_a^{\pi_1}(t_N) \otimes C_b^{\pi_2}(t_N) \cdots \otimes C_d^{\pi_N}(t_N) \otimes \mathbb{I}^{\omega/\pi_1\pi_2\cdots\pi_N}])$$

$$(11.23)$$

and by the relations (11.9) and (11.10) this joint probability is equal to the Born probability (11.21).

A second case generalises aspects of the case sketched above. The starting point is the idea that in physics one does not directly observe properties possessed at different instants but that one can only observe properties possessed at the same instant. This means that if one has two measurement outcomes which come into existence at different times, say, at t_1 and at t_2, then one cannot directly compare these outcomes. Instead one has to have some kind of record of the first outcome which still exists at t_2. Then at t_2 one can only compare this record of the first outcome with the second outcome. This simple fact reveals the presupposition that records of measurements earlier performed are supposed to be faithful in the sense that they do not change in time. Records, in the language of Chapter 8, thus have to evolve deterministically. To illustrate this, consider two spin measurements. At t_1 one measures the spin of a first particle and at t_2 one measures the spin of a second particle. Both measurements produce a black spot on a photographic plate. If one now concludes that the two measurement outcomes were sequentially 'up' because one has at t_2 two 'up' black spots on the photographic plate, one implicitly assumes that the black spot created at t_1 evolved with probability 1 to the black spot present at t_2. One thus assumes that the spot evolved deterministically.

The second case is therefore as follows. Assume explicitly that the records $\{R_j^{\pi_1}\}_j$, $\{R_k^{\pi_2}\}_k$, ... of the outcomes ascribed at $t_1, t_2, \ldots$, respectively, evolve deterministically. So, if at t_1 the system π_1 possesses outcome $R_j^{\pi_1}$, then π_1 also possesses $R_j^{\pi_1}$ at any time t later than t_1 and *mutatis mutandis* for π_2, etc. Assume, furthermore, that the records, once they have been established, are so-called conserved quantities with regard to the evolution of ω. This implies that after t_1 the Hamiltonian H^ω of ω commutes with $R_j^{\pi_1} \otimes \mathbb{I}^{\omega/\pi_1}$, etc. A consequence of this second assumption is that if, for instance, the state of π_1 at t_1 is given by $W^{\pi_1}(t_1) = R_j^{\pi_1}/\mathrm{Tr}^{\pi_1}(R_j^{\pi_1})$, then this state remains constant after t_1.

With these two assumptions one can again derive that the predictions of the spectral and atomic modal interpretations agree with those of the standard formulation. From the first assumption one can derive (11.22) and thus (11.23). From the second assumption it follows that $R_j^{\pi_1} \otimes \mathbb{I}^{\omega/\pi_1}$ commutes with $U^\omega(t, t_1)$ for all $t > t_1$, that $R_k^{\pi_2} \otimes \mathbb{I}^{\omega/\pi_2}$ commutes with $U^\omega(t, t_2)$ for all $t > t_2$, etc., so one can derive (11.21) from (11.19). The rest is similar to the first case.

To sum up, the standard formulation of quantum mechanics gives by the Born rule empirically adequate predictions about the probabilities with which single measurements have outcomes. If the bi, spectral and atomic modal interpretations solve the measurement problem by ascribing outcomes after measurements, then these three interpretations reproduce these predicted probabilities for single measurements. Furthermore, the standard formulation gives by the Born rule empirically adequate correlations between the outcomes of different measurements. The spectral and atomic modal interpretations reproduce these correlations for the case of simultaneously performed measurements. And, in so far as it is possible to calculate the correlations for the case of sequentially performed measurements, the spectral and atomic modal interpretations reproduce the predicted correlations as well.

11.3 Correlations between preparations and measurements

The above proofs have one drawback, and that is that it is assumed that the standard formulation and the modal interpretations assign the same states to systems before the first measurement. In the proof for sequentially established outcomes, for instance, I have calculated the correlations between the outcomes by assuming that the standard formulation and the modal interpretations all assign the same initial state $W^\omega(t_1)$ to ω. The possibility that they disagree about this initial state is thus ignored.[128]

If one holds the position that a system has its state as a kind of intrinsic feature which pertains inalienably to that system, then this assumption seems warranted. One can then argue that at the Big Bang the universe was born in an initial state $W^{\text{Universe}}(t_0)$. One can then argue that, in order to be empirically adequate, the standard formulation and the modal interpretations must all agree that this initial state of the universe is given by $W^{\text{Universe}}(t_0)$. The standard formulation and the modal interpretations then also agree that the state of the universe evolves unitarily to the state $W^{\text{Universe}}(t) = U^{\text{Universe}}(t, t_0) W^{\text{Universe}}(t_0) U^{\text{Universe}}(t_0, t)$ up to the instant t_1 when the first measurement is performed. Only after this first measurement do the standard formulation and the modal interpretations start to disagree about the state of the universe and about the states of the systems within the universe.

Within the context of modal interpretations, this position seems very natural. In modal interpretations the state of a system represents the properties possibly possessed by the system. And since it seems natural to take these

---

[128] In the proof for sequentially established outcomes, I do not ignore that the standard formulation and the modal interpretations disagree about the state of ω after the first measurement. According to the standard formulation the state collapses and I calculated its predictions from this collapsing state; and according to the modal interpretations the state of ω evolves unitarily and I calculated their predictions from this unitarily evolving state.

possibly possessed properties as intrinsic to the system, it also seems natural to take the state as intrinsic to the system.

Within the context of the standard formulation, however, other more instrumental views are possible as well. One can, for instance, hold the position that *we, the people*, assign states to systems in order that *we* can calculate the correlations between the outcomes of different measurements. Paradigmatically, one then assigns a state to a system only after a first measurement on that system. The first measurement thus functions as a preparation of the state and this state is then used to calculate the statistics for the outcomes of subsequent measurements. On this instrumental view, states assigned to systems are thus not intrinsic to those systems but rather bookkeeping devices for us which codify information about the outcome of the first preparational measurement. If one accepts this view, it is meaningless to speak about the state of a system before any preparational measurement takes place. And, as a consequence, the above proofs that the standard formulation and the modal interpretations yield the same predictions about measurement outcomes break down.

In this section I discuss an example of a series of two measurements performed on a system σ in which the first counts as a preparation of the state of σ. I show that the predictions by the standard formulation in the sketched instrumental view agree with the predictions by the modal interpretations, regardless of whether the standard formulation and the modal interpretations disagree about the state of σ. The discussion is based on the one given by Bacciagaluppi and Hemmo (1998, Sect. 3).

The example consists of two consecutive measurements of the second kind. The first measurement measures $A^\sigma = \sum_{q=1} a_q |a_q^\sigma\rangle\langle a_q^\sigma|$ by means of the interaction

$$
\begin{aligned}
|\Psi^{\sigma\mu}\rangle &= \sum_q c_q |a_q^\sigma\rangle \otimes |D_0^{\bar\mu}\rangle \otimes |R_0^\pi\rangle \longmapsto |\tilde\Psi^{\sigma\mu}\rangle \\
&= \sum_q c_q |\tilde a_q^\sigma\rangle \otimes |D_q^{\bar\mu}\rangle \otimes |R_q^\pi\rangle.
\end{aligned}
\tag{11.24}
$$

The second measures a magnitude $B^\sigma = \sum_{r=1} b_r |b_r^\sigma\rangle\langle b_r^\sigma|$ by means of the interaction

$$
\begin{aligned}
|\Psi^{\sigma\nu}\rangle &= \sum_r d_r |b_r^\sigma\rangle \otimes |D_0^{\bar\nu}\rangle \otimes |R_0^\rho\rangle \longmapsto |\tilde\Psi^{\sigma\nu}\rangle \\
&= \sum_r d_r |\tilde b_r^\sigma\rangle \otimes |D_r^{\bar\nu}\rangle \otimes |R_r^\rho\rangle.
\end{aligned}
\tag{11.25}
$$

The systems $\bar\mu$ and $\bar\nu$ are the mechanisms of the two measurement devices μ

and v, respectively. The sets $\{|D_q^{\bar{\mu}}\rangle\}_q$ and $\{|D_r^{\bar{v}}\rangle\}_r$ are sets of pair-wise orthogonal vectors such that both measurements satisfy the sufficient condition (10.42), given in Section 10.4, for solving the measurement problem with the spectral modal interpretation. The systems π and ρ are the pointers of the devices μ and v, respectively. And the pointer reading vectors $\{|R_q^{\pi}\rangle\}_q$ and $\{|R_r^{\rho}\rangle\}_r$ both satisfy the sufficient conditions (10.15) and (10.19), given in Section 10.3, for solving the measurement problem with the atomic modal interpretation.

According to the instrumental approach to the standard formulation, the first measurement on σ counts as a preparation of the state of σ: if the outcome of this measurement is $|R_j^{\pi}\rangle\langle R_j^{\pi}|$, the state of σ is $|\tilde{a}_j^{\sigma}\rangle\langle\tilde{a}_j^{\sigma}|$. This state can then be used to calculate the probabilities of finding outcomes at the end of the second measurement: given that the first measurement indeed yields the outcome $|R_j^{\pi}\rangle\langle R_j^{\pi}|$ and given that σ evolves freely during the time interval between the two measurements (say the first measurement ends at t_1 and the second starts at t_2), the probability of finding (say at t_3) the outcome $|R_k^{\rho}\rangle\langle R_k^{\rho}|$ corresponding to the eigenstate $|b_k^{\sigma}\rangle\langle b_k^{\sigma}|$, is by the Born rule equal to

$$p_{\text{Born}}(|R_k^{\rho}\rangle\langle R_k^{\rho}| \text{ at } t_3 / |R_j^{\pi}\rangle\langle R_j^{\pi}| \text{ at } t_1) = |\langle b_k^{\sigma}|e^{\frac{t_2-t_1}{i\hbar}H^{\sigma}}|\tilde{a}_j^{\sigma}\rangle|^2. \qquad (11.26)$$

Note that this conditional probability is calculated without assuming that σ has a state before the first measurement.

If one applies the spectral and atomic modal interpretations to these two measurements, one needs to assume that σ has a state before the first measurement. Let the first measurement begin at t_0 and assume that σ has at t_0 the (unknown) state $W^{\sigma}(t_0)$. The measurement devices are initially in their 'ready-to-measure' states so the whole lot is initially in the state

$$W^{\sigma\mu v}(t_0) = W^{\sigma}(t_0) \otimes |D_0^{\bar{\mu}}\rangle\langle D_0^{\bar{\mu}}| \otimes |R_0^{\pi}\rangle\langle R_0^{\pi}| \otimes |D_0^{\bar{v}}\rangle\langle D_0^{\bar{v}}| \otimes |R_0^{\rho}\rangle\langle R_0^{\rho}|. \qquad (11.27)$$

The system σ then interacts from t_0 to t_1 with the first device μ. This interaction is given by (11.24). From t_1 to t_2, σ evolves freely and with $U^{\sigma}(t_2,t_1) = \exp([(t_2-t_1)/i\hbar]\,H^{\sigma})$ and from t_2 to t_3, σ interacts with the second device v according to (11.25). If one now assumes that the first measurement device evolves freely from t_1 to t_3 with $U^{\mu}(t_3,t_1) = \mathbb{I}^{\mu}$ and that the second device evolves freely from t_0 to t_2 with $U^{v}(t_2,t_0) = \mathbb{I}^{v}$, one can calculate the final state of $\sigma\mu v$ after the second measurement. The result is

$$W^{\sigma\mu v}(t_3) = \sum_{q,q',r,r'} \langle b_r^{\sigma}|e^{\frac{t_2-t_1}{i\hbar}H^{\sigma}}|\tilde{a}_q^{\sigma}\rangle \langle a_q^{\sigma}|W^{\sigma}(t_0)|a_{q'}^{\sigma}\rangle \langle\tilde{a}_{q'}^{\sigma}|e^{-\frac{t_2-t_1}{i\hbar}H^{\sigma}}|b_{r'}^{\sigma}\rangle$$

$$\times |\tilde{b}_r^{\sigma}\rangle\langle\tilde{b}_{r'}^{\sigma}| \otimes |D_q^{\bar{\mu}}\rangle\langle D_{q'}^{\bar{\mu}}| \otimes |R_q^{\pi}\rangle\langle R_{q'}^{\pi}| \otimes |D_r^{\bar{v}}\rangle\langle D_{r'}^{\bar{v}}| \otimes |R_r^{\rho}\rangle\langle R_{r'}^{\rho}|. \qquad (11.28)$$

The spectral and atomic modal interpretations predicts that π possesses at t_1 one of the readings $\{|R_q^\pi\rangle\langle R_q^\pi|\}_q$ and that ρ possesses at t_3 one of the readings $\{|R_r^\rho\rangle\langle R_r^\rho|\}_r$. Moreover, if π evolves freely from t_1 to t_3 with $U^\pi(t_3, t_1) = \mathbb{I}^\pi$, the actual outcome $|R_j^\pi\rangle\langle R_j^\pi|$ at t_1 evolves deterministically to itself at t_3. Hence, the predicted joint probability for the outcomes of the two measurements is

$$p(|R_j^\pi\rangle\langle R_j^\pi| \text{ at } t_1, |R_k^\rho\rangle\langle R_k^\rho| \text{ at } t_3) = p(|R_j^\pi\rangle\langle R_j^\pi| \text{ at } t_3, |R_k^\rho\rangle\langle R_k^\rho| \text{ at } t_3) =$$

$$\text{Tr}^{\pi\rho}(W^{\pi\rho}(t_3)\left[|R_j^\pi\rangle\langle R_j^\pi| \otimes |R_k^\rho\rangle\langle R_k^\rho|\right]) = |\langle b_k^\sigma|e^{\frac{t_2-t_1}{i\hbar}H^\sigma}|\tilde{a}_j^\sigma\rangle|^2 \langle a_j^\sigma|W^\sigma(t_0)|a_j^\sigma\rangle$$

$$(11.29)$$

and, since

$$p(|R_j^\pi\rangle\langle R_j^\pi| \text{ at } t_1) = \sum_k p(|R_j^\pi\rangle\langle R_j^\pi| \text{ at } t_1, |R_k^\rho\rangle\langle R_k^\rho| \text{ at } t_3) = \langle a_j^\sigma|W^\sigma(t_0)|a_j^\sigma\rangle,$$

$$(11.30)$$

it follows that the spectral and atomic modal interpretations yield the same conditional probability (11.26) as the Born rule.

Hence, the spectral and atomic modal interpretations can also (at least in this example) reproduce the predictions by the standard formulation if one takes the view that the standard formulation assigns states to systems only after preparational measurements.

Part three
Philosophy

In this final part I analyse the modal interpretations from a more philosophical point of view.

In Chapter 12 I start by arguing that modal interpretations describe noumenal states of affairs and that therefore metaphysically tenable interpretations need only to meet the criteria of Consistency and Internal Completeness. Then I analyse the relations between properties, states and outcomes of measurements in the modal description of reality. I end by discussing how modal interpretations, when restricted to the description of measurement outcomes, recover the standard formulation of quantum mechanics.

Chapter 13 concerns the relations between the properties ascribed to composite systems and subsystems. I show that the property ascriptions of the bi and spectral modal interpretations can be characterised as holistic and non-reductionistic, whereas the property ascription of the atomic modal interpretation is non-holistic and to a large extent reductionistic. I argue that notwithstanding the lack of reductionism, the bi and spectral modal interpretations are empirically adequate and metaphysically tenable. I also discuss the possibility of saving the metaphysical tenability of the atomic modal interpretation by taking holistic properties as dispositional properties.

In Chapter 14 I give a survey of the possibilities and impossibilities of modal interpretations and in Chapter 15 I end with general conclusions.

12

Properties, states, measurement outcomes and effective states

Starting with a more philosophical review, I discuss how modal interpretations describe reality. Firstly, it is shown that they describe states of affairs which need not be observable. Secondly, I underpin my position that in order to be metaphysically tenable, interpretations need only to satisfy the criteria of Consistency and Internal Completeness. Thirdly, I analyse how properties, states and measurement outcomes are related to one another within the modal descriptions of reality. Finally, I show how modal interpretations recover the standard formulation of quantum mechanics if one defines, in addition to the true states, so-called effective states of systems.

12.1 Noumenal states of affairs

The aim of an interpretation of quantum mechanics is, as I said in the introduction, to provide a description of what reality would be like if quantum mechanics were true. This formulation underlines two aspects of interpretations. Firstly, interpretations intend to construe quantum mechanics in terms of a description of reality and not merely in terms of the outcomes of measurements. Secondly, this description need not be correct. The objective is only to prove that there *exists* a construal of quantum mechanics which yields an acceptable description of reality. And because it is already difficult enough to give such a proof of existence, it is not (yet) the aim of interpretations to provide the one and only correct description of reality.

The above formulation, however, disregards another aspect of interpretations of quantum mechanics, namely that they describe states of affairs which need not always be observable. The project of interpreting quantum mechanics is thus in principle different from, say, a historian's attempt to give a description of what Athens would have been like at the time Socrates lived, or from a child's dream of imagining what life would be like if one were as small as, say, a smurf.[129] These latter two projects concern a reality which can, in principle, be observed, namely by, respectively, Socrates and the smurfs. They thus aim at describing states of affairs which are, in principle, observable and which can be designated in that sense as *phenomenal* states of affairs. In contrast, when interpreting quantum mechanics, one not only aims at describing phenomenal states of affairs, such as, for instance, the properties possessed by a measurement device, but also at describing states of affairs which are, in principle, not observable.

Consider, for instance, the description of the properties possessed by two elementary particles α and β and their composite $\alpha\beta$. The spectral modal interpretation says that α, β and $\alpha\beta$ simultaneously possess, say, the respective core properties P_a^α, P_b^β and $P_c^{\alpha\beta}$. If we now want to observe the joint possession of these properties, one has to measure them simultaneously. However, because these properties P_a^α, P_b^β and $P_c^{\alpha\beta}$ are not, in general, comeasurable (the projections $P_a^\alpha \otimes \mathbb{I}^\beta$, $\mathbb{I}^\alpha \otimes P_b^\beta$ and $P_c^{\alpha\beta}$ in general do not commute), quantum mechanics rules out such a simultaneous measurement. Hence, the state of affairs for α, β and $\alpha\beta$ given by the spectral modal interpretation cannot be observed. Or consider the transition probabilities for the actually possessed properties of an atom α during the time interval that α evolves freely and that the composite state of α and a second atom β is constantly given by the pure state $|\Psi^{\alpha\beta}\rangle = \sum_j c_j |c_j^\alpha\rangle \otimes |c_j^\beta\rangle$. The bi, spectral and atomic modal interpretations then all yield that these transition probabilities are $p(|c_b^\alpha\rangle\langle c_b^\alpha|$ at $t / |c_a^\alpha\rangle\langle c_a^\alpha|$ at $s) = \delta_{ab}$ (see Section 8.2). But this can never be observed, not by us nor by any other being. To prove this, assume that there exist beings, say quantum smurfs, which can, in accordance with the rules of quantum mechanics, directly monitor the properties of individual atoms (like we can monitor the properties of macroscopic systems). Then an observation of these transition probabilities for α would consist of the observation that if α possesses the property $|c_a^\alpha\rangle\langle c_a^\alpha|$ at a first instant s, then α possesses with probability 1 the same property at a later instant t while the state of $\alpha\beta$ remains constantly equal to $|\Psi^{\alpha\beta}\rangle$. However, even a quantum smurf cannot observe that α possesses $|c_a^\alpha\rangle\langle c_a^\alpha|$ at s without disturbing the pure state of $\alpha\beta$ at s because any non-trivial correlation between α and the smurf makes the state of α become entangled with the state of the smurf. This entanglement is ruled out because the state of $\alpha\beta$ at s is pure.[130] This conclusion holds for every possible observer, so, according to quantum

---

[129] See Peyo (1967).

[130] Let σ be the smurf. Then, because the state of $\alpha\beta$ is pure at instant s, it follows (see footnote 11) that the state of $\sigma\alpha\beta$ at s is equal to $W^{\sigma\alpha\beta}(s) = W^\sigma(s) \otimes |\Psi^{\alpha\beta}\rangle\langle\Psi^{\alpha\beta}|$. The state of $\sigma\alpha$ thus factorises, that is, $W^{\sigma\alpha}(s) = W^\sigma(s) \otimes W^\alpha(s)$, ruling out any entanglement between the states of σ and of α.

mechanics, no being can observe the transition probabilities for a freely evolving atom α part of a composite $\alpha\beta$ with a pure state.

Thus, to conclude, a third aspect of interpretations of quantum mechanics is that they aim not only at describing phenomenal states of affairs but also at describing states of affairs which cannot be observed. Interpretations of quantum mechanics thus exceed the realm of the phenomena and are in that sense aiming at describing *noumenal* states of affairs, where 'noumenal' should be taken in its literal meaning as 'known by the mind as against the senses.'[131]

Let's now return to the discussion in Section 3.3 of the demands an interpretation of quantum mechanics should meet. I listed three: an interpretation should yield a description of reality which is well developed, empirically adequate and metaphysically tenable. Consider, firstly, the last demand. It is beyond controversy to require that a metaphysically tenable interpretation should yield a consistent description of reality and that such an interpretation should deliver all that it promises to deliver. That is, the description should be free of contradiction and be complete with regard to the standards set by the interpretation itself. The spectral and atomic modal interpretations, for instance, aim at probabilistically ascribing properties to systems and at giving correlations between these properties. Given this, it is warranted to demand that these interpretations consistently correlate the properties in the case of both the phenomenal and the noumenal states of affairs. In Section 3.3 I have called these two criteria, respectively, Consistency and Internal Completeness. On the other hand, I believe that one should be careful about accepting further 'natural' criteria for metaphysically tenable interpretations. Natural criteria usually capture our intuitions about what we observe, but the analysis which I have given above proves that interpretations also describe noumenal states of affairs. Interpretations of quantum mechanics thus enter domains beyond what we observe. And it seems incorrect to me to demand that the description of noumenal states of affairs should conform to our intuitions about the phenomena. For instance, Clifton (1996) has argued that a metaphysically tenable interpretation should satisfy the condition that if a system α possesses a property Q^{α}, then any composite $\alpha\beta$ should simultaneously possess the property $Q^{\alpha} \otimes \mathbb{I}^{\beta}$.[132] I agree that on the

---

[131] I am well aware that the term 'noumenon' has a strong Kantian connotation. However, according to two handbooks of philosophy, Lacey (1976, page 145) and Honderich (1995, pages 657-658), 'noumenon' also has a literal pre-Kantian meaning of 'thing known by the mind as against the senses,' respectively 'things that are thought.' When characterising interpretations of quantum mechanics as (also) aiming at describing noumenal states of affairs, I refer to this literal meaning. I thus do not mean that interpretations intend to describe the realm of the Kantian *Ding-an-sich*.

[132] The condition that if α possesses Q^{α}, then $\alpha\beta$ should simultaneously possess $Q^{\alpha} \otimes \mathbb{I}^{\beta}$, is called the Property Composition condition, and is extensively discussed in the next chapter.

basis of our observations of properties of systems and composite systems, one can make a case for this condition. However, our observations yield by definition information about only phenomenal states of affairs and not about the noumenal states of affairs. Hence, there is no reason whatsoever to also assume that these noumenal states of affairs should satisfy the condition that if α possesses Q^α, then $\alpha\beta$ possesses $Q^\alpha \otimes \mathbb{I}^\beta$. Put more generally, I take the somewhat liberal position that the descriptions of the noumenal states of affairs by modal interpretations need only to satisfy Consistency and Internal Completeness, and that the descriptions may for the rest defy all our common intuitions.

With regard to the descriptions of the phenomena, one should of course expect more. One should demand that interpretations of quantum mechanics should yield empirically adequate descriptions of phenomenal states of affairs. Hence, a third criterion for interpretations is that their descriptions of the phenomenal states of affairs satisfy Empirical Adequacy.

One can now go one step further and also formulate every-day intuitions about the phenomena and demand that an interpretation is well developed only if its descriptions of the phenomenal states of affairs satisfy these every-day intuitions as well. Examples of such intuitions are Instantaneous and Dynamical Autonomy as presented in Section 3.3, and the conditions on full property ascriptions as discussed in Section 5.3. I think, however, that one should not impose these intuitions as criteria on interpretations.[133] As I said in Section 3.3, there is not yet consensus about which conditions should be imposed on well-developed interpretations. I therefore do not impose further criteria on modal interpretations apart from Consistency and Internal Completeness with regard to the phenomenal and the noumenal states of affairs, and Empirical Adequacy with regard to the phenomenal states of affairs. The route I follow in this book is firstly to develop the modal descriptions of reality and afterwards to assess (and to leave it to the judgement of the reader) whether these developed descriptions are acceptable (see Chapters 14 and 15).

12.2 Relations between properties, states and measurement outcomes

Quantum mechanics in the standard formulation can be understood as an interplay between states and outcomes of measurements. By means of states of systems quantum mechanics yields predictions about outcomes of measurements performed on those systems. Conversely, outcomes of so-called

---

[133] In Section 3.3 I argued that Instantaneous and Dynamical Autonomy are reasonable *assumptions* about interpretations; I did not present them as generally valid truths.

preparational measurements (see Section 11.3) determine, via the projection postulate, the states of systems.

Within modal interpretations this picture changes, yielding a different understanding of quantum mechanics. A first change is induced by the introduction of properties of quantum systems: quantum mechanics becomes therefore an interplay between states, measurement outcomes *and* properties. A second change is due to the rejection of the projection postulate: states do not collapse during measurements blocking the possibility of preparing a state by means of a measurement. In this section I analyse this new picture of quantum mechanics by discussing how properties, states and measurement outcomes interlock.

FROM STATES TO PROPERTIES: Consider, firstly, how the state of a system fixes the properties of the system. Recapitulating Chapters 4 and 5, modal interpretations determine a core property ascription $\{\langle p_j, C_j^{\alpha}\rangle\}_j$ by means of the state W^{α} of a system α. This core property ascription says that α possesses with probability $p_k = p(C_k^{\alpha})$ the core property C_k^{α} and induces a full property ascription $\langle p_k, \mathscr{DP}_k, [.]_k \rangle$ to α given by

$$\left\langle p_k = p(C_k^{\alpha}), \ \mathscr{DP}_k = \mathscr{F}(C_k^{\alpha}), \ [Q^{\alpha}]_k = \frac{\mathrm{Tr}^{\alpha}(C_k^{\alpha}Q^{\alpha})}{\mathrm{Tr}^{\alpha}(C_k^{\alpha})} \right\rangle. \tag{12.1}$$

This full property ascription implies that it is with probability $p(C_k^{\alpha})$ the case that α actually possesses the properties $Q^{\alpha} \in \mathscr{DP}_k$ with $[Q^{\alpha}]_k = 1$, that α actually does not possess the properties $Q^{\alpha} \in \mathscr{DP}_k$ with $[Q^{\alpha}]_k = 0$, and that the properties $Q^{\alpha} \notin \mathscr{DP}_k$ are actually indefinite. Hence, the state of a system determines probabilistically the actual properties of the system.

FROM STATES TO OUTCOMES: If modal interpretations solve the measurement problem, then states determine in the usual quantum mechanical way the outcomes of measurements. That is, if at the end of a measurement the state of a pointer π is W^{π}, then (see Section 11.1) this pointer possesses the outcome R_k^{π} with the Born probability $\mathrm{Tr}^{\pi}(W^{\pi}R_k^{\pi})$.

FROM PROPERTIES TO STATES: Suppose we know those properties which are actually possessed by an individual system α and those which are actually not possessed. What does this then say about the state of the system? Given that in modal interpretations the actual properties of α form a faux-Boolean algebra $\mathscr{F}(C_k^{\alpha})$, one can easily fix the core property C_k^{α} of α: it is simply the property actually possessed by α that is represented by the lowest-dimensional projection. This core property C_k^{α} now yields some information about the state of α but does not fix it: in the bi and spectral modal interpretations C_k^{α} is an eigenprojection of the state W^{α}. So, the spectral resolution $W^{\alpha} = \sum_j w_j P_j^{\alpha}$ must contain a term $w_k C_k^{\alpha}$. However, the corresponding eigenvalue w_k can have any value strictly larger than 0 and the other eigenprojections $\{P_j^\alpha\}_{j\neq k}$ can be any set of projections orthogonal to C_k^α (and to one another). In the atomic modal interpretation C_k^α is a product $P_a^{\beta_1} \otimes P_b^{\beta_2} \otimes \cdots$ of eigenprojections of the states of the atoms in α. In this case C_k^α thus says even less about the state of α. Hence, if one knows the actual properties of a system, one can determine the core property of the system but one cannot determine the state of that system.

In contrast, if one considers within the bi and spectral modal interpretations an ensemble of N systems which are all in the same state W^α, one can obtain more conclusive information about this state. Let $\#(C_j^\alpha)$ denote the number of systems in this ensemble which actually possess the core property C_j^α. If one then accepts a frequency interpretation of probabilities, it is clear that the state defined by $\widetilde{W}^\alpha = \sum_j (\#(C_j^\alpha)/[N \dim(C_j^\alpha)]) \, C_j^\alpha$ converges to W^α in some way if N goes to infinity. Within the atomic modal interpretation one can reach the same conclusion only for ensembles of atoms. (For ensembles of molecules ω there exist mutually different states W_1^ω and W_2^ω which yield the same statistics for the core properties of ω.[134] So, if one knows these statistics, one still cannot uniquely fix the state of the molecules.) Hence, the actually possessed properties for an ensemble of systems with the same state, determine the core property ascription to those systems. And this core property ascription determines in the bi and spectral modal interpretations in some approximation the state of the systems. If these systems are atoms, the same conclusion holds in the atomic modal interpretation.

FROM PROPERTIES TO OUTCOMES: Consider how the actual properties of a system are related to the outcomes of a measurement performed on that system. Take a system α which actually possesses the core property C_j^α at some initial instant and perform a measurement on α yielding an outcome R_k^π possessed by a pointer π at a final instant. Let the measured magnitude be given by $A^\alpha = \sum_p a_p \, |a_p^\alpha\rangle\langle a_p^\alpha|$ and let R_k^π be the outcome that corresponds to the eigenvalue a_k of A^α. Assume, moreover, that there exists a snooper system σ for α at the initial instant and that this snooper evolves freely during the measurement. Then it can be proved for the perfect measurements discussed in Sections 10.3 and 10.4 that the spectral and atomic modal

---

[134] Consider a system consisting of two atoms α and β. The states

$$W_1^{\alpha\beta} = \sum_{j,j'} c_j \bar{c}_{j'} \, |e_j^\alpha\rangle\langle e_{j'}^\alpha| \otimes |e_j^\beta\rangle\langle e_{j'}^\beta| \quad \text{and} \quad W_2^{\alpha\beta} = \sum_j |c_j|^2 \, |e_j^\alpha\rangle\langle e_j^\alpha| \otimes |e_j^\beta\rangle\langle e_j^\beta|$$

then both yield that the core property of $\alpha\beta$ is $|e_k^\alpha\rangle\langle e_k^\alpha| \otimes |e_k^\beta\rangle\langle e_k^\beta|$ with probability $|c_k|^2$.

interpretations[135] yield the following conditional probabilities between the initially possessed core projection of α and the finally possessed outcomes:

$$p(R_k^\pi/C_j^\alpha) = \text{Tr}^\alpha\left(\frac{1}{\dim(C_j^\alpha)}C_j^\alpha\,|a_k^\alpha\rangle\langle a_k^\alpha|\right). \tag{12.2}$$

This result is rather attractive for if these conditional probabilities are generally valid, it proves that one can use, in addition to the state of a system, the actually possessed core property of the system to give predictions about the outcomes of a measurement. Furthermore, the actually possessed core property determines these predictions for outcomes independently of the specific state of the system. Finally, the way in which the actual core property C_j^α determines these predictions appears to be quantum mechanical in the sense that (12.2) can be taken as the Born probability of finding the outcome R_k^π, given that the state of α is $C_j^\alpha/\dim(C_j^\alpha)$. So, it seems that before the measurement α effectively has a state equal to the (normalised) actually possessed core property.

Unfortunately, the conditional probabilities (12.2) are not generally valid. In Section 9.2 it was proved that the spectral and atomic modal interpretations do not satisfy the stronger assumption of Dynamical Autonomy for measurements (see page 168). That is, one can construct two identical measurement situations (given by, respectively, an object system α interacting with a pointer π and an object system α' interacting with a pointer π', where α and α' initially possess the same core property and where the composites $\alpha\pi$ and $\alpha'\pi'$ initially have the same composite state which evolves with the same Hamiltonian) for which $p(R_k^\pi/C_j^\alpha)$ and $p(R_k^{\pi'}/C_j^{\alpha'})$ are nevertheless different. Thus, the nice conditional probabilities (12.2) are not generally valid.

Hence, there do exist conditional probabilities for the initially possessed core property of a system and the final outcomes of a measurement performed on that system, but these conditional probabilities are neither unique functions of the initially possessed core property nor unique functions of the initial core property and the composite state of the system and the measurement device. Thus there does not exist a unique general rule with which one can predict measurement outcomes from the initially possessed core property of a system.

FROM OUTCOMES TO PROPERTIES: As mentioned at the beginning of this section, within the standard formulation of quantum mechanics one can use an outcome R_j^π of a so-called preparational measurement to fix the (final) state of the system on which the measurement was performed. This follows

---

[135] Analogous results cannot be derived for the bi modal interpretation since this interpretation does not in general correlate the properties of the snooper and of the pointer after the measurement.

because one has the projection postulate within the standard formulation. Modal interpretations, however, reject this projection postulate and therefore outcomes of preparational measurements do not yield such determinate information about the object system. To demonstrate this let's consider a preparational measurement and see what the outcome reveals about the object system. I start by discussing the relation between the outcome and the properties possessed by the object system. After that, I focus on the relation between the outcome and the final state of the object system.

As I said in Section 11.3, measurements of the second kind count in the standard formulation as preparational measurements. So, take a measurement of the second kind on a system α measuring the magnitude $A^\alpha = \sum_p a_p |a_p^\alpha\rangle\langle a_p^\alpha|$. The measurement interaction yields

$$|a_p^\alpha\rangle \otimes |D_0^{\bar\mu}\rangle \otimes |R_0^\pi\rangle \longmapsto |\tilde a_p^\alpha\rangle \otimes |D_p^{\bar\mu}\rangle \otimes |R_p^\pi\rangle \qquad (12.3)$$

and, given the initial state $W_0^\alpha \otimes |D_0^{\bar\mu}\rangle\langle D_0^{\bar\mu}| \otimes |R_0^\pi\rangle\langle R_0^\pi|$, the final state of $\alpha\mu$ is

$$W^{\alpha\mu} = \sum_{p,q} \langle a_p^\alpha|W_0^\alpha|a_q^\alpha\rangle \, |\tilde a_p^\alpha\rangle\langle \tilde a_q^\alpha| \otimes |D_p^{\bar\mu}\rangle\langle D_q^{\bar\mu}| \otimes |R_p^\pi\rangle\langle R_q^\pi|. \qquad (12.4)$$

The state of α after the measurement is thus equal to

$$W^\alpha = \sum_p \langle a_p^\alpha|W_0^\alpha|a_p^\alpha\rangle \, |\tilde a_p^\alpha\rangle\langle \tilde a_p^\alpha| \qquad (12.5)$$

and due to the rejection of the projection postulate, this remains the state of α. It can now be proved that the set of core properties assigned to α by the spectral and atomic modal interpretations after the measurement depends on the initial state of α.[136] Hence, a first conclusion is that after a preparational

---

[136] Let α be an atom defined on a two-dimensional Hilbert space. Let $\{|e_1^\alpha\rangle, |e_2^\alpha\rangle\}$ be a basis for this Hilbert space and assume that $|\tilde a_1^\alpha\rangle = |e_1^\alpha\rangle$ and $|\tilde a_2^\alpha\rangle = \frac{1}{2}\sqrt 2 |e_1^\alpha\rangle + \frac{1}{2}\sqrt 2 |e_2^\alpha\rangle$. Let the initial state of α be given by $W_0^\alpha = w_1 |a_1^\alpha\rangle\langle a_1^\alpha| + w_2 |a_2^\alpha\rangle\langle a_2^\alpha|$. Its final state (12.5) is then equal to

$$W^\alpha = \begin{pmatrix} w_1 + \frac{w_2}{2} & \frac{w_2}{2} \\ \frac{w_2}{2} & \frac{w_2}{2} \end{pmatrix}$$

with respect to the basis $\{|e_1^\alpha\rangle, |e_2^\alpha\rangle\}$. Consider the case that $w_1 = \frac{3}{7}$ and $w_2 = \frac{4}{7}$. The eigenprojections of W^α are then given by the matrices

$$P_1^\alpha = \tfrac{1}{5}\begin{pmatrix} 4 & 2 \\ 2 & 1 \end{pmatrix}, \qquad P_2^\alpha = \tfrac{1}{5}\begin{pmatrix} 1 & -2 \\ -2 & 4 \end{pmatrix}.$$

Consider, secondly, the case that $w_1 = \frac{4}{7}$ and $w_2 = \frac{3}{7}$. The eigenprojections of W^α are then

$$P_1^\alpha = \tfrac{1}{10}\begin{pmatrix} 9 & 3 \\ 3 & 1 \end{pmatrix}, \qquad P_2^\alpha = \tfrac{1}{10}\begin{pmatrix} 1 & -3 \\ -3 & 9 \end{pmatrix}.$$

Hence, the core properties which are ascribed to α by the spectral and atomic modal interpretations after the measurement depend on the state of α before the measurement.

measurement the finally possessed core properties of the object system are not in general uniquely fixed.

Assume that the measurement interaction (12.3) satisfies the conditions laid down in the Sections 10.3 and 10.4 such that the spectral and atomic modal interpretations ascribe after the measurement the outcomes $\{|\mathbb{R}_j^\pi\rangle\langle\mathbb{R}_j^\pi|\}_j$ to the pointer. Let $\{C_k^\alpha\}_k$ be the core properties ascribed to α after the measurement. The conditional probability that α actually possesses the core property C_k^α given that π possesses outcome $|\mathbb{R}_j^\pi\rangle\langle\mathbb{R}_j^\pi|$ is then in the spectral and atomic modal interpretations equal to

$$p(C_k^\alpha/|\mathbb{R}_j^\pi\rangle\langle\mathbb{R}_j^\pi|) = \frac{\mathrm{Tr}^{\alpha\mu}(W^{\alpha\mu}[C_k^\alpha \otimes \mathbb{I}^{\bar\mu} \otimes |\mathbb{R}_j^\pi\rangle\langle\mathbb{R}_j^\pi|])}{\mathrm{Tr}^\pi(W^\pi|\mathbb{R}_j^\pi\rangle\langle\mathbb{R}_j^\pi|)} = \mathrm{Tr}^\alpha(|\tilde{a}_j^\alpha\rangle\langle\tilde{a}_j^\alpha|\,C_k^\alpha). \quad (12.6)$$

So, a second conclusion is that after a preparational measurement the probabilities for the finally possessed core properties of the object system, conditional on the measurement outcomes, are fixed independently of the initial state of the object system.

An interesting special case is given by a preparational measurement (12.3), where the vectors $\{|\tilde{a}_p^\alpha\rangle\}_p$ are pair-wise orthogonal. The core properties of α are then in the spectral modal interpretation given by $\{|\tilde{a}_p^\alpha\rangle\langle\tilde{a}_p^\alpha|\}_p$ (barring degeneracies) independently of the initial state W_0^α, and the conditional probabilities (12.6) become equal to δ_{jk}. Within the spectral modal interpretation an outcome $|\mathbb{R}_j^\pi\rangle\langle\mathbb{R}_j^\pi|$ of such a special preparational measurement thus determines with probability 1 that the actually possessed core property of α is given by $|\tilde{a}_j^\alpha\rangle\langle\tilde{a}_j^\alpha|$. For the atomic modal interpretation such special measurements also exist. If the vectors $\{|\tilde{a}_p^\alpha\rangle\}_p$ are pair-wise orthogonal and if they are given by $|\tilde{a}_p^\alpha\rangle = |e_{p1}^{\beta_1}\rangle \otimes |e_{p2}^{\beta_2}\rangle \cdots$, where the systems $\{\beta_i\}_i$ are the atoms in α and the sets of vectors $\{|e_{pi}^{\beta_i}\rangle\}_{pi}$ are sets of pair-wise orthogonal vectors, then an outcome $|\mathbb{R}_j^\pi\rangle\langle\mathbb{R}_j^\pi|$ also reveals with probability 1 that α possesses the core property $|\tilde{a}_j^\alpha\rangle\langle\tilde{a}_j^\alpha|$. Hence, a final conclusion is that in special cases, the outcome of a preparational measurement does uniquely fix the actually possessed core property of the object system. That is, the outcome of such a special measurement determines with probability 1 which core property is actually possessed by the object system after the measurement, independently of the initial state of that object system.

FROM OUTCOMES TO STATES: After the preparational measurement (12.3), the object system α remains in the state (12.5) because modal interpretations reject the projection postulate. Moreover, this final state depends on the initial state W_0^α of the object system. Hence, if one does not know this initial state, one cannot determine the state (12.5) of the object system after this measurement, regardless of whether one knows which outcome the measurement has.

An exception to this conclusion is given by the case in which the preparational measurements are of the special kind which determine with probability 1 the actually possessed core property of the object system after the measurement. If one performs N such special measurements on an ensemble of N systems which all have the same initial state W_0^α, one can determine the final state of these systems. Let $\#(|R_j^\pi\rangle\langle R_j^\pi|)$ denote the number of times that the measurements yield the outcome $|R_j^\pi\rangle\langle R_j^\pi|$. Then one can conclude that $\#(|R_j^\pi\rangle\langle R_j^\pi|)$ object systems actually possess the core property $|\tilde{a}_j^\alpha\rangle\langle\tilde{a}_j^\alpha|$ after the measurements. Then using how the actually possessed core properties of systems determine the states of those systems (see FROM PROPERTIES TO STATES), one can conclude that the state $\widetilde{W}^\alpha = \sum_j(\#(|R_j^\pi\rangle\langle R_j^\pi|)/N)\,|\tilde{a}_j^\alpha\rangle\langle\tilde{a}_j^\alpha|$ converges in some way to the final state (12.5) of α. (This holds in the spectral modal interpretation for every ensemble of systems and in the atomic modal interpretation only for ensembles of atoms.)

When reviewing all these relations between properties, states and measurement outcomes, the following picture arises. The state of a system probabilistically determines the actually possessed properties of the system and probabilistically determines the outcomes of measurements performed on that system. Conversely, neither the actual core property of an individual system nor the outcome of a preparational measurement performed on an individual system yields conclusive information about the state of that system. The state of a system is thus rather elusive from the perspective of properties and measurement outcomes. Furthermore, the initially possessed core property of a system probabilistically determines the outcomes of a measurement performed on the system. But, since this probabilistic relation is not unique and is only derivable in special cases, it is not of much use. Finally, there exists a special class of preparational measurements for which the outcome determines with probability 1 the actually possessed core property of the object system after the measurement.

12.3 States and effective states

The aim of an interpretation is to provide a description of what reality would be like if quantum mechanics were true. And modal interpretations aim to furnish this description by ascribing properties to quantum systems on the basis of the states of those systems. However, given the analysis of the previous section, one now has to conclude that is quite hard to obtain definite knowlegde of this modal description of reality because it is often impossible to determine the states of systems. Consider, for instance, an incoming muon reaching earth from outer space. This muon has some unknown state and a (preparational) measurement on this muon will neither disclose this state nor alter it to a known one. At best a preparational measurement can determine with probability 1 the final actually possessed property of the muon. On the other hand, in some special cases it is imaginable that one can obtain definite knowledge about states of systems. Examples are processes for which one can decide on theoretical grounds what the final state of a system is (this might be the case for particles emitted during a well-defined atomic de-excitation). Or one can think of processes which yield an ensemble of systems which are always in the same unknown final state. A series of measurements can then determine this state. However, in general it is difficult to know the states of systems. Hence, within modal interpretations states have, as already noted, epistemologically a rather elusive status. And because modal interpretations describe reality on the basis of these states, these descriptions are elusive as well: modal interpretations ascribe properties to systems but it is, in general, impossible to know these properties.

Given that one indeed cannot determine the states of most systems, one faces a second problem, namely, how to properly generate predictions about measurement outcomes with modal interpretations. In Chapter 11 I gave the expressions which modal interpretations yield for the probabilities and the correlations of measurement outcomes and these probabilities and correlations (formulae (11.3), (11.12) and (11.23)) are all functions of the states of systems. So, if one, in general, cannot know states, how can one then still determine these probabilities and correlations? This second problem can, however, be addressed and this brings us back to the collapsed states assigned to systems by the standard formulation.

Consider a measurement performed on a system α from t_1 to t_2 and assume that one knows the state of α. One can then calculate within the spectral and atomic modal interpretations the probabilities for the outcomes by applying the Born rule. That is,

$$p(R_j^\pi \text{ at } t_2) = \text{Tr}^{\alpha\mu}(U^{\alpha\mu}(t_2, t_1) [W^\alpha(t_1) \otimes W^\mu(t_1)] U^{\alpha\mu}(t_1, t_2) [\mathbb{I}^{\alpha\bar{\mu}} \otimes R_j^\pi]), \quad (12.7)$$

where $W^\mu(t_1)$ is the initial state of the measurement device and $U^{\alpha\mu}(t_2, t_1)$ is the measurement interaction. If, on the other hand, one does not know the state of α, the probabilities for the outcomes are still given by (12.7) but one can obviously not calculate them. Return now for a moment to the stardard formulation of quantum mechanics and assume that before t_1 a preparational measurement was performed on α yielding at t_0 an outcome R_i^ρ possessed by some pointer ρ. One can then determine the state of α:

conditional on the outcome R_i^ρ, the state of α at t_1 is within the standard formulation the Schrödinger evolute $\widehat{W}^\alpha(t_1)$ of the collapsed state $\widehat{W}^\alpha(t_0)$ assigned to α after the preparational measurement. With this state one can then calculate the conditional probabilities as

$$p_{\text{Born}}(R_j^\pi \text{ at } t_2 / R_i^\rho \text{ at } t_0) =$$

$$\text{Tr}^{\alpha\mu}(U^{\alpha\mu}(t_2, t_1) [\widehat{W}^\alpha(t_1) \otimes W^\mu(t_1)] U^{\alpha\mu}(t_1, t_2) [\mathbb{I}^{\alpha\bar\mu} \otimes R_j^\pi]). \qquad (12.8)$$

In Section 11.3 it was proved that, if the state of the pointer ρ evolves freely from t_0 to t_2, these last conditional probabilities are also correct within the spectral and atomic modal interpretations regardless of whether or not one knows the true state of α.

The similarity of (12.7) and (12.8) now suggests the following solution to our second problem concerning the prediction of measurement outcomes. Say, one cannot determine the state of a system α but one does know that a preparational measurement has been performed on that system. Then, in order to calculate the probabilities for the outcomes of a second measurement within the spectral and atomic modal interpretations, one can pretend that the collapsed state $\widehat{W}^\alpha(t)$, assigned by the standard formulation, is the state of α. This collapsed state is, of course, not the real state of the system, so the collapsed state does not fix the properties of α, but is merely a bookkeeping device to calculate the statistics for future measurements.

I end this chapter by discussing a train of thought which leads to the conclusion that this collapsed state assigned by the standard formulation plays within modal interpretations a rôle more significant than that of a bookkeeping device. I start by considering a discussion by Bacciagaluppi and Hemmo (1998, Sects. 2 and 3) of what they call a selective preparational measurement.[137] In such a measurement one begins with an ensemble of systems α, then performs a preparational measurement on each member of the ensemble and finally constructs a subensemble by selecting all those systems for which the measurements yield the same outcome. Suppose the measurement is a measurement of the second kind given by (12.3) and say one selects the systems which produce the outcome $|R_j^\pi\rangle\langle R_j^\pi|$. This selected subensemble is homogeneous according to the standard formulation: all its members have the same collapsed state $|\tilde{a}_j^\alpha\rangle\langle\tilde{a}_j^\alpha|$. This collapsed state is thus a proper characterisation of the subensemble.

Within modal interpretations it is less obvious how to characterise the selected subensemble. The states of the systems are in any case not suitable. Assume, for instance, that all the systems in the ensemble with which one

---

[137] See also Section 3.2.3 of Bacciagaluppi (1996b).

started have the same initial state. Then all the systems also have the same state after the measurement. Hence, non-trivial subensembles cannot be distinguished by means of the state of the systems in the ensemble. The actually possessed properties of the systems also do not yield a proper characterisation. The possible possessed properties of a system α after a preparational measurement in general depend, as was proved in the previous section, on the initial state W_0^α. So, if the selected subensemble is inhomogeneous with regard to this initial state, the subensemble can also be inhomogeneous with regard to the set of possible properties possessed after the measurement and thus with regard to the actually possessed properties.

The question is thus how to characterise the selected subensemble within modal interpretations. Bacciagaluppi and Hemmo observed that all members of this subensemble have in common that the past preparational measurements yield the same outcome $|R_j^\pi\rangle\langle R_j^\pi|$. Hence, they propose characterising the subensemble by means of the outcome $|R_j^\pi\rangle\langle R_j^\pi|$ possessed by the pointers of the measurement devices.

Secondly, Bacciagaluppi and Hemmo discuss the question of how to predict the probabilities for the outcomes of future measurements performed on the selected subensemble. Repeating the discussion of Section 11.3, one predicts these probabilities within the standard formulation by applying the Born rule to the collapsed state $|\tilde{a}_j^\alpha\rangle\langle\tilde{a}_j^\alpha|$, and one predicts these probabilities within modal interpretations by calculating the joint probabilities $p(|R_j^\pi\rangle\langle R_j^\pi|$ at $t_3, |R_k^\rho\rangle\langle R_k^\rho|$ at $t_3)$.[138] Bacciagaluppi and Hemmo therefore conclude that within modal interpretations the measurement outcome $|R_j^\pi\rangle\langle R_j^\pi|$ of the preparational measurement is in two respects a proper description of the selected subensemble: it characterises the subensemble and it determines the probabilities for outcomes of future measurements on this ensemble.

I believe one can improve on this answer in two respects. Firstly, quantum mechanics is within modal interpretations a theory about properties of *individual* systems and not (merely) a theory about ensembles of systems. I therefore think that one should take the outcome $|R_j^\pi\rangle\langle R_j^\pi|$ as a characterisation of each individual system in the selected ensemble and not as a characterisation of that ensemble as a whole. Secondly, it is rather unusual to represent the future behaviour of a system α (or, if one doesn't like the first improvement, of an ensemble of systems) by means of a feature pertaining to a second system. More precisely, the assignment of the outcome $|R_j^\pi\rangle\langle R_j^\pi|$ to α to characterise its behaviour during future measurements, means that

---

[138] It is here assumed that the outcome $|R_j^\pi\rangle\langle R_j^\pi|$ of the preparational measurement evolves deterministically to the instant t_3 at which the future measurement produces its outcome $|R_k^\rho\rangle\langle R_k^\rho|$.

The effective state

After a preparational measurement performed on a system α yielding outcome R_j^π at t, one may assign an effective state $W_{R_j^\pi}^\alpha(t)$ to α at t. This effective state is equal to the collapsed state which the standard formulation of quantum mechanics assigns to α after the preparational measurement, and is, in general, different from the true state $W^\alpha(t)$ of α. The effective state evolves according to the Schrödinger equation and generates via the Born rule the probabilities for outcomes of future measurements on α conditional on the outcome R_j^π of the preparational measurement.

---

a disposition of α is represented by an operator defined on a Hilbert space $\mathscr{H}^\pi$ which is not associated with α. Instead, it is usual in quantum theory to represent a feature or a disposition of a system by means of an operator defined on the Hilbert space associated with the system itself.

So, in order to improve on the answer of Bacciagaluppi and Hemmo, I propose the following. If a preparational measurement is performed on a system α yielding an outcome represented by, say, the projection R_j^π, then one may assign, within modal interpretations, the collapsed state $\widehat{W}^\alpha(t_1)$ to α to describe its behaviour during future measurements. I propose to call this assigned state the **effective state** of α. And because an effective state is assigned conditional on an outcome R_j^π of a preparational measurement, I denote it by $W_{R_j^\pi}^\alpha(t)$. This effective state is assigned to individual systems and is by definition defined on $\mathscr{H}^\alpha$.[139] (As an illustration: in the case of the preparational measurement (12.3), the effective state $W_{|R_j^\pi\rangle\langle R_j^\pi|}^\alpha(t_1)$ of α is $|\tilde{a}_j^\alpha\rangle\langle\tilde{a}_j^\alpha|$.)

With this last addition, the modal picture of quantum mechanics becomes as follows. A system has a state which determines the possible possessed properties of the system and which determines via the Born rule the (unconditional) probabilities of finding outcomes after measurements. The dynamics of the state is given by the Schrödinger equation. In addition to this state, one can assign after a preparational measurement an effective state to a system which determines via the Born rule the probabilities for outcomes of future measurements conditional on the outcome of the preparational meas-

---

[139] This proposal to assign the state $W_{R_j^\pi}^\alpha(t)$ to α after a preparational measurement can also be found in Healey (1989) ($W_{R_j^\pi}^\alpha(t)$ is for Healey the *quantum state* of α). The motivation to call it an effective state is given by the analogy between $W_{R_j^\pi}^\alpha(t)$ and the effective state in Bohmian mechanics.

urement. The dynamics of this effective state is also given by the Schrödinger equation. A system always has a state which need not be known and a system sometimes has an effective state (only if a preparational measurement has been performed) which is then always known. During a (preparational) measurement neither the state nor the possible effective states collapse; the outcome of this measurement at best allows one to define a new effective state of the system.

By means of this picture one can now reconstruct the standard formulation of quantum mechanics as that part of the modal theory with which one can best predict the outcomes of measurements in the case that preparational measurements are performed. To see this, consider a system α on which a preparational measurement is performed at time t_1. It then follows within modal interpretations that the best predictions about the outcomes of a future measurement on α at $t > t_1$ are obtained by applying the Born rule to the Schrödinger evolute of the effective state $W^{\alpha}_{R^{\pi_1}_a}(t_1)$ assigned to α at t_1. If now a second preparational measurement is performed on α at $t_2 > t_1$, one can improve on our theory about measurement outcomes because then the best predictions for a future measurement on α at $t > t_2$ are no longer generated by applying the Born rule to the Schrödinger evolute of $W^{\alpha}_{R^{\pi_1}_a}(t_1)$ but by applying this rule to the Schrödinger evolute of the new effective state $W^{\alpha}_{R^{\pi_2}_b}(t_2)$ assigned to α at t_2. And if a third preparational measurement is performed at $t_3 > t_2$, one can again improve and predict the outcomes of future measurements most accurately by using the evolute to the effective state $W^{\alpha}_{R^{\pi_3}_c}(t_3)$, etc. The succession in time of these effective states with which one can best predict the outcomes of future measurements on α establishes exactly the collapsing dynamics of the state of α in the standard formulation of quantum mechanics.

13

Holism versus reductionism

In quantum mechanics the properties of a composite system can be divided into properties which are reducible to the properties of the subsystems of the composite and properties which are irreducible to those subsystem properties. The irreducible properties are called holistic and emerge when one describes the system as a whole.

In this chapter I show that the bi and spectral modal interpretations can ascribe the holistic properties to composites but, in general, fail to reproduce the relations between the reducible properties of a composite and the properties of the subsystems. The atomic modal interpretation, on the other hand, fails to ascribe the holistic properties but is much more successful in reproducing the relations between the reducible properties and the subsystem properties. I discuss whether these failures harm the empirical adequacy and the metaphysical tenability of modal interpretations.

13.1 Holistic properties of composite systems

The basic principle of reductionism, that all the properties of composite systems are amalgamates of the properties of the subsystems of these composites, has triumphed so much in classical physics that I find it difficult to come up with a clear example of a classical property that cannot be taken as a complex of the subsystem properties. So, in order to make a case for the idea that composites can also possess, in addition to reducible properties, so-called holistic properties which cannot be reduced to the properties of the subsystems, one has to move outside physics or to draw upon some counterfactual development of physics. In the first case, one may claim that many of the (mental) properties which we ascribe in psychology or sociology to humans cannot be reduced to the properties of the atoms and molecules of those humans. And, in the second case, one may argue that if it had been proved that it is impossible to reduce thermodynamics to classical statistical mechanics, then thermodynamical properties like pressure, temperature and entropy would have been taken as holistic properties of, say, a gas. For then these properties would not have been reducible to the properties of the particles of the gas. However, I guess such examples do not do much good to the idea of the existence of holistic properties in physics. Firstly, the claim that the mental properties of humans cannot be reduced to the properties of the particles of humans is not generally accepted. Moreover, even if it is the case that the properties ascribed by psychology or sociology to humans cannot be reduced to the properties ascribed by physics to particles, this does not yet prove that the properties ascribed by physics itself to composites like humans cannot be reduced to the properties ascribed to the particles. Secondly, in physics, properties like the pressure, temperature and entropy of a gas are generally taken as reducible to the properties of the particles of the gas. So, even though we might imagine that these properties are holistic, in actual physics this is not the case.

One of the peculiarities of quantum mechanics is now that it is highly holistic. And this holism is present within quantum mechanics itself and is thus not something which arises because quantum mechanics and some other physical theory cannot be reduced to one another (analogously to that thermodynamical properties would be holistic in the counterfactual case that thermodynamics could not be reduced to statistical mechanics). That is, holism arises because the quantum mechanical properties of the composites cannot be reduced to the quantum mechanical properties of the subsystems of those composites. Hence, although reductionism has triumphed in classical physics, holism is strongly present in quantum physics.

Before proving this quantum mechanical holism, I lay down more explicitly what I mean by reducible and holistic properties of composites. Consider two particles α and β which both have two energy levels: E_1^α and $E_2^\alpha > E_1^\alpha$, and E_1^β and $E_2^\beta > E_1^\beta$, respectively. Then the property of the composite $\alpha\beta$ that these two particles have together the energy $E_1^\alpha + E_1^\beta$ is clearly reducible to a pair of properties of α and β. Namely, $\alpha\beta$ possesses the property that the joint energy is $E_1^\alpha + E_1^\beta$ if and only if it is simultaneously the case that α possesses the property that its energy is E_1^α and that β possesses the property that its energy is E_1^β. The property of $\alpha\beta$ that the joint energy is strictly larger than $E_1^\alpha + E_1^\beta$ is in its turn reducible to a number of pairs of properties of α and β: the composite $\alpha\beta$ possesses this second property if and only if it is the case that α has energy E_1^α and β has energy E_2^β, *or* α has energy E_2^α and β has energy E_1^β, *or* α has energy E_2^α and β has energy

E_2^β. More generally, a property $Q^{\alpha\beta}$ of a composite $\alpha\beta$ is reducible to the pairs of properties $\{\langle Q_j^\alpha, Q_j^\beta \rangle\}_j$ in the case that $\alpha\beta$ possesses $Q^{\alpha\beta}$ if and only if α and β possesses simultaneously Q^α and Q^β, respectively, for some pair $\langle Q^\alpha, Q^\beta \rangle \in \{\langle Q_j^\alpha, Q_j^\beta \rangle\}_j$. In contrast, all the properties $Q^{\alpha\beta}$ for which there do not exist such pairs $\{\langle Q_j^\alpha, Q_j^\beta \rangle\}_j$ to which it can be reduced are irreducible holistic properties of $\alpha\beta$. Let's now generalise this to composites containing more than two systems. One then arrives at the following definitions.

Reducible and holistic properties

A property Q^ω of $\omega = \alpha\beta\cdots$ is reducible to the properties $\{\langle Q_j^\alpha, Q_j^\beta, \ldots \rangle\}_j$ of α, β, etc., if and only if

$$[Q^\omega] = 1 \iff \exists \langle Q^\alpha, Q^\beta, \ldots \rangle \in \{\langle Q_j^\alpha, Q_j^\beta, \ldots \rangle\}_j : [Q^\alpha] = [Q^\beta] = \cdots = 1. \quad (13.1)$$

A property Q^ω of $\omega = \alpha\beta\cdots$ is holistic if and only if it is not reducible to the properties of any set of the subsystems of ω.

Let's consider whether quantum mechanics is holistic in the sense of these definitions. A first remark is that quantum mechanics in its standard formulation can be neither reductionistic nor holistic: the standard formulation does not speak about properties, so one cannot apply the above definitions. Hence, it is only meaningful in the context of an interpretation to consider the question of whether quantum mechanics is holistic. It is now possible to argue that any interpretation of quantum mechanics which reproduces the predictions of the standard formulation with regard to measurement outcomes is necessarily holistic. The argument goes as follows.

If a property $Q^{\alpha\beta}$ of the composite $\alpha\beta$ is reducible to the properties of α and β, then there should exist at least one pair $\langle Q^\alpha, Q^\beta \rangle$ of properties for which it holds that if α and β possess Q^α and Q^β, respectively, then $\alpha\beta$ possesses $Q^{\alpha\beta}$. Or, in terms of measurement outcomes, if $Q^{\alpha\beta}$ is reducible, then there should exist at least one pair $\langle Q^\alpha, Q^\beta \rangle$ such that if two ideal measurements performed on α and on β, respectively, reveal with probability 1 that Q^α and Q^β are possessed, then a subsequent measurement on $\alpha\beta$ yields with probability 1 that $\alpha\beta$ possesses $Q^{\alpha\beta}$. This latter consequence is now violated by the predictions of the standard formulation of quantum mechancs and is therefore violated by any interpretation which reproduces these predictions.

Assume that α and β are each defined on two-dimensional Hilbert spaces, $\mathscr{H}^\alpha$ and $\mathscr{H}^\beta$, with the orthonormal bases $\{|e_1^\alpha\rangle, |e_2^\alpha\rangle\}$ and $\{|f_1^\beta\rangle, |f_2^\beta\rangle\}$, respectively. Consider a property of $\alpha\beta$ represented by the projection $Q^{\alpha\beta} = |\Psi^{\alpha\beta}\rangle\langle\Psi^{\alpha\beta}|$, where $|\Psi^{\alpha\beta}\rangle$ is a non-trivially entangled vector

$|\Psi^{\alpha\beta}\rangle = \sum_{j=1}^{2} c_j |e_j^{\alpha}\rangle \otimes |f_j^{\beta}\rangle$. If this property is reducible, there exist properties $\langle Q^{\alpha}, Q^{\beta} \rangle$ such that if two ideal measurements on α and on β, respectively, reveal with probability 1 that these properties are possessed, then a subsequent measurement on $\alpha\beta$ reveals with probability 1 that $Q^{\alpha\beta}$ is possessed as well. However, such properties $\langle Q^{\alpha}, Q^{\beta} \rangle$ do not exist:

Because α and β are defined on two-dimensional Hilbert spaces, Q^{α} and Q^{β} are zero-, one- or two-dimensional projections. The possibility that Q^{α} or Q^{β} is a zero-dimensional projection falls away; it makes no sense to reduce a property of a composite to a property of a part for which it holds that a measurement always reveals with probability 1 that it is not possessed. So, Q^{α} and Q^{β} are either one- or two-dimensional. Consider, firstly, the case that they are both one-dimensional, say $Q^{\alpha} = |a^{\alpha}\rangle\langle a^{\alpha}|$ and $Q^{\beta} = |b^{\beta}\rangle\langle b^{\beta}|$. If the state of $\alpha\beta$ is given by $W^{\alpha\beta} = |a^{\alpha}\rangle\langle a^{\alpha}| \otimes |b^{\beta}\rangle\langle b^{\beta}|$, then the measurements on α and β to check whether Q^{α} and Q^{β}, respectively, are possessed yield with probability 1 positive outcomes. But a subsequent measurement of $Q^{\alpha\beta}$ yields a positive outcome with a probability $|\langle\Psi^{\alpha\beta}|(|a^{\alpha}\rangle\otimes|b^{\beta}\rangle)|^2$, and this probability is strictly smaller than 1 because $|\Psi^{\alpha\beta}\rangle$ is non-trivially entangled. So, $Q^{\alpha\beta}$ cannot be reduced to properties $\langle Q^{\alpha}, Q^{\beta} \rangle$ represented by one-dimensional projections.

Consider, secondly, the case that Q^{α} is one-dimensional ($Q^{\alpha} = |a^{\alpha}\rangle\langle a^{\alpha}|$) and that Q^{β} is two-dimensional ($Q^{\beta} = \mathbb{I}^{\beta}$). If the initial state of $\alpha\beta$ is again $|a^{\alpha}\rangle\langle a^{\alpha}| \otimes |b^{\beta}\rangle\langle b^{\beta}|$, the measurements on α and β again yield that Q^{α} and Q^{β} are possessed with probability 1. However, the subsequent measurement of $Q^{\alpha\beta}$ still yields a positive outcome with a probability $|\langle\Psi^{\alpha\beta}|(|a^{\alpha}\rangle\otimes|b^{\beta}\rangle)|^2 < 1$. The same conclusion holds if Q^{α} is two-dimensional and Q^{β} one-dimensional and if Q^{α} and Q^{β} are both two-dimensional. So, $Q^{\alpha\beta}$ is not reducible to properties $\langle Q^{\alpha}, Q^{\beta} \rangle$ of α and β.[140]

Hence, in any interpretation of quantum mechanics that reproduces the predictions of the standard formulation, one must take the property $Q^{\alpha\beta} =$

---

[140] The probability $|\langle\Psi^{\alpha\beta}|(|a^{\alpha}\rangle \otimes |b^{\beta}\rangle)|^2$ is calculated by the following measurements. In the case of a one-dimensional projection $Q^X = |q^X\rangle\langle q^X|$, where $|q^X\rangle$ can be $|a^{\alpha}\rangle$, $|b^{\beta}\rangle$ or $|\Psi^{\alpha\beta}\rangle$, the measurement interaction is

$$|q^X\rangle \otimes |D_0^{\tilde{\mu}}\rangle \otimes |R_0^{\pi}\rangle \longmapsto |q^X\rangle \otimes |D_1^{\tilde{\mu}}\rangle \otimes |R_1^{\pi}\rangle,$$

$$|\tilde{q}^X\rangle \otimes |D_0^{\tilde{\mu}}\rangle \otimes |R_0^{\pi}\rangle \longmapsto |\tilde{q}^X\rangle \otimes |D_2^{\tilde{\mu}}\rangle \otimes |R_2^{\pi}\rangle, \qquad \text{for all } |\tilde{q}^X\rangle \text{ with } \langle q^X|\tilde{q}^X\rangle = 0.$$

In the case of a two-dimensional projection $Q^X = \mathbb{I}^X$, where X can be α or β, the measurement interaction is

$$|q^X\rangle \otimes |D_0^{\tilde{\mu}}\rangle \otimes |R_0^{\pi}\rangle \longmapsto |q^X\rangle \otimes |D_1^{\tilde{\mu}}\rangle \otimes |R_1^{\pi}\rangle, \qquad \text{for all } |q^X\rangle.$$

All these measurements are ideal in the sense that they establish strict correlations between the eigenstates $\{|q^X\rangle\}$ of Q^X and the outcome $|R_1^{\pi}\rangle$, and in the sense that they do not perturb these eigenstates $\{|q^X\rangle\}$ to states $\{|\tilde{q}^X\rangle\}$ different to $\{|q^X\rangle\}$.

$|\Psi^{\alpha\beta}\rangle\langle\Psi^{\alpha\beta}|$ with $|\Psi^{\alpha\beta}\rangle = \sum_{j=1}^{2} c_j |e_j^{\alpha}\rangle \otimes |f_j^{\beta}\rangle$, as a holistic property. Or, more generally, in any such interpretation the properties of a composite ω represented by projections $Q^{\omega} = |\Psi^{\omega}\rangle\langle\Psi^{\omega}|$ with $|\Psi^{\omega}\rangle$ a non-trivially entangled vector are holistic properties of ω. That such properties exist in quantum mechanics is proved by a composite of two spin $\frac{1}{2}$-particles σ and τ. The spin of the composite $\sigma\tau$ in the $\vec{v}$ direction is represented by the operator $S_{\vec{v}}^{\sigma\tau} = S_{\vec{v}}^{\sigma} \otimes \mathbb{I}^{\tau} + \mathbb{I}^{\sigma} \otimes S_{\vec{v}}^{\tau}$, where $S_{\vec{v}}^{\sigma} = (\hbar/2)|u_{\vec{v}}^{\sigma}\rangle\langle u_{\vec{v}}^{\sigma}| - (\hbar/2)|d_{\vec{v}}^{\sigma}\rangle\langle d_{\vec{v}}^{\sigma}|$ and $S_{\vec{v}}^{\tau} = (\hbar/2)|u_{\vec{v}}^{\tau}\rangle\langle u_{\vec{v}}^{\tau}| - (\hbar/2)|d_{\vec{v}}^{\tau}\rangle\langle d_{\vec{v}}^{\tau}|$. And the squared total spin of $\sigma\tau$ is given by $S_{\vec{v}}^{\sigma\tau} \cdot S_{\vec{v}}^{\sigma\tau} = [S_{\vec{x}}^{\sigma\tau}]^2 + [S_{\vec{y}}^{\sigma\tau}]^2 + [S_{\vec{z}}^{\sigma\tau}]^2$. The property '$[S_{\vec{z}}^{\sigma\tau}] = 0$ and $[S_{\vec{v}}^{\sigma\tau} \cdot S_{\vec{v}}^{\sigma\tau}] = 0$' now has a clear physical meaning and is represented by a holistic property $Q^{\sigma\tau} = |\Psi^{\sigma\tau}\rangle\langle\Psi^{\sigma\tau}|$ because $|\Psi^{\sigma\tau}\rangle$ is the singlet state $\frac{1}{2}\sqrt{2}(|u_{\vec{v}}^{\sigma}\rangle \otimes |d_{\vec{v}}^{\tau}\rangle - |d_{\vec{v}}^{\sigma}\rangle \otimes |u_{\vec{v}}^{\tau}\rangle)$.

Quantum mechanical properties of composites are, of course, not all holistic. Return, for a moment, to the example of the two particles α and β with the two energy levels E_1^{α} and $E_2^{\alpha} > E_1^{\alpha}$, and E_1^{β} and $E_2^{\beta} > E_1^{\beta}$. Let the projection Q_j^{α} represent the property that α has energy E_j^{α} (the Hamiltonian of α is then $H^{\alpha} = E_1^{\alpha} Q_1^{\alpha} + E_2^{\alpha} Q_2^{\alpha}$) and let Q_k^{β} represent the property that β has energy E_k^{β}. It is then standard in quantum mechanics to assume that the property of the composite $\alpha\beta$ that α and β have together the energy $E_j^{\alpha} + E_k^{\beta}$ is represented by the projection $Q_j^{\alpha} \otimes Q_k^{\beta}$.[141] Consider now the property that α and β have together the energy $E_1^{\alpha} + E_1^{\beta}$. This property is by the above assumption represented by the projection $Q_1^{\alpha} \otimes Q_1^{\beta}$. And since this property is reducible to the properties that α has energy E_1^{α} and that β has energy E_1^{β}, it follows that the property $Q_1^{\alpha} \otimes Q_1^{\beta}$ of $\alpha\beta$ is reducible to the pair $\langle Q_1^{\alpha}, Q_1^{\beta}\rangle$ of properties of α and β. More generally, the above assumption implies that $Q_j^{\alpha} \otimes Q_k^{\beta}$ is reducible to $\langle Q_j^{\alpha}, Q_k^{\beta}\rangle$. It should thus be the case that $\alpha\beta$ possesses $Q_j^{\alpha} \otimes Q_k^{\beta}$ if and only if α possesses Q_j^{α} and β possesses Q_k^{β}.

Consider, secondly, the property of $\alpha\beta$ that the joint energy of α and β is strictly larger than $E_1^{\alpha} + E_1^{\beta}$. This property is, as I said above, reducible to a number of pairs of properties of α and β. Namely, $\alpha\beta$ possesses this property if and only if α possesses Q_1^{α} and β possesses Q_2^{β}, or α possesses Q_2^{α} and β possesses Q_1^{β}, or α possesses Q_2^{α} and β possesses Q_2^{β}. Because, by the above assumption, $\alpha\beta$ possesses $Q_j^{\alpha} \otimes Q_k^{\beta}$ if and only if α possesses Q_j^{α} and β possesses Q_k^{β}, it follows that this second property is possessed by $\alpha\beta$, if and only if $\alpha\beta$ possesses $Q_1^{\alpha} \otimes Q_2^{\beta}$ or $Q_2^{\alpha} \otimes Q_1^{\beta}$ or $Q_2^{\alpha} \otimes Q_2^{\beta}$. Using the Weakening condition, one can thus conclude that $\alpha\beta$ possesses the property

---

[141] Assume, here, that the sums $\{E_j^{\alpha} + E_k^{\beta}\}_{j,k}$ are not degenerate, that is, $E_1^{\alpha} + E_2^{\beta} \neq E_2^{\alpha} + E_1^{\beta}$.

that the joint energy of α and β is strictly larger than $E_1^\alpha + E_1^\beta$ if and only if $\alpha\beta$ possesses the property $Q_1^\alpha \otimes Q_2^\beta + Q_2^\alpha \otimes Q_1^\beta + Q_2^\alpha \otimes Q_2^\beta$. Hence, one may conclude that the property represented by $Q_1^\alpha \otimes Q_2^\beta + Q_2^\alpha \otimes Q_1^\beta + Q_2^\alpha \otimes Q_2^\beta$ should be reducible to the pairs of properties $\{\langle Q_1^\alpha, Q_2^\beta \rangle, \langle Q_2^\alpha, Q_1^\beta \rangle, \langle Q_2^\alpha, Q_2^\beta \rangle\}$.

Moveover, if one generalises this to arbitrary composites $\omega = \alpha\beta\cdots$, one may take any property $Q^\omega = \sum_j Q_j^\alpha \otimes Q_j^\beta \otimes \cdots$, represented by a projection which is a sum of products of projections defined on $\mathscr{H}^\alpha$, $\mathscr{H}^\beta$, etc., as a property that is reducible to the strings of properties $\{\langle Q_j^\alpha, Q_j^\beta, \ldots \rangle\}_j$.[142]

One can now raise two questions about an interpretation of quantum mechanics with regard to the reducible and the holistic properties. The first is whether the property ascription of an interpretation respects the relations between the reducible properties and the properties to which they are reducible. That is, does it hold that ω possesses a reducible property Q^ω if and only if α, β, etc., possess the properties $\langle Q^\alpha, Q^\beta, \ldots \rangle \in \{\langle Q_j^\alpha, Q_j^\beta, \ldots \rangle\}_j$. Let's call an interpretation which respects this relation a *reductionistic* interpretation. The second question is whether the property ascription of an interpretation can ascribe the holistic properties to composite systems and let's call an interpretation which can a *holistic* interpretation. Ideally an interpretation is both reductionistic and holistic. In the next section I show that modal interpretations do not meet this ideal.

13.2 The violations of holism and of reductionism

The bi and spectral modal interpretations are without restriction holistic because they can ascribe any holistic (or reducible) property Q^ω to any composite ω: if ω has the state $W^\omega = Q^\omega / \dim(Q^\omega)$ and, in the case of the bi modal interpretation, if ω is also a part of a larger composite in a pure state, then ω possesses Q^ω with probability 1.

The bi and spectral modal interpretations are, however, not reductionistic. Consider the composite $\alpha\beta$ and assume that α and β are both defined on the two-dimensional Hilbert spaces $\mathscr{H}^\alpha$ and $\mathscr{H}^\beta$ with the orthonormal bases $\{|e_1^\alpha\rangle, |e_2^\alpha\rangle\}$ and $\{|f_1^\beta\rangle, |f_2^\beta\rangle\}$, respectively. Take then the reducible property $Q^{\alpha\beta} = |e_1^\alpha\rangle\langle e_1^\alpha| \otimes |f_1^\beta\rangle\langle f_1^\beta|$ and assume that $\alpha\beta$ has the state vector $|\Psi^{\alpha\beta}\rangle = \frac{\sqrt{3}}{2}|e_1^\alpha\rangle \otimes |f_1^\beta\rangle + \frac{1}{2}|e_2^\alpha\rangle \otimes |f_2^\beta\rangle$. The bi and spectral modal interpretations then

---

[142] Note that these properties satisfy the necessary condition for reducibility: if a series of ideal measurements on α, β, etc., yields that they possess the properties $\langle Q^\alpha, Q^\beta, \ldots \rangle \in \{\langle Q_j^\alpha, Q_j^\beta, \ldots \rangle\}_j$, then a subsequent measurement of $Q^\omega = \sum_j Q_j^\alpha \otimes Q_j^\beta \otimes \cdots$ has with probability 1 a positive outcome.
Note also that I have not given a clear demarcation between holistic and reducible properties; I have only argued that such a demarcation should yield that $Q^\omega = |\Psi^\omega\rangle\langle\Psi^\omega|$, with $|\Psi^\omega\rangle$ a non-trivially entangled vector, is a holistic property and that $Q^\omega = \sum_j Q_j^\alpha \otimes Q_j^\beta \otimes \cdots$ is a reducible property.

yield that with probability $\frac{3}{4}$ the systems α and β possess simultaneously the properties $|e_1^\alpha\rangle\langle e_1^\alpha|$ and $|f_1^\beta\rangle\langle f_1^\beta|$, respectively. If these two interpretations are reductionistic, then they should yield that $\alpha\beta$ possesses the property $Q^{\alpha\beta}$ with a probability equal to or larger than $\frac{3}{4}$. However, if one applies these interpretations to $\alpha\beta$, one obtains (assuming in the case of the bi modal interpretation that $\alpha\beta$ is a part of a larger composite with a pure state) that its core property is $|\Psi^{\alpha\beta}\rangle\langle\Psi^{\alpha\beta}|$ with probability 1. And since $Q^{\alpha\beta}|\Psi^{\alpha\beta}\rangle$ is equal to neither $|\Psi^{\alpha\beta}\rangle$ nor 0, it follows that the full property ascription (5.12) to $\alpha\beta$ leaves $Q^{\alpha\beta}$ indefinite with probability 1. Hence, the bi and spectral modal interpretations are not reductionistic.

For the atomic modal interpretation the situation is more or less the other way round: its property ascription is not holistic but is to a large extent reductionistic. To see that it is not holistic, take a composite of two atoms α and β and consider a property of $\alpha\beta$ represented by the projection $|\Psi^{\alpha\beta}\rangle\langle\Psi^{\alpha\beta}|$, where $|\Psi^{\alpha\beta}\rangle$ is a non-trivial entangled vector $|\Psi^{\alpha\beta}\rangle = \sum_j c_j |e_j^\alpha\rangle \otimes |f_j^\beta\rangle$. If the atomic modal interpretation can ascribe this property to $\alpha\beta$, there must exist, according to the full property ascription (5.12), a core property $C^{\alpha\beta}$ such that $|\Psi^{\alpha\beta}\rangle\langle\Psi^{\alpha\beta}|\, C^{\alpha\beta} = C^{\alpha\beta}$. Since $|\Psi^{\alpha\beta}\rangle\langle\Psi^{\alpha\beta}|$ is a one-dimensional projection, this can only be the case if $C^{\alpha\beta} = |\Psi^{\alpha\beta}\rangle\langle\Psi^{\alpha\beta}|$. However, in the atomic modal interpretation the core property of $\alpha\beta$ is always a product $P_a^\alpha \otimes P_b^\beta$, so $C^{\alpha\beta}$ can never be equal to $|\Psi^{\alpha\beta}\rangle\langle\Psi^{\alpha\beta}|$ and it follows that the holistic property $|\Psi^{\alpha\beta}\rangle\langle\Psi^{\alpha\beta}|$ cannot be ascribed to $\alpha\beta$.

The atomic modal interpretation is partly reductionistic in the sense that it reproduces the relations between a reducible property $Q^\omega = \sum_j Q_j^\alpha \otimes Q_j^\beta \otimes \cdots$ of a composite ω and the properties $\{\langle Q_j^\alpha, Q_j^\beta, \ldots\rangle\}_j$ of the subsystems α, β, etc., only if the sets $\{Q_j^\alpha\}_j$, $\{Q_j^\beta\}_j$, etc., are all sets of pair-wise orthogonal projections.

Consider, for instance, the property $Q^\alpha \otimes Q^\beta$ of the composite $\alpha\beta$. Let the atoms that make up α be $\sigma_1,\ldots\sigma_m$, and let the atoms that make up β be $\sigma_{m+1},\ldots\sigma_n$ (because α and β are disjoint they do not share atoms). Consider any simultaneous core property ascription to the atoms $\sigma_1,\ldots\sigma_n$, for instance, $[P_d^{\sigma_1}] = 1,\ldots[P_d^{\sigma_m}] = 1$, and $[P_e^{\sigma_{m+1}}] = 1,\ldots[P_h^{\sigma_n}] = 1$. Then the core property ascriptions to α, β and $\alpha\beta$ are according to the joint probability (6.35) simultaneously given by

$$
\left.
\begin{aligned}
[P_{a\cdots d}^\alpha] &= [P_a^{\sigma_1} \otimes \cdots \otimes P_d^{\sigma_m}] = 1, \\
[P_{e\cdots h}^\beta] &= [P_e^{\sigma_{m+1}} \otimes \cdots \otimes P_h^{\sigma_n}] = 1, \\
[P_{a\cdots de\cdots h}^{\alpha\beta}] &= [P_{a\cdots d}^\alpha \otimes P_{e\cdots h}^\beta] = 1.
\end{aligned}
\right\} \qquad (13.2)
$$

Assume that $[Q^\alpha \otimes Q^\beta] = 1$. It then follows that

$$(Q^\alpha \otimes Q^\beta) P^{\alpha\beta}_{a\cdots de\cdots h} = P^{\alpha\beta}_{a\cdots de\cdots h}. \tag{13.3}$$

Because $P^{\alpha\beta}_{a\cdots de\cdots h}$ is equal to $P^\alpha_{a\cdots d} \otimes P^\beta_{e\cdots h}$, it follows that

$$Q^\alpha P^\alpha_{a\cdots d} = P^\alpha_{a\cdots d} \quad \text{and} \quad Q^\beta P^\beta_{e\cdots h} = P^\beta_{e\cdots h}. \tag{13.4}$$

And given the simultaneous core property ascription (13.2), one can conclude that $[Q^\alpha] = 1$ and $[Q^\beta] = 1$. In a similar way one can prove that $[Q^\alpha] = 1$ and $[Q^\beta] = 1$ implies that $[Q^\alpha \otimes Q^\beta] = 1$.

In the MATHEMATICS it is proved that the atomic modal interpretation is also reductionistic with regard to the reducible properties $Q^\omega = \sum_j Q^\alpha_j \otimes Q^\beta_j \otimes \cdots$ if the sets $\{Q^\alpha_j\}_j$, $\{Q^\beta_j\}_j$, etc., are all sets of pair-wise orthogonal projections.

However, the atomic modal interpretation ceases to be reductionistic for more general reducible properties. Return for a second time to the example of the two particles α and β with the two energy levels E^α_1 and $E^\alpha_2 > E^\alpha_1$, and E^β_1 and $E^\beta_2 > E^\beta_1$, and let these particles be atoms. The property that $\alpha\beta$ has an energy strictly larger than $E^\alpha_1 + E^\beta_1$ is represented by $Q^{\alpha\beta} = Q^\alpha_1 \otimes Q^\beta_2 + Q^\alpha_2 \otimes Q^\beta_1 + Q^\alpha_2 \otimes Q^\beta_2$ and should be reducible to the pairs of properties $\{\langle Q^\alpha_1, Q^\beta_2 \rangle, \langle Q^\alpha_2, Q^\beta_1 \rangle, \langle Q^\alpha_2, Q^\beta_2 \rangle\}$. The projections Q^α_1 and Q^α_2 are orthogonal and Q^β_1 and Q^β_2 are orthogonal (they are the pair-wise orthogonal eigenprojections of the Hamiltonians of α and β, respectively). Hence $Q^{\alpha\beta}$ is reducible to three pairs $\{\langle Q^\alpha_j, Q^\beta_j \rangle\}^3_{j=1}$ for which it holds that the projections $\{Q^\alpha_j\}^3_{j=1}$ and $\{Q^\beta_j\}^3_{j=1}$ are not pair-wise orthogonal. Let the state of $\alpha\beta$ now be given by $|\Psi^{\alpha\beta}\rangle = |E^\alpha_2\rangle \otimes |\psi^\beta\rangle$, where $|\psi^\beta\rangle = c_1 |E^\beta_1\rangle + c_2 |E^\beta_2\rangle$ and where $Q^\alpha_2 |E^\alpha_2\rangle = |E^\alpha_2\rangle$, $Q^\beta_1 |E^\beta_1\rangle = |E^\beta_1\rangle$ and $Q^\beta_2 |E^\beta_2\rangle = |E^\beta_2\rangle$. The atomic core property ascription to $\alpha\beta$ then yields that $[|\Psi^{\alpha\beta}\rangle\langle\Psi^{\alpha\beta}|] = 1$ with probability 1. Since $Q^{\alpha\beta} |\Psi^{\alpha\beta}\rangle\langle\Psi^{\alpha\beta}| = |\Psi^{\alpha\beta}\rangle\langle\Psi^{\alpha\beta}|$, this yields that $\alpha\beta$ possesses $Q^{\alpha\beta}$ with probability 1. The core property ascription to β yields, however, that $[|\psi^\beta\rangle\langle\psi^\beta|] = 1$ with probability 1. And since it is generally not the case that $Q^\beta_1 |\psi^\beta\rangle\langle\psi^\beta| = |\psi^\beta\rangle\langle\psi^\beta|$ or that $Q^\beta_2 |\psi^\beta\rangle\langle\psi^\beta| = |\psi^\beta\rangle\langle\psi^\beta|$, it follows that β possesses neither Q^β_1 nor Q^β_2. Hence, the atomic modal interpretation does not (always) confirm that if $[Q^{\alpha\beta}] = 1$ with $Q^{\alpha\beta} = Q^\alpha_1 \otimes Q^\beta_2 + Q^\alpha_2 \otimes Q^\beta_1 + Q^\alpha_2 \otimes Q^\beta_2$, there exists a pair $\langle Q^\alpha, Q^\beta \rangle \in \{\langle Q^\alpha_1, Q^\beta_2 \rangle, \langle Q^\alpha_2, Q^\beta_1 \rangle, \langle Q^\alpha_2, Q^\beta_2 \rangle\}$ with $[Q^\alpha] = 1$ and $[Q^\beta] = 1$.

In the next sections I discuss the consequences of the violations of reductionism and holism. In the case of the bi and spectral modal interpretations I do so not by focusing directly on reductionism, but by focusing on the conditions of **Property Composition and Property Decomposition**. Property

---

Property Composition and Property Decomposition

An interpretation satisfies Property Composition if $[Q^\alpha] = x$ implies $[Q^\alpha \otimes \mathbb{I}^{\omega/\alpha}] = x$ (with x either 1 or 0) for any system α and any composite ω which contains α.

An interpretation satisfies Property Decomposition if $[Q^\alpha \otimes \mathbb{I}^{\omega/\alpha}] = x$ implies $[Q^\alpha] = x$ (with x either 1 or 0) for any system α and any composite ω which contains α.

Composition says that if a system α possesses a property Q^α, then any system ω containing α as a subsystem, simultaneously possesses the property represented by $Q^\alpha \otimes \mathbb{I}^{\omega/\alpha}$. And if α does not possess Q^α, then such a system ω simultaneously does not possess $Q^\alpha \otimes \mathbb{I}^{\omega/\alpha}$. This condition follows from two assumptions. The first is that (A) a property of α is also a property of a composite ω that contains α, and the property of α represented by the projection Q^α is, as a property of ω, represented by the projection $Q^\alpha \otimes \mathbb{I}^{\omega/\alpha}$. The second assumption is that (B) the property ascription to α is re-endorsed by the property ascription to ω. The condition of Property Decomposition says the converse, namely that if the composite ω possesses $Q^\alpha \otimes \mathbb{I}^{\omega/\alpha}$, then α simultaneously possesses Q^α. And if ω does not possess $Q^\alpha \otimes \mathbb{I}^{\omega/\alpha}$, then α simultaneously does not possess Q^α. This second condition follows from the assumption (A) and the converse of (B).

It can now be proved that a modal interpretation satisfies the conditions of Property Composition and Decomposition if and only if it is reductionistic with regard to the reducible properties $Q^\omega = Q^\alpha \otimes Q^\beta \otimes \cdots$ (see the MATHEMATICS). Hence, the bi and spectral modal interpretations violate Property Composition and Decomposition and the atomic modal interpretation satisfies these two conditions.

A last remark is that it may seem simple to the reader to define a modal interpretation which is both holistic and reductionistic. On the one hand, just ascribe to systems all the properties the bi or spectral modal interpretation ascribes such that one obtains a holistic interpretation, and then, on the other hand, add enough extra properties that it also becomes reductionistic.[143] Unfortunately, Bacciagaluppi (1995) and Clifton (1996)

---

[143] Healey's (1989) modal interpretation comes close to this outlined interpretation. Healey's property ascription satisfies by construction what he calls the Composition, the System Representative and the Subspace Decomposition conditions. By the Composition condition his interpretation always satisfies what I call Property Composition. And by the System Representative and Subspace Decomposition conditions it often ascribes the properties the bi modal interpretation ascribes (see Healey (1989, Sect. 2.2)). However, because the System Representative condition contains a proviso, it can happen that Healey's interpretation ascribes less holistic properties than the bi modal interpretation.

have proved that such a super modal interpretation yields an improper property ascription. Briefly summarised, Bacciagaluppi took a composite ω defined on a nine-dimensional Hilbert space and considered a number of factorisations $\omega = \alpha_i\beta_i$, $i = 1, 2, \ldots$, of ω in subsystems $\{\alpha_i\}_i$ and $\{\beta_i\}_i$ defined on three-dimensional Hilbert spaces. Then he considered the core properties $\{P_j^{\alpha_i}\}_{i,j}$ and $\{P_k^{\beta_i}\}_{i,k}$ ascribed to these subsystems by means of the bi modal interpretation. Next he ascribed these properties via Property Composition to ω (such that the property ascription is reducible with regard to the properties $\{P_j^{\alpha_i} \otimes P_k^{\beta_i}\}_{i,j,k}$ of ω). Finally Bacciagaluppi proved that all the properties ascribed to ω include the set of properties for which Kochen and Specker (1967) have shown that it does not allow a homomorphism to the set of values $\{0, 1\}$. Hence, the envisaged super modal interpretation ascribes a set of properties to ω which does not allow a proper value assignment.

One can block Bacciagaluppi's no-go result by denying that a composite ω can be freely factorised into pairs of subsystems. For instance, if one takes an atomistic view, it makes sense only to factorise ω into its atoms. Clifton (1996) has proved, however, that even then the super modal interpretation does not yield a proper property ascription. Clifton considered a composite ω defined on a 64-dimensional Hilbert space which is factorisable into only two subsystems α and β defined on eight-dimensional Hilbert spaces. By ascribing properties with the bi modal interpretation to α, β and ω and by employing Property Composition, he derived that ω possesses a set of properties similar to a set for which Kernaghan (1994) proved that it does not allow a homomorphism to $\{0, 1\}$.[144]

So, to conclude, there does not exist a super modal interpretation which integrates the holistic and reductionistic features of the bi, spectral and atomic modal interpretations. Instead one has to choose and settle for either a holistic or a reductionistic interpretation. In the next sections I discuss the consequences of this choice.

MATHEMATICS

To prove that the atomic modal interpretation is reductionistic with regard to the reducible properties $Q^\omega = \sum_j Q_j^\alpha \otimes Q_j^\beta \otimes \cdots$ if the sets $\{Q_j^\alpha\}_j$, $\{Q_j^\beta\}_j$, etc., are all sets of pair-wise orthogonal projections, I need:

Theorem 13.1
Let $\{\alpha, \beta, \gamma, \ldots\}$ be N mutually disjoint systems and let $\{Q_j^\alpha\}_j$, $\{Q_j^\beta\}_j$, $\{Q_j^\gamma\}_j$,

---

[144] See Bacciagaluppi and Vermaas (1999) and Bacciagaluppi (2000) for a more extensive discussion of the no-go results by Bacciagaluppi and by Clifton.

etc., be sets of pair-wise orthogonal projections. Then one has that

$$\left(\sum_j Q_j^\alpha \otimes Q_j^\beta \otimes \cdots\right)(\widehat{Q}^\alpha \otimes \widehat{Q}^\beta \otimes \cdots) = \widehat{Q}^\alpha \otimes \widehat{Q}^\beta \otimes \cdots \tag{13.5}$$

if and only if

$$\exists !\langle Q_k^\alpha, Q_k^\beta, \ldots\rangle \in \{\langle Q_j^\alpha, Q_j^\beta, \ldots\rangle\}_j : \quad Q_k^\alpha \widehat{Q}^\alpha = \widehat{Q}^\alpha, \quad Q_k^\beta \widehat{Q}^\beta = \widehat{Q}^\beta, \text{ etc.} \tag{13.6}$$

Proof: The 'if' part is trivial, so consider the 'only if' part. Take, firstly, the case that $N = 2$, so assume that

$$\left(\sum_j Q_j^\alpha \otimes Q_j^\beta\right)(\widehat{Q}^\alpha \otimes \widehat{Q}^\beta) = \widehat{Q}^\alpha \otimes \widehat{Q}^\beta. \tag{13.7}$$

Multiply this relation from the left-hand side with $\mathbb{I}^\alpha \otimes Q_k^\beta$. Because the projections $\{Q_j^\beta\}_j$ are pair-wise orthogonal, one obtains

$$Q_k^\alpha \widehat{Q}^\alpha \otimes Q_k^\beta \widehat{Q}^\beta = \widehat{Q}^\alpha \otimes Q_k^\beta \widehat{Q}^\beta, \tag{13.8}$$

from which it follows that $(Q_k^\alpha \widehat{Q}^\alpha - \widehat{Q}^\alpha) \otimes Q_k^\beta \widehat{Q}^\beta = 0$ and that

$$Q_k^\alpha \widehat{Q}^\alpha = \widehat{Q}^\alpha \quad \text{or} \quad Q_k^\beta \widehat{Q}^\beta = 0. \tag{13.9}$$

And if one multiplies (13.7) from the left-hand side with $Q_k^\alpha \otimes \mathbb{I}^\beta$, one obtains similarly

$$Q_k^\alpha \widehat{Q}^\alpha = 0 \quad \text{or} \quad Q_k^\beta \widehat{Q}^\beta = \widehat{Q}^\beta. \tag{13.10}$$

One can show by these results that the pair $\langle Q_k^\alpha, Q_k^\beta \rangle$ satisfies

$$[Q_k^\alpha \widehat{Q}^\alpha = \widehat{Q}^\alpha \text{ and } Q_k^\beta \widehat{Q}^\beta = \widehat{Q}^\beta] \quad \text{or} \quad [Q_k^\alpha \widehat{Q}^\alpha = 0 \text{ and } Q_k^\beta \widehat{Q}^\beta = 0]. \tag{13.11}$$

Assume, firstly, that $Q_k^\alpha \widehat{Q}^\alpha = \widehat{Q}^\alpha$. By (13.10) it then follows that $Q_k^\beta \widehat{Q}^\beta = \widehat{Q}^\beta$. Assume, secondly, that $Q_k^\alpha \widehat{Q}^\alpha = 0$. By (13.9) it then follows that $Q_k^\beta \widehat{Q}^\beta = 0$. Assume, finally, that $0 \neq Q_k^\alpha \widehat{Q}^\alpha \neq \widehat{Q}^\alpha$. By (13.9) it then follows that $Q_k^\beta \widehat{Q}^\beta = 0$, and by (13.10) it follows that $Q_k^\beta \widehat{Q}^\beta = \widehat{Q}^\beta$. This is contradictory, so $Q_k^\alpha \widehat{Q}^\alpha$ is equal to either 0 or $\widehat{Q}^\alpha$.

Return to relation (13.7) and assume that there does not exist a single pair $\langle Q_k^\alpha, Q_k^\beta \rangle \in \{\langle Q_j^\alpha, Q_j^\beta \rangle\}_j$ for which it holds that $Q_k^\alpha \widehat{Q}^\alpha = \widehat{Q}^\alpha$ and $Q_k^\beta \widehat{Q}^\beta = \widehat{Q}^\beta$. This implies that the left-hand side of (13.7) is equal to 0, which contradicts (13.7). Assume, secondly, that there is exactly one pair $\langle Q_k^\alpha, Q_k^\beta \rangle$ for which it holds that $Q_k^\alpha \widehat{Q}^\alpha = \widehat{Q}^\alpha$ and $Q_k^\beta \widehat{Q}^\beta = \widehat{Q}^\beta$. This is consistent with (13.7). And assume, finally, that there exist two or more pairs $\langle Q_k^\alpha, Q_k^\beta \rangle$ with $Q_k^\alpha \widehat{Q}^\alpha = \widehat{Q}^\alpha$ and $Q_k^\beta \widehat{Q}^\beta = \widehat{Q}^\beta$. The left-hand side of (13.7) is then equal to $2\widehat{Q}^\alpha \otimes \widehat{Q}^\beta$, which again contradicts (13.7). Hence, given relation (13.7), one has that

$$\exists! \langle Q_k^\alpha, Q_k^\beta \rangle \in \{ \langle Q_j^\alpha, Q_j^\beta \rangle \}_j : \quad Q_k^\alpha \hat{Q}^\alpha = \hat{Q}^\alpha \text{ and } Q_k^\beta \hat{Q}^\beta = \hat{Q}^\beta. \tag{13.12}$$

Take, secondly, the case with $N > 2$. The projections $\{ Q_j^\beta \otimes Q_j^\gamma \otimes \cdots \}_j$ in (13.5) are pair-wise orthogonal, so using that (13.7) implies (13.12), one can conclude that (13.5) implies

$$\exists! \langle Q_k^\alpha, Q_k^\beta \otimes Q_k^\gamma \otimes \cdots \rangle \in \{ \langle Q_j^\alpha, Q_j^\beta \otimes Q_j^\gamma \otimes \cdots \rangle \}_j :$$
$$Q_k^\alpha \hat{Q}^\alpha = \hat{Q}^\alpha \text{ and } (Q_k^\beta \otimes Q_k^\gamma \otimes \cdots)(\hat{Q}^\beta \otimes \hat{Q}^\gamma \otimes \cdots) = \hat{Q}^\beta \otimes \hat{Q}^\gamma \otimes \cdots. \tag{13.13}$$

The relation $(Q_k^\beta \otimes Q_k^\gamma \otimes \cdots)(\hat{Q}^\beta \otimes \hat{Q}^\gamma \otimes \cdots) = \hat{Q}^\beta \otimes \hat{Q}^\gamma \otimes \cdots$ implies in its turn that $Q_k^\beta \hat{Q}^\beta = \hat{Q}^\beta$, $Q_k^\gamma \hat{Q}^\gamma = \hat{Q}^\gamma$, etc., and so one obtains (13.6). $\qquad \square$

Let's now apply the atomic modal interpretation to the property $Q^\omega = \sum_j Q_j^\alpha \otimes Q_j^\beta \otimes \cdots$, where the sets $\{Q_j^\alpha\}_j$, $\{Q_j^\beta\}_j$, etc., are all sets of pair-wise orthogonal projections. Let ω consist of the atoms $\{\sigma_{pq}\}_{p,q}$, let α consist of the atoms $\{\sigma_{1q}\}_q$ and let β consist of the atoms $\{\sigma_{2q}\}_q$, etc. Consider any simultaneous core property ascription to the atoms $\{\sigma_{pq}\}_{p,q}$. For instance, $[P_{a_{pq}}^{\sigma_{pq}}] = 1$ for all p, q. Then the core property ascriptions to α, β, ..., and ω are, according to the joint probability (6.35), simultaneously given by, respectively,

$$\left.\begin{aligned} [P_{a_{11}a_{12}a_{13}\cdots}^\alpha] &= [P_{a_{11}}^{\sigma_{11}} \otimes P_{a_{12}}^{\sigma_{12}} \otimes P_{a_{13}}^{\sigma_{13}} \otimes \cdots] = 1, \\ [P_{a_{21}a_{22}a_{23}\cdots}^\beta] &= [P_{a_{21}}^{\sigma_{21}} \otimes P_{a_{22}}^{\sigma_{22}} \otimes P_{a_{23}}^{\sigma_{23}} \otimes \cdots] = 1, \text{ etc.,} \\ [P_{a_{11}a_{12}a_{13}\cdots a_{21}a_{22}a_{23}\cdots}^\omega] &= [P_{a_{11}a_{12}a_{13}}^\alpha \otimes P_{a_{21}a_{22}a_{23}\cdots}^\beta \otimes \cdots] = 1. \end{aligned}\right\} \tag{13.14}$$

Assume that $Q^\omega = \sum_j Q_j^\alpha \otimes Q_j^\beta \otimes \cdots$ is assigned value 1. It then follows that

$$\left(\sum_j Q_j^\alpha \otimes Q_j^\beta \otimes \cdots\right) P_{a_{11}a_{12}\cdots a_{21}a_{22}\cdots}^\omega = P_{a_{11}a_{12}\cdots a_{21}a_{22}\cdots}^\omega. \tag{13.15}$$

Since $P_{a_{11}a_{12}\cdots a_{21}a_{22}\cdots}^\omega$ is equal to $P_{a_{11}a_{12}\cdots}^\alpha \otimes P_{a_{21}a_{22}\cdots}^\beta \otimes \cdots$, one has by Theorem 13.1 and (13.15) that there exists exactly one set of properties $\langle Q_k^\alpha \otimes Q_k^\beta \otimes \cdots \rangle \in \{ \langle Q_j^\alpha \otimes Q_j^\beta \otimes \cdots \rangle \}_j$ for which it holds that

$$Q_k^\alpha P_{a_{11}a_{12}\cdots}^\alpha = P_{a_{11}a_{12}\cdots}^\alpha, \quad Q_k^\beta P_{a_{21}a_{22}\cdots}^\beta = P_{a_{21}a_{22}\cdots}^\beta, \quad \text{etc.} \tag{13.16}$$

Given the simultaneous core property ascription (13.14), one can conclude that $[Q_k^\alpha] = 1$ and $[Q_k^\beta] = 1$, etc. Hence, if $Q^\omega = \sum_j Q_j^\alpha \otimes Q_j^\beta \otimes \cdots$ is assigned value 1, there exist properties $\langle Q_k^\alpha \otimes Q_k^\beta \otimes \cdots \rangle \in \{ \langle Q_j^\alpha \otimes Q_j^\beta \otimes \cdots \rangle \}_j$ with $[Q_k^\alpha] = 1$ and $[Q_k^\beta] = 1$, etc.

Conversely, if there exists properties $\langle Q_k^\alpha \otimes Q_k^\beta \otimes \cdots \rangle \in \{\langle Q_j^\alpha \otimes Q_j^\beta \otimes \cdots \rangle\}_j$ with $[Q_k^\alpha] = 1$ and $[Q_k^\beta] = 1$, etc., then one can prove by means of the simultaneous core property ascription (13.14) that $Q^\omega = \sum_j Q_j^\alpha \otimes Q_j^\beta \otimes \cdots$ is assigned value 1. Hence, the atomic modal interpretation is reductionistic with regard to $Q^\omega = \sum_j Q_j^\alpha \otimes Q_j^\beta \otimes \cdots$, if $\{Q_j^\alpha\}_j$, $\{Q_j^\beta\}_j$, etc., are sets of pair-wise orthogonal projections.

Finally I prove that modal interpretations satisfy Property Composition and Decomposition if and only if they are reductionistic with regard to the properties $Q^\omega = Q^\alpha \otimes Q^\beta \otimes \cdots$.

Proof: Take a modal interpretation that satisfies Property Composition and Decomposition and consider a property $Q^{\alpha\beta} = Q^\alpha \otimes Q^\beta$ (I consider only properties with two factors; the proof can easily be generalised to properties with three or more factors). If $\alpha\beta$ possesses $Q^{\alpha\beta}$, it also possesses $Q^\alpha \otimes \mathbb{I}^\beta$ and $\mathbb{I}^\alpha \otimes Q^\beta$ since for modal interpretations the full property ascription satisfies the Weakening condition (see page 72). Application of Property Decomposition yields that α possesses Q^α and β possesses Q^β. Conversely, if α possesses Q^α and β possesses Q^β, application of Property Composition yields that $\alpha\beta$ possesses $Q^\alpha \otimes \mathbb{I}^\beta$ and $\mathbb{I}^\alpha \otimes Q^\beta$. And because the full property ascription satisfies the Closure condition (see page 68), $\alpha\beta$ also possesses the conjunction $Q^\alpha \otimes \mathbb{I}^\beta \wedge \mathbb{I}^\alpha \otimes Q^\beta = Q^{\alpha\beta}$. Hence, if a modal interpretation satisfies Property Composition and Decomposition, it is reductionistic with regard to $Q^{\alpha\beta} = Q^\alpha \otimes Q^\beta$.

Take, secondly, a modal interpretation which is reductionistic with regard to $Q^\omega = Q^\alpha \otimes Q^\beta \otimes \cdots$. Let α possess property Q^α and consider any composite ω which comprises α, say $\omega = \alpha\beta$. According to modal interpretations β possesses the property $\mathbb{I}^\beta$. By using reductionism, one can thus infer that $\alpha\beta$ possesses $Q^\alpha \otimes \mathbb{I}^\beta$. So, $[Q^\alpha] = 1$ implies $[Q^\alpha \otimes \mathbb{I}^\beta] = 1$. Let α now not possess Q^α. Since the full property ascription satisfies the Closure condition, α does possess the negation $\neg Q^\alpha = \mathbb{I}^\alpha - Q^\alpha$ and one can conclude that $\alpha\beta$ possesses $(\mathbb{I}^\alpha - Q^\alpha) \otimes \mathbb{I}^\beta$. The full property ascription to $\alpha\beta$ also satisfies Closure, so $\alpha\beta$ does not possess the negation of $(\mathbb{I}^\alpha - Q^\alpha) \otimes \mathbb{I}^\beta$. Hence, $[Q^\alpha] = 0$ implies $[Q^\alpha \otimes \mathbb{I}^\beta] = 0$. A modal interpretation which is reductionistic with regard to $Q^\omega = Q^\alpha \otimes Q^\beta \otimes \cdots$ thus satisfies Property Composition.

Now let $\alpha\beta$ possess $Q^\alpha \otimes \mathbb{I}^\beta$. Reductionism then immediately yields that α possesses Q^α, so $[Q^\alpha \otimes \mathbb{I}^\beta] = 1$ implies $[Q^\alpha] = 1$. Let, finally, $\alpha\beta$ not possess $Q^\alpha \otimes \mathbb{I}^\beta$. By using Closure it follows that $\alpha\beta$ possesses $(\mathbb{I}^\alpha - Q^\alpha) \otimes \mathbb{I}^\beta$, by using reductionism it follows that α possesses $\mathbb{I}^\alpha - Q^\alpha$ and by again using Closure, it follows that α does not possess Q^α. So, $[Q^\alpha \otimes \mathbb{I}^\beta] = 0$ implies

$[Q^\alpha] = 0$. A modal interpretation which is reductionistic with regard to $Q^\omega = Q^\alpha \otimes Q^\beta \otimes \cdots$, thus also satisfies Property Decomposition. $\qquad\square$

13.3 Holism with observational reductionism

If one accepts the bi or spectral modal interpretations, one does not have reductionism and, consequently, one does not have Property Composition and Decomposition. The question is how one should judge this absence. In the literature Arntzenius (1990, 1998) and Clifton (1995c, 1996) especially took the position that this absence makes the bi and spectral modal interpretations metaphysically untenable. In Vermaas (1998c) I tried to counter this conclusion. The following discussion focuses on the violations of Property Composition and Decomposition.

So, consider the property ascription to a system α and to a composite $\alpha\beta$ and let's try to make sense of the fact that the bi and spectral modal interpretations sometimes assign different values to the projections Q^α and $Q^\alpha \otimes \mathbb{I}^\beta$. As I said in the previous section, Property Composition follows from two assumptions, namely that (A) a property of α is also a property of a composite $\alpha\beta$, where the property of α represented by Q^α is, as a property of $\alpha\beta$, represented by $Q^\alpha \otimes \mathbb{I}^\beta$, and that (B) the property ascription to α is re-endorsed by the property ascription to $\alpha\beta$. Property Decomposition follows from assumption (A) and from the converse of (B), namely that (B') the property ascription to $\alpha\beta$ is re-endorsed by the property ascription to α. If one clings to these assumptions, the assignment of different values to Q^α and $Q^\alpha \otimes \mathbb{I}^\beta$ leads straightforwardly to a contradiction: Let the property ascription to α yield that Q^α has value 1 and let the property ascription to $\alpha\beta$ yield that $Q^\alpha \otimes \mathbb{I}^\beta$ has no definite value. Then α possesses the property Q^α and from (A) it follows that $\alpha\beta$ possesses the (same) property $Q^\alpha \otimes \mathbb{I}^\beta$. Hence, $[Q^\alpha \otimes \mathbb{I}^\beta] = 1$ which, according to (B), should be re-endorsed by the property ascription to $\alpha\beta$. However, this contradicts that the property ascripiton to $\alpha\beta$ yields that $Q^\alpha \otimes \mathbb{I}^\beta$ has no definite value. Therefore, to save the bi and spectral modal interpretations from being inconsistent, one should reject (A), reject (B) and (B'), or reject all three assumptions.

If one adopts perspectivalism (see Sections 4.3 and 4.4), then it is possible to reject assumptions (B) and (B'), for then one accepts that the ascription of properties to systems depends on the perspective with which one considers those systems. And since there does not exist a perspective with which one can simultaneously ascribe properties to both α and $\alpha\beta$ (a division of the universe ω into disjoint systems cannot yield both the systems α and $\alpha\beta$), (B) and (B') relate property ascriptions defined from different perspectives.

However, perspectivalism implies that property ascriptions defined from different perspectives are not related to one another, so one can in a natural way deny (B) and (B′). And if one does so, the fact that Q^α and $Q^\alpha \otimes \mathbb{I}^\beta$ are sometimes assigned different values only reflects that if one compares the properties ascribed from different perspectives, one gets different results. Let the property ascription to α yield that $[Q^\alpha] = 1$ and let the property ascription to $\alpha\beta$ yield that $Q^\alpha \otimes \mathbb{I}^\beta$ has no definite value. Then from the perspective 'ω divided by $\alpha, \beta, \gamma, \ldots$,' from which the property ascription to α is defined, one can say that α possesses Q^α and, by assumption (A), that $\alpha\beta$ possesses $Q^\alpha \otimes \mathbb{I}^\beta$. And from the perspective 'ω divided by $\alpha\beta, \gamma, \ldots$' one can say that the property $Q^\alpha \otimes \mathbb{I}^\beta$ is indefinite for $\alpha\beta$ and, by (A), that Q^α is indefinite for α. But these different property ascriptions do not lead to a contradiction; they only confirm that the property ascription depends on perspectives.[145]

If one does not adopt perspectivalism, it is much harder to deny (B) and (B′) because then one has to deal with two different sources to ascribe properties to, say, $\alpha\beta$. The first is the property ascription of the bi and spectral modal interpretations directly applied to $\alpha\beta$ and the second is the property ascription applied to α and via assumption (A) mapped to $\alpha\beta$. One might now take the position that these two sources are independent sources of information about the properties of $\alpha\beta$ which need not re-endorse one another. However, a consequence of this position is that $\alpha\beta$ possesses both the properties ascribed directly to $\alpha\beta$, and the properties obtained by applying assumption (A) to the properties ascribed to α. This integrated property ascription leads, however, to the super modal interpretation discussed in the previous section and Bacciagaluppi and Clifton proved that that modal interpretation does not allow a proper property ascription.

So, if one does not adopt perspectivalism, one has to deny assumption (A), that is, one has to deny that the projections Q^α and $Q^\alpha \otimes \mathbb{I}^\beta$ represent the same property. However, such a denial brings one into immediate conflict with what is seen as a basic tenet of quantum mechanics, namely that the operators A^α and $A^\alpha \otimes \mathbb{I}^\beta$ represent the same magnitude (see, for instance, Healey (1989, pages 231–232)). The question thus becomes whether it is

---

[145] Note that if one adopts perspectivalism, denies (B) and (B′) but accepts (A), then not only the correlations between properties but also the properties of systems themselves are perspective-dependent. For in this case, as shown above, one ascribes from the perspective 'ω divided by $\alpha, \beta, \gamma, \ldots$' other properties to, say, α than from the perspective 'ω divided by $\alpha\beta, \gamma, \ldots$' (from the first perspective α possesses Q^α whereas from the second perspective Q^α is indefinite). If one wants to avoid such perspective-dependence of the properties themselves, one should not only deny (B) and (B′) but also (A). In that case α can only be ascribed properties from the perspectives that divide ω into α and other systems disjoint from α. From all these perspectives α has the same set of possible possessed properties (see also the discussion in Section 14.2).

possible to reject this tenet. There are various arguments in support of the tenet which roughly say that one cannot distinguish the magnitudes represented by A^α and by $A^\alpha \otimes \mathbb{I}^\beta$ by means of measurements.[146] Firstly, quantum mechanics in the standard formulation predicts that the possible outcomes of measurements of A^α and $A^\alpha \otimes \mathbb{I}^\beta$ correspond to the same values (A^α and $A^\alpha \otimes \mathbb{I}^\beta$ have the same eigenvalues $\{a_j\}_j$). Secondly, for every state of the universe ω it holds that the Born probability $p_{\text{Born}}(a_a)$ of finding an outcome corresponding to the eigenvalue a_a after an A^α measurement is equal to the Born probability $p_{\text{Born}}(a_a)$ of finding an outcome corresponding to that same eigenvalue after an $A^\alpha \otimes \mathbb{I}^\beta$ measurement (if P_a^α is the eigenprojection of A^α corresponding to a_a, then $P_a^\alpha \otimes \mathbb{I}^\beta$ is the eigenprojection of $A^\alpha \otimes \mathbb{I}^\beta$ corresponding to a_a; the respective Born probabilities are thus $\text{Tr}^\alpha(\text{Tr}^{\omega/\alpha}(W^\omega) P_a^\alpha)$ and $\text{Tr}^{\alpha\beta}(\text{Tr}^{\omega/\alpha\beta}(W^\omega) [P_a^\alpha \otimes \mathbb{I}^\beta])$, which are obviously equal). And, thirdly, it is assumed that any model of a measurement of A^α also counts as a model of a measurement of $A^\alpha \otimes \mathbb{I}^\beta$ and *vice versa*.

Now, I think that it is possible to question the assumption that any model of an $A^\alpha \otimes \mathbb{I}^\beta$ measurement is also a model of an A^α measurement (see the MATHEMATICS). But the above points certainly make a strong case that at the level of observation (that is, at the level of measurement outcomes) one cannot notice a difference between the magnitudes represented by A^α and by $A^\alpha \otimes \mathbb{I}^\beta$. In addition, the spectral modal interpretation[147] confirms this observational indistinguishability of A^α and $A^\alpha \otimes \mathbb{I}^\beta$. Consider for simplicity the magnitudes represented by the projections Q^α and $Q^\alpha \otimes \mathbb{I}^\beta$ and measure them by means of perfect Von Neumann measurements in an arbitrary order. Assuming that the state of β does not evolve during the Q^α measurement, one can prove the following proposition within the spectral modal interpretation.

Proposition 13.1
If a Q^α measurement has a positive or a negative outcome, then a subsequent $Q^\alpha \otimes \mathbb{I}^\beta$ measurement has with probability 1 also a positive or a negative outcome.

If a $Q^\alpha \otimes \mathbb{I}^\beta$ measurement has a positive or a negative outcome, then a subsequent Q^α measurement has with probability 1 also a positive or a negative outcome.

If one accepts the assumption that a measurement of A^α is automatically

---

[146] This discussion is not intended to do justice to all the arguments in favour of the tenet; I am only exploring whether it is possible to distinguish the magnitudes represented by A^α and by $A^\alpha \otimes \mathbb{I}^\beta$.

[147] It is unclear whether what follows also holds for the bi modal interpretation; the truth of Proposition 13.1 cannot be established within the bi modal interpretation because this interpretation does not, in general, correlate the outcomes of two different measurements.

a measurement of $A^\alpha \otimes \mathbb{I}^\beta$ and *vice versa*, then this proposition is trivial since Von Neumann measurements, when repeated, yield with probability 1 the same result. However, if one rejects this assumption, Proposition 13.1 becomes less trivial but still holds.[148]

However, this observational indistinguishability of the magnitudes represented by A^α and by $A^\alpha \otimes \mathbb{I}^\beta$ does not necessarily force one also to take these magnitudes from a theoretical point of view as indistinguishable. In fact quantum mechanics makes a clear distinction between these two magnitudes. The operator A^α is defined on the Hilbert space $\mathcal{H}^\alpha$ associated with the system α so the magnitude represented by A^α is according to quantum mechanics, a magnitude pertaining to α. And $A^\alpha \otimes \mathbb{I}^\beta$ is defined on the Hilbert space $\mathcal{H}^{\alpha\beta}$ associated with $\alpha\beta$ and thus represents a magnitude of $\alpha\beta$. The operators A^α and $A^\alpha \otimes \mathbb{I}^\beta$ thus represent magnitudes pertaining to different systems so that they cannot, strictly speaking, represent the same magnitude. Hence, it follows that the tenet that A^α and $A^\alpha \otimes \mathbb{I}^\beta$ represent the same magnitude can be viewed as an addition to quantum mechanics, which can be accepted but which can also be denied as, for instance, Van Fraassen (1991, Sect. 9.4) does.

Let's return to assumption (A). As I have argued, one can deny that A^α and $A^\alpha \otimes \mathbb{I}^\beta$ represent the same magnitude so that one can deny also that Q^α and $Q^\alpha \otimes \mathbb{I}^\beta$ represent the same property. On the other hand, if one indeed takes such a position, one should acknowledge that at the level of observation one still cannot distinguish the properties Q^α and $Q^\alpha \otimes \mathbb{I}^\beta$. It is, for instance, quite impossible to assume that Q^α represents the energy of α and that $Q^\alpha \otimes \mathbb{I}^\beta$ represents the position of $\alpha\beta$. Instead one should demand that the properties Q^α and $Q^\alpha \otimes \mathbb{I}^\beta$ are both different from a theoretical point of view and indistinguishable from an observational point of view.

Arntzenius (1990, page 245), when discussing the violation of Property Composition in the bi modal interpretation, put forward a description of the properties represented by Q^α and $Q^\alpha \otimes \mathbb{I}^\beta$. He considered the left-hand side of a table (system α) and the table as whole (system $\alpha\beta$) and he let Q^α represent

---

[148] In Vermaas (1998c, App. A) one can find an explicit proof of Proposition 13.1. The Von Neumann measurements of Q^α and $Q^\alpha \otimes \mathbb{I}^\beta$ are given by the respective interactions:

$$|q_{jk}^\alpha\rangle \otimes |\mathrm{D}_0^{\bar\mu}\rangle \otimes |\mathrm{R}_0^\pi\rangle \longmapsto |q_{jk}^\alpha\rangle \otimes |\mathrm{D}_{jk}^{\bar\mu}\rangle \otimes |\mathrm{R}_{jk}^\pi\rangle,$$

$$|\bar q_{lm}^{\alpha\beta}\rangle \otimes |\mathrm{D}_0^{\bar\mu}\rangle \otimes |\mathrm{R}_0^\pi\rangle \longmapsto |\bar q_{lm}^{\alpha\beta}\rangle \otimes |\mathrm{D}_{lm}^{\bar\mu}\rangle \otimes |\mathrm{R}_{lm}^\pi\rangle.$$

Here, the vectors $\{|q_{jk}^\alpha\rangle\}_{j,k}$ are given by the spectral resolution $Q^\alpha = \sum_{j=1}^2 \sum_k q_j^\alpha |q_{jk}^\alpha\rangle\langle q_{jk}^\alpha|$, where $q_1^\alpha = 1$ and $q_2^\alpha = 0$. And the vectors $\{|\bar q_{lm}^{\alpha\beta}\rangle\}_{l,m}$ are given by the spectral resolution $Q^\alpha \otimes \mathbb{I}^\beta = \sum_{l=1}^2 \sum_m \bar q_l^{\alpha\beta} |\bar q_{lm}^{\alpha\beta}\rangle\langle \bar q_{lm}^{\alpha\beta}|$, where $\bar q_1^{\alpha\beta} = 1$ and $\bar q_2^{\alpha\beta} = 0$.

More briefly one can note that since Proposition 13.1 is definitely true in the standard formulation, this proposition is by the results of Chapter 11 also true in the spectral modal interpretation.

the property 'greenness' of the object 'the left-hand side of the table' and $Q^\alpha \otimes \mathbb{I}^\beta$ represent the property 'greenness of the left-hand side' of the object 'table'. This description of the properties Q^α and $Q^\alpha \otimes \mathbb{I}^\beta$ meets the demands of being different and of being observationally indistinguishable: From a logical point of view Q^α represents the proposition 'The left-hand side of the table is green' and $Q^\alpha \otimes \mathbb{I}^\beta$ represents the proposition 'The table is green at the left-hand side.' And these propositions can be analysed as predicating two different predicates, 'green' and 'green at the left-hand side,' respectively, to two different individuals, 'the left-hand side of the table' and the 'the table as a whole,' respectively. Hence, logically speaking (the propositions represented by) Q^α and $Q^\alpha \otimes \mathbb{I}^\beta$ can be taken as different. On the other hand, in daily life one normally does not distinguish between these propositions.

If one indeed accepts that Q^α and $Q^\alpha \otimes \mathbb{I}^\beta$ represent different properties, then the fact that these projections are sometimes assigned different values reflects (trivially) that they indeed do represent different properties.

To sum up, one can save the bi and spectral modal interpretations from being inconsistent by either adopting perspectivalism and rejecting that the property ascriptions to α and $\alpha\beta$ re-endorse one another, or rejecting that the properties Q^α and $Q^\alpha \otimes \mathbb{I}^\beta$ represent the same property. This, however, still leaves us with the fact that the bi and spectral modal interpretations violate Property Composition and Decomposition. Both Arntzenius and Clifton argued that these violations themselves challenge the tenability of these two interpretations.

Arntzenius (1990, page 245) discussed the violation of Property Composition and judged this violation to be *bizarre* since it assigns different truth values to propositions like 'the left-hand side of a table is green' and 'the table has a green left-hand side' which are normally not distinguished. I agree with Arntzenius: the property ascription by the bi and spectral modal interpretations is certainly at odds with our every-day notions about properties. However, it is still open to debate whether such bizarreness makes these two interpretations untenable.

In Section 12.1 I have argued that an interpretation should meet the criteria of Consistency and Internal Completeness in order to be metaphysically tenable, and that an interpretation should meet the criterion of Empirical Adequacy with regard to the phenomena in order to be empirically adequate. As discussed, the violations of Property Composition and Decomposition need not reveal an inconsistency. Furthermore, the bi and spectral modal interpretations never claimed to deliver Property Composition and Decomposition, so the violations also do not reveal an internal incompleteness. That leaves us with the criterion of Empirical Adequacy.

I believe it is indeed possible to take Property Composition and Decomposition as instances of the criterion of Empirical Adequacy because we usually observe that composites and their parts simultaneously possess Q^α and $Q^\alpha \otimes \mathbb{I}^\beta$. But because Proposition 13.1 holds for at least the spectral modal interpretation, Property Composition and Decomposition are not violated at the level of observation (taken in the limited sense of the observation of outcomes of measurements). Hence, one can conclude that the spectral modal interpretation *observationally* satisfies Property Composition and Decomposition and that it is consequently *observationally reductionistic* (with regard to the reducible properties $Q^\omega = Q^\alpha \otimes Q^\beta \otimes \cdots$).

Hence, given my liberal conception of tenable interpretations of quantum mechanics, the violations of Property Composition and Decomposition do not prove that the spectral modal interpretation is metaphysically untenable nor that it is empirically inadequate (whether this also holds for the bi modal interpretation is difficult to say since this interpretation is silent about Proposition 13.1). Still I agree with Arntzenius when he judges that these violations are bizarre. But if indeed these violations never manifest themselves at the level of observation, this bizarreness is restricted only to the description of non-observable or noumenal states of affairs. And I can live with the conclusion that the description of noumenal states of affairs is on occasions bizarre in the sense of abnormal when compared with the description of phenomenal states of affairs. By definition we have never observed a noumenal state of affairs, so I do not see it as a problem to accept that the description of such a state of affairs deviates from the description of the phenomenal states of affairs which we can observe.

Clifton (1996, Sect. 2.3), however, developed a telling example in which it seems that the violations of Property Composition and Decomposition do have implications which exceed the realm of the noumenal states of affairs. The example is a plane with a possibly warped left-hand wing: α is the left-hand wing and $\alpha\beta$ is the plane as a whole (a Boeing 747). The projection Q^α represents the wing property of being warped and $Q^\alpha \otimes \mathbb{I}^\beta$ represents the plane property of the left-hand wing being warped. The implication of a violation of Property Composition ($[Q^\alpha] = 1$ and $[Q^\alpha \otimes \mathbb{I}^\beta] \neq 1$) is according to Clifton now that '*a pilot could still be confident flying in the 747 despite the fault in its left-hand wing.*' If, on the other hand, Property Decomposition fails ($[Q^\alpha \otimes \mathbb{I}^\beta] = 1$ and $[Q^\alpha] \neq 1$) the implication is '*certainly no one would fly in the 747; but, then again, a mechanic would be hard-pressed to locate any flaw in its left-hand wing.*'[149] It can, however, be argued that the bi

---

[149] Quotations from Clifton (1996, page 385).

and spectral modal interpretations are not necessarily committed to these implications. The argument goes as follows.

Let's assume that either the bi or the spectral modal interpretation is the correct interpretation of quantum mechanics and let's assume that assumption (A) does not hold (if one prefers to reject (B) and (B'), the following holds *mutatis mutandis*). That is, assume that A^α and $A^\alpha \otimes \mathbb{I}^\beta$ do not represent the same magnitudes and assume that reality satisfies the description given by one of these interpretations. Then it follows, that we, the inhabitants of this 'bi' or 'spectral' world, will have become accustomed to the fact that in our world the conditions of Property Composition and Decomposition hold for observed systems, but that theoretical descriptions of non-observed systems have proved that these conditions do not hold in general. Hence, the inhabitants of the 'bi' or 'spectral' world, will reasonably use Property Composition and Decomposition not as a generally valid law about their world, but as a rule which applies to observations only. Consequently, they will take the questions of whether α and $\alpha\beta$ possess A^α and $A^\alpha \otimes \mathbb{I}^\beta$, respectively, as two separate questions; only the questions about the outcomes of measurements of A^α and $A^\alpha \otimes \mathbb{I}^\beta$ will be regarded as equivalent questions. Clifton's alleged implications of the violation of Property Composition and Decomposition now only arise because he assumes that people still take Property Composition and Decomposition as general truths. For instance, the confidence of Clifton's pilot that, despite the fault, the left-hand wing is fine rests on a deduction by means of Property Decomposition: because it holds for the plane as a whole that $[Q^\alpha \otimes \mathbb{I}^\beta] \neq 1$, Clifton's pilot concludes that $[Q^\alpha] \neq 1$ for the left-hand wing. And that is exactly what a pilot would not do if the bi and spectral modal interpretations correctly describe reality. A pilot who is well trained in 'bi' or 'spectral' ontology acknowledges that the properties of the plane as a whole do not reveal information about the properties of the wings and *vice versa*, and therefore will check the properties of the wings independently of the property ascription to the plane as a whole. And even in our world most cockpits are actually equipped with instruments which directly reveal the properties of the parts of the plane.

Consider, secondly, the mechanic in Clifton's example. The pilot notices that the plane as a whole possesses the property $[Q^\alpha \otimes \mathbb{I}^\beta] = 1$ and concludes (incorrectly, since the conclusion is reached on the basis of Property Decomposition) that the left-hand wing is warped, that is, that $[Q^\alpha] = 1$. The mechanic is sent to fix the wing but according to Clifton cannot locate the flaw because the wing does not possess the property Q^α. But if the mechanic is a skilled 'bi' or 'spectral' mechanic, he or she knows how to handle the pilot's report. Within at least the spectral modal interpretation it can be proved, given an assumption,[150] that for, for instance, measurements of the second kind, the following proposition holds

Proposition 13.2
If the property ascription to $\alpha\beta$ yields that $[Q^\alpha \otimes \mathbb{I}^\beta] = 1$, then a Q^α measurement yields with probability 1 a positive outcome.

(See the MATHEMATICS for a proof.) From this proposition and the information given by the pilot, the mechanic deduces that it takes only one test (good mechanics do perform tests) to reveal with probability 1 that the wing is indeed warped. Hence, the mechanic is not at all hard pressed to find the flaw but just performs a measurement.

So, if one accepts the bi or spectral modal interpretation, the implications put forward by Clifton need not arise. The 747 example is begging the question for it proves the truism that, given an interpretation that violates Property Composition and Decomposition, one gets into trouble as soon as one reasons as if Property Composition and Decomposition hold.

Arntzenius (1998, Sect. 5) agrees that the violations of Property Composition and Decomposition do not lead to problems at the level of observation. Instead he thinks the problems are of a metaphysical nature. Elaborating on Clifton's example Arntzenius writes: '*One should view the mechanic as having a list of all the definite properties of the left-hand wing handed to him, e.g. by God, while the pilot has handed to him a list of all the definite properties of the entire plane. The pilot says to the mechanic: "Hm, the left-hand wing is warped, that's a problem". The mechanic responds: "No, I've got all the properties of the left-hand wing, and nowhere it is listed that it is warped". This seems bizarre.*'

It is indeed bizarre that Property Composition and Decomposition are sometimes violated. But, as illustrated by Arntzenius' example, these violations can appear only if one considers the properties of systems from a theoretical (or theological) point of view: if the pilot and the mechanic leave their armchairs and go to their work, the violations of Property Composition and Decomposition do not occur. The bizarreness related to these violations is thus metaphysical in the sense that one is confronted with it not when one considers the phenomena but only if one considers noumenal states of affairs. Pilots and mechanics thus need not worry, only philosophers are confronted with violations of Property Composition and Decomposition when they are studying noumenal states of affairs as given by the spectral modal interpretation. And, where Arntzenius takes the position that these violations

---

[150] The assumption is that there exists a snooper system for the composite $\alpha\beta$.

damage the metaphysical tenability of the spectral modal interpretation, I take, on the basis of my liberal criteria for tenable interpretations, a more light-hearted view.

MATHEMATICS

Consider the assumption that any model of a measurement of a magnitude A^α pertaining to α counts as a measurement of the magnitude $A^\alpha \otimes \mathbb{I}^\beta$ pertaining to $\alpha\beta$ and *vice versa*. This assumption can be questioned if one adopts the view that a measurement of a magnitude pertaining to α is an interaction between *only* α and a measurement device. For then it follows that a measurement of A^α should keep the state of any system β disjoint from α, in principle, constant. Consequently, if an $A^\alpha \otimes \mathbb{I}^\beta$ measurement counts as an A^α measurement, any measurement of $A^\alpha \otimes \mathbb{I}^\beta$ should also keep the state of β constant. This consequence need, however, not be true.

Take a magnitude of α represented by the operator $Q^\alpha = |q_1^\alpha\rangle\langle q_1^\alpha| + |q_2^\alpha\rangle\langle q_2^\alpha|$ (a two-dimensional projection) and consider the eigenvectors $\{|\tilde{q}_j^{\alpha\beta}\rangle\}_j$ of $Q^\alpha \otimes \mathbb{I}^\beta$. If $\mathscr{H}^\alpha$ is three-dimensional and $\mathscr{H}^\beta$ is two-dimensional, these eigenvectors are, for instance,

$$
\left.
\begin{aligned}
|\tilde{q}_1^{\alpha\beta}\rangle &= |q_1^\alpha\rangle \otimes |e_1^\beta\rangle, &\qquad |\tilde{q}_4^{\alpha\beta}\rangle &= |q_2^\alpha\rangle \otimes |e_2^\beta\rangle, \\
|\tilde{q}_2^{\alpha\beta}\rangle &= \tfrac{1}{2}\sqrt{2}(|q_1^\alpha\rangle \otimes |e_2^\beta\rangle + |q_2^\alpha\rangle \otimes |e_1^\beta\rangle), &\qquad |\tilde{q}_5^{\alpha\beta}\rangle &= |q_3^\alpha\rangle \otimes |e_1^\beta\rangle, \\
|\tilde{q}_3^{\alpha\beta}\rangle &= \tfrac{1}{2}\sqrt{2}(|q_1^\alpha\rangle \otimes |e_2^\beta\rangle - |q_2^\alpha\rangle \otimes |e_1^\beta\rangle), &\qquad |\tilde{q}_6^{\alpha\beta}\rangle &= |q_3^\alpha\rangle \otimes |e_2^\beta\rangle,
\end{aligned}
\right\}
\quad (13.17)
$$

where $|e_1^\beta\rangle$ and $|e_2^\beta\rangle$ are mutually orthogonal vectors. The eigenvectors $|\tilde{q}_1^{\alpha\beta}\rangle$, $|\tilde{q}_2^{\alpha\beta}\rangle$, $|\tilde{q}_3^{\alpha\beta}\rangle$ and $|\tilde{q}_4^{\alpha\beta}\rangle$ correspond to the eigenvalue 1 of $Q^\alpha \otimes \mathbb{I}^\beta$, the eigenvectors $|\tilde{q}_5^{\alpha\beta}\rangle$ and $|\tilde{q}_6^{\alpha\beta}\rangle$ correspond to the eigenvalue 0.

The measurement interaction

$$
|\Psi^{\alpha\beta\mu}\rangle = \sum_j c_j |\tilde{q}_j^{\alpha\beta}\rangle \otimes |D_0^{\bar\mu}\rangle \otimes |R_0^\pi\rangle \longmapsto |\widetilde{\Psi}^{\alpha\beta\mu}\rangle = \sum_j c_j |\tilde{q}_j^{\alpha\beta}\rangle \otimes |D_j^{\bar\mu}\rangle \otimes |R_j^\pi\rangle
$$

$$(13.18)$$

can then be taken as a measurement of $Q^\alpha \otimes \mathbb{I}^\beta$ because the interaction correlates the eigenvectors of $Q^\alpha \otimes \mathbb{I}^\beta$ to the pointer readings $\{|R_j^\pi\rangle\langle R_j^\pi|\}_j$. The outcome $|R_a^\pi\rangle\langle R_a^\pi|$ can thus be taken as an outcome corresponding to an eigenvalue of $Q^\alpha \otimes \mathbb{I}^\beta$.

This measurement of $Q^\alpha \otimes \mathbb{I}^\beta$ does not keep the state of β constant. Start with the state $W^{\alpha\beta} = |q_1^\alpha\rangle\langle q_1^\alpha| \otimes |e_2^\beta\rangle\langle e_2^\beta|$ such that the initial state of β is $W^\beta = |e_2^\beta\rangle\langle e_2^\beta|$. Then, after the measurement, $\widetilde{W}^{\alpha\beta}$ is equal to $\tfrac{1}{2}|\tilde{q}_2^{\alpha\beta}\rangle\langle \tilde{q}_2^{\alpha\beta}| + \tfrac{1}{2}|\tilde{q}_3^{\alpha\beta}\rangle\langle \tilde{q}_3^{\alpha\beta}|$ so that the final state of β is $\widetilde{W}^\beta = \tfrac{1}{2}|e_1^\beta\rangle\langle e_1^\beta| + \tfrac{1}{2}|e_2^\beta\rangle\langle e_2^\beta|$. Hence, if a measurement of Q^α keeps the state of β constant, then the above measurement of $Q^\alpha \otimes \mathbb{I}^\beta$ is not a measurement of Q^α. (This argument can be blocked by taking the view that the above measurement is actually a measurement of the non-degenerate operator $A^{\alpha\beta} = \sum_j a_j^{\alpha\beta} |\tilde{q}_j^{\alpha\beta}\rangle\langle\tilde{q}_j^{\alpha\beta}|$ and not a measurement of $Q^\alpha \otimes \mathbb{I}^\beta$.)

Proposition 13.2 can be proved for the spectral modal interpretation for measurements of the second kind provided that there exists a snooper system for $\alpha\beta$ at the beginning of the Q^α measurement.

Proof: Let $Q^\alpha = \sum_{j=1}^2 \sum_k q_j^\alpha |q_{jk}^\alpha\rangle\langle q_{jk}^\alpha|$ be a spectral decomposition ($q_1^\alpha = 1$ and $q_2^\alpha = 0$) and let the measurement interaction be

$$|q_{jk}^\alpha\rangle \otimes |D_0^{\bar{\mu}}\rangle \otimes |R_0^\pi\rangle \longmapsto |\tilde{q}_{jk}^\alpha\rangle \otimes |D_{jk}^{\bar{\mu}}\rangle \otimes |R_{jk}^\pi\rangle. \tag{13.19}$$

Let $\{|R_{jk}^\pi\rangle\langle R_{jk}^\pi|\}_{j,k}$ represent the measurement outcomes and let $[|R_{jk}^\pi\rangle\langle R_{jk}^\pi|] = 1$ correspond to a positive outcome of the Q^α measurement if $j = 1$ and to a negative outcome if $j = 2$. Let $\{|D_{jk}^{\bar{\mu}}\rangle\}_{j,k}$ be a set of pair-wise orthogonal vectors such that (see Section 10.4) the spectral modal interpretation indeed ascribes outcomes to the pointer after the measurement.

Take now a composite $\alpha\beta$ and assume that $[Q^\alpha \otimes \mathbb{I}^\beta] = 1$ at $t = 0$. The core property ascription to $\alpha\beta$ is then given by $[P_x^{\alpha\beta}] = 1$, where $P_x^{\alpha\beta}$ is an eigenprojection of $W^{\alpha\beta}$ which satisfies $[Q^\alpha \otimes \mathbb{I}^\beta] P_x^{\alpha\beta} = P_x^{\alpha\beta}$. Assume, furthermore, that there exists a snooper σ for $\alpha\beta$ at $t = 0$. So, σ possesses P_x^σ with probability 1 if and only if α possesses $P_x^{\alpha\beta}$ and the state $\sigma\alpha\beta$ thus satisfies

$$\mathrm{Tr}^{\sigma\alpha\beta}(W^{\sigma\alpha\beta}(0)\,[P_a^\sigma \otimes P_b^{\alpha\beta}]) = \delta_{ab}\,\mathrm{Tr}^{\sigma\alpha\beta}(W^{\sigma\alpha\beta}(0)\,[P_a^\sigma \otimes P_a^{\alpha\beta}]). \tag{13.20}$$

Then perform the Q^α measurement and assume that during this measurement the states of σ and β evolve freely and remain constant. The state of $\sigma\alpha\beta\bar{\mu}\pi$ after the measurement, say at $t = 1$, can be calculated (a rather long expression) and it follows that the state of snooper and pointer becomes

$$W^{\sigma\pi}(1) = \sum_{a,b,c,j,k} \langle e_a^\sigma| \otimes \langle q_{jk}^\alpha| \otimes \langle e_b^\beta| W^{\sigma\alpha\beta}(0)|e_c^\sigma\rangle \otimes |q_{jk}^\alpha\rangle \otimes |e_b^\beta\rangle \, |e_a^\sigma\rangle\langle e_c^\sigma| \otimes |R_{jk}^\pi\rangle\langle R_{jk}^\pi|,$$

$$\tag{13.21}$$

where $\{|e_a^\sigma\rangle\}_a$ and $\{|e_b^\beta\rangle\}_b$ are arbitrary orthonormal bases for, respectively, $\mathcal{H}^\sigma$ and $\mathcal{H}^\beta$.

We are interested in the conditional probabilities

$$p(|R_{2y}^\pi\rangle\langle R_{2y}^\pi| \text{ at } t = 1/P_x^{\alpha\beta} \text{ at } t = 0) \tag{13.22}$$

since they give the probabilities with which one obtains a negative outcome after the Q^α measurement given that $[Q^\alpha \otimes \mathbb{I}^\beta] = 1$. By using the one-to-one correlations between the core properties of $\alpha\beta$ and σ at $t = 0$ and by using the deterministic evolution of the core properties of σ from $t = 0$ to $t = 1$, one can derive that these conditional probabilities are equal to

$$\frac{p(P_x^\sigma \text{ at } t = 1, |R_{2y}^\pi\rangle\langle R_{2y}^\pi| \text{ at } t = 1)}{p(P_x^\sigma \text{ at } t = 1)}. \tag{13.23}$$

The numerator is equal to $\text{Tr}^{\sigma\pi}(W^{\sigma\pi}(1)\,[P_x^\sigma \otimes |R_{2y}^\pi\rangle\langle R_{2y}^\pi|])$ for the spectral modal interpretation and by substituting the state $W^{\sigma\pi}(1)$, it becomes

$$\text{Tr}^{\sigma\alpha\beta}(W^{\sigma\alpha\beta}(0)\,[P_x^\sigma \otimes |q_{2y}^\alpha\rangle\langle q_{2y}^\alpha| \otimes \mathbb{I}^\beta]) =$$
$$\text{Tr}^{\alpha\beta}(\text{Tr}^\sigma(W^{\sigma\alpha\beta}(0)\,[P_x^\sigma \otimes \mathbb{I}^{\alpha\beta}])\,|q_{2y}^\alpha\rangle\langle q_{2y}^\alpha| \otimes \mathbb{I}^\beta). \tag{13.24}$$

Since $W^{\sigma\alpha\beta}(0)$ satisfies (13.20) one can apply Theorem 6.1 (see page 91) and rewrite the numerator as

$$\text{Tr}^{\alpha\beta}(W^{\alpha\beta}(0)\,P_x^{\alpha\beta}\,[|q_{2y}^\alpha\rangle\langle q_{2y}^\alpha| \otimes \mathbb{I}^\beta]) = w_x^{\alpha\beta}\,\text{Tr}^{\alpha\beta}(P_x^{\alpha\beta}\,[|q_{2y}^\alpha\rangle\langle q_{2y}^\alpha| \otimes \mathbb{I}^\beta]), \tag{13.25}$$

where $P_x^{\alpha\beta}$ is an eigenprojection of $W^{\alpha\beta}$ corresponding to the eigenvalue $w_x^{\alpha\beta}$. Since $[Q^\alpha \otimes \mathbb{I}^\beta]\,P_x^{\alpha\beta} = P_x^{\alpha\beta}$, it follows that $[|q_{2y}^\alpha\rangle\langle q_{2y}^\alpha| \otimes \mathbb{I}^\beta]\,P_x^{\alpha\beta} = 0$ (multiply the first relation at the left-hand side with $|q_{2y}^\alpha\rangle\langle q_{2y}^\alpha| \otimes \mathbb{I}^\beta$). Hence, the numerator is zero such that the conditional probability of finding a negative outcome after the Q^α measurement if $[Q^\alpha \otimes \mathbb{I}^\beta] = 1$, is zero. $\qquad\square$

13.4 Reductionism with dispositional holism

If, instead of the bi and spectral modal interpretations, one accepts the atomic modal interpretation, one has to a large extent reductionism but now holism is lacking. And again one can raise the question of whether this makes the atomic modal interpretation unacceptable.

The fact that the atomic modal interpretation does not ascribe holistic properties to composites makes it in a sense a quite classical interpretation of quantum mechanics: it is generally assumed that in classical physics the properties of composites are all reducible to the properties of their subsystems and the atomic modal interpretation now proves that one can also interpret quantum mechanics in terms of solely reducible properties. On the other hand, it is quite bold not to ascribe the holistic properties. Quantum mechanics in the standard formulation says that one can measure holistic properties and that they can have a clear physical meaning. Consider again the example of the property '$[S_{\vec{z}}^{\sigma\tau}] = 0$ and $[S_{\vec{v}}^{\sigma\tau} \cdot S_{\vec{v}}^{\sigma\tau}] = 0$' of two spin $\frac{1}{2}$-particles. The atomic modal interpretation denies that $\sigma\tau$ ever possesses this property, even though one can easily measure it. However, I believe that this bold denial leads to neither an inconsistency nor an internal incompleteness in the atomic property ascription. Hence, given my criteria, the atomic modal interpretation is not metaphysically untenable due to its lack of holism. So, if the lack of holism is to cause problems, I suspect that it will be because this lack in some way violates the criterion of Empirical Adequacy. Such a violation would clearly arise if observations yield that systems sometimes actually possess the holistic properties that the atomic modal interpretation cannot ascribe. In particular, if there exist measurements with outcomes represented by pointer readings $\{R_j^{\pi}\}_j$ which are such holistic properties, then the atomic modal interpretation would fail to be empirically adequate since then it denies that the pointer ever can possess its readings. I return to this latter possible threat at the end of this section.

A further question one can raise when assessing the atomic modal interpretation is concerned with the ontological status of holistic properties. As said, the atomic modal interpretation does not ascribe every holistic property. However, composite systems can be in states such that a measurement of such a non-ascribable holistic property has with probability 1 a positive outcome. If, for instance, the composite $\sigma\tau$ of the two spin $\frac{1}{2}$-particles is in the singlet state $|\Psi^{\sigma\tau}\rangle = \frac{1}{2}\sqrt{2}(|u_{\tilde{z}}^{\sigma}\rangle \otimes |d_{\tilde{z}}^{\tau}\rangle - |d_{\tilde{z}}^{\sigma}\rangle \otimes |u_{\tilde{z}}^{\tau}\rangle)$, a measurement of the property '$[S_{\tilde{z}}^{\sigma\tau}] = 0$ and $[S_{\tilde{v}}^{\sigma\tau} \cdot S_{\tilde{v}}^{\sigma\tau}] = 0$' has with certainty a positive outcome. The question is now what such a measurement reveals according to the atomic modal interpretation. And this question also arose at the end of Section 5.5 when discussing the full property ascription. As proved in the MATHEMATICS of Section 5.4, the atomic modal interpretation does not satisfy the Certainty condition. That is, it is sometimes the case that properties (which we can now identify as holistic properties) have a Born probability 1 even though they are not possessed according to the atomic modal interpretation. The measurement of such a property has with probability 1 a positive outcome and the question is thus what does this outcome reveal since it cannot reveal that the measured property was initially possessed.

Clifton (1996, Sect. 2) has given a first answer to this question. If a measurement of a property Q^{α} has with probability 1 a positive outcome and α does not possess Q^{α}, then the measurement reveals that α has a *disposition* to yield with probability 1 a positive outcome. By this answer, Clifton doubles the description of reality. Firstly, a system α has actually possessed properties which are really there. These actually possessed properties are given by the property ascription of the atomic modal interpretation. Secondly, a system α has dispositions which, when measured, elicit that the measurement device possesses with probability 1 a positive outcome as an actual property at the end of the measurement. Such dispositional properties are represented by projections Q^α with Born probability 1 and do not need to correspond to the actually possessed properties.

Dieks (1998b, Sect. 4) has given a more extensive answer. If a measurement of Q^α has with probability 1 a positive outcome and α does not possess Q^α, then the measurement reveals that α has a *collective dynamical effect* on the measurement device. The idea behind this second answer is that in quantum mechanics the atoms in a molecular system α can interact as a collective with the environment. Dieks illustrates this idea with an ammonia molecule which interacts with an electric field. Such a molecule consists of three hydrogen atoms and one nitrogen atom but behaves essentially like a two-state system which emits and absorbs energy quanta. The interaction between the molecule and the field is, of course, the result of the interaction between the field and the individual atoms in the molecule. However, this interaction can also be considered as an interaction between the field and the molecule taken as one collective. And with regard to the effects the molecule has on the state of the field, such a description of the molecule as a collective suffices in the sense that if one considers the ammonia molecule as a system which emits and absorbs quanta, one need not focus on the individual atoms anymore. Hence, generalising this, the idea is that composite systems, when interacting with their environment, can behave as collectives, screening off the contributions of the individual atoms. This collective behaviour of the composite is called by Dieks a collective dynamical effect of the composite on the environment.

In measurements molecules can exhibit such collective effects. Let the state of a system α, which consists of the atoms σ and τ, be given by an entangled state vector $|\Psi^{\sigma\tau}\rangle$. Then a measurement of a property Q^α, which satisfies $Q^\alpha |\Psi^{\sigma\tau}\rangle = |\Psi^{\sigma\tau}\rangle$, yields with probability 1 a positive outcome because the two atoms interact collectively by means of their composite state with the device. This collective effect of $\sigma\tau$ is now represented by the projection $|\Psi^{\sigma\tau}\rangle\langle\Psi^{\sigma\tau}|$.

According to Dieks, the collective effects of a molecule α are related to a *coarse-grained* description in which α is taken *as if* it is an atomic system. There is still only one fundamental description of the properties of systems and that is the one given by the atomic modal interpretation. Molecular systems consist of atoms, these atoms are ascribed properties and the properties of the molecular systems are built up from these atomic properties. However, in physics one often replaces this fundamental *fine-grained* description by a coarse-grained description in which collections of atoms are taken as the building blocks of nature. If one does so and applies the property ascription of the atomic modal interpretation to such an *as if* atom α, one obtains not the definite properties of α, but the collective effects of α. In the case of our molecule $\sigma\tau$ with the state $|\Psi^{\sigma\tau}\rangle$ coarse-graining implies taking $\sigma\tau$ as if it is an atom. The property ascription of the atomic modal interpretation then indeed yields that the property $|\Psi^{\sigma\tau}\rangle\langle\Psi^{\sigma\tau}|$ is a collective effect of $\sigma\tau$.

The answers by Clifton and Dieks reintroduce holism into the atomic modal interpretation since any holistic property Q^α can be ascribed to any system α as a disposition or as a collective effect: just assume that the state of α is given by $W^\alpha = Q^\alpha / \dim(Q^\alpha)$. The clever thing in this reintroduction is that now the atomic modal interpretation can be said to be both reductionistic and holistic without falling prey to the no-go theorems of Bacciagaluppi (1995) and Clifton (1996), or to the no-go theorem discussed in Section 6.3. To see how these theorems are evaded, suppose that systems possess their dispositions and collective effects unconditionally. One is then immediately led back to the problem of defining a joint value assignment to all the properties, dispositions and collective effects of a composite, as well as to the problem of defining a joint probability that different systems α, β, ... possess simultaneously their properties, dispositions and collective effects. However, dispositions and collective effects can be understood as *contextual* features of systems: conditional on a specific interaction with an environment, a system exhibits a corresponding disposition or collective effect. And given that there do not exist interactions between α, β, ... in which all the dispositions and collective effects can be exhibited simultaneously, the task of defining joint value assignments and joint probabilities is evaded.

A further and slightly more critical comment is the following. As I have said, the introduction of dispositions (I concentrate here on dispositions, but the following holds *mutatis mutandis* for collective effects) doubles the description of reality: a system possesses not only real properties, but also dispositional properties. This doubling is, however, unproblematic because both the real properties and the dispositions have a meaning in terms of the actual properties: The real properties of a system are the properties which it actually possesses, and the dispositions are features of the system which manifest themselves in measurement contexts by eliciting that the pointer of the measurement device actually possesses an outcome at the end of the measurement. So, although the description of reality is doubled, both descriptions still refer to one single realm of actual properties.

This last conclusion can now be endangered depending on the precise definition of dispositional properties. If dispositions are only those properties of systems which are revealed by measurements which have with certainty an actual outcome, then the above conclusion is fine: all dispositions then elicit by definition actually possessed outcomes. If, however, dispositions are defined as those properties Q^α of systems which have a Born probability $\text{Tr}^\alpha(W^\alpha Q^\alpha)$ equal to 1, the above conclusion should be weakened. The reason for this is that it need not always be the case that a measurement of a property with Born probability 1 yields an actual outcome. To see this, consider the possibility that the atomic modal interpretation does not always solve the measurement problem. Assume, for instance, that the already mentioned threat that there exist pointer readings which are holistic properties themselves becomes true. Then, since the atomic modal interpretation cannot ascribe such holistic readings as actually possessed properties, a measurement of a property with Born probability 1 need not always yield an actual outcome.

Therefore, if one defines dispositions of systems as those properties which have a Born probability 1, then dispositions are features which only elicit in some measurement contexts that the pointer of the measurement device actually possesses an outcome. On the other hand, since a measurement of a disposition always yields that the final outcome has a Born probability equal to 1, this final outcome is itself a disposition of the pointer after the measurement. So, dispositions are features which in any case elicit in measurement contexts that the pointer of the measurement device possesses an outcome as a disposition. Hence, the conclusion is now that the introduction of dispositions doubles the description of reality and that the description in terms of dispositions only refers in some cases to the realm of the actual properties. In the worst case (if we, for instance, only have measurement devices with holistic pointer readings at our disposal) dispositions manifest themselves only in terms of other dispositions such that they describe a realm of reality which is completely cut off from the realm of the actually possessed properties. Dispositions (and collective effects) then describe a realm which, in the words of Guido Bacciagaluppi in Bacciagaluppi and Vermaas (1999), can be characterised as a *virtual reality*.

So, to conclude, the lack of holism does not affect the metaphysical tenability of the atomic modal interpretation. And, provided that the readings of the pointers of measurement devices are never holistic properties, this lack also need not affect its empirical adequacy. For if pointer readings are not holistic, it is still possible to solve the measurement problem by ascribing them after measurements as actually possessed properties. Moreover, if the atomic modal interpretation indeed solves the measurement problem in this way, one can in a sensible way ascribe holistic properties to composites as dispositions or as collective effects.

14

Possibilities and impossibilities

As a last step towards general conclusions, I collect and assess the more important results obtained in the previous chapters. In the next and final chapter I then consider the question of whether the modal descriptions of reality meet the demands of being well developed, empirically adequate and metaphysically tenable.

14.1 Indefinite properties and inexact magnitudes

The bi, spectral and atomic modal interpretations do manage to interpret quantum mechanics by providing a description of what reality could be like in terms of systems possessing properties. Moreover, if one accepts the full property ascription (5.12) as developed in Section 5.4, this description of reality satisfies a number of desirable conditions. However, these modal interpretations do not simultaneously ascribe all the possible properties to a system. Instead they select at each instant a specific subset of the properties pertaining to a system and ascribe only these. And the properties which are not selected are taken to be indefinite.

This possibility that properties of systems are sometimes not definite is in itself not that problematic for in some cases it is perfectly understandable what this means. Consider again the example of a tossing coin. During the toss the properties 'heads' and 'tails' can be said to be indefinite and this can be understood as meaning that these two properties are applicable only to coins which lie on surfaces. However, in other cases it is less clear what is going on. Suppose that during the toss the property 'the centre of mass of the coin has position $\vec{r}$' becomes indefinite for every vector $\vec{r}$, and that nevertheless the (coarse-grained) property 'the centre of mass of the coin is in this room' is possessed at all times. Such cases can occur according to the bi, spectral and atomic modal interpretations. (For macroscopic coins which interact continuously with their environment one will not easily encounter such cases but for more microscopic systems one can.) And it is now much harder to understand what happens with the coin: should this description be understood as that the coin disperses in some kind of cloud located within the room when it is tossed? And does it then again 'materialise' at a specific position when it comes down after the toss? However, even though this description of a coin or of any other system may seem strange to us who are living at the dawn of the third millennium, I guess that the judgement of whether or not such a description is understandable is in the end not that self-evident. This judgement depends, among other things, on what one is used to. If an experimentalist manages to demonstrate that objects can indeed disperse when they are tossed, we would soon consider this behaviour as an acceptable phenomenon — we presently accept that a fluid like superfluid helium can flow up the sides of its container and over the brim, something which must have been very strange to the physicists who first observed this. Nevertheless, the fact that the bi, spectral and atomic modal interpretations sometimes leave properties like position indefinite makes their descriptions of reality at the least rather 'futuristic' when compared to the descriptions of systems in daily life. (Our attempts to interpret quantum mechanics indeed made us part of the adventures of Captain Kirk in *Star Trek*!)

On the other hand, the properties which modal interpretations do ascribe to systems satisfy the Closure condition, the Exclusion condition, the Weakening condition and in the case of the bi and spectral modal interpretations, also the Certainty condition (see Section 5.5). The atomic modal interpretation does not satisfy Certainty but, as discussed in Section 13.4, if the atomic modal interpretation solves the measurement problem, one can introduce dispositional properties in a meaningful way. And by means of these dispositions, the atomic modal interpretation satisfies the Certainty condition at least in spirit. Hence, with regard to the definite properties, the property ascriptions of the bi, spectral and atomic modal interpretations are quite satisfactory. The set of ascribed properties is closed under negation, conjunction and disjunction, and satisfies the rules of classical logic. If a system α possesses a property represented by the projection Q^α, it also possesses all the weaker properties represented by the projections $\tilde{Q}^\alpha$ with $\tilde{Q}^\alpha Q^\alpha = Q^\alpha$. Moreover, if a measurement of the property Q^α has with certainty a positive outcome, then that property Q^α was indeed possessed before the measurement (as a disposition in the case of the atomic modal interpretation).

The property ascription to a system can be transposed to a value assignment to the magnitudes pertaining to that system. According to the discussion in Section 5.6, the choice of rules which govern this transposition confronts one with a dilemma: either one assigns exact values only to a specific set of magnitudes represented by operators with a discrete spectral resolution, or one assigns genuinely inexact values to all magnitudes pertaining to a system. Either way it follows that there exist some magnitudes with exact values. Consider, for instance, a spin $\frac{1}{2}$-particle with a state W^σ which is diagonal with respect to the projections $|u_{\tilde{x}}^\sigma\rangle\langle u_{\tilde{x}}^\sigma|$ and $|d_{\tilde{x}}^\sigma\rangle\langle d_{\tilde{x}}^\sigma|$. The bi, spectral and atomic modal interpretations then all yield that the magnitude 'spin in $\tilde{x}$ direction' represented by the operator $S_{\tilde{x}}^\sigma$ has a precise value, namely $[S_{\tilde{x}}^\sigma]$ is $-\frac{1}{2}\hbar$ or $+\frac{1}{2}\hbar$. If one now chooses to assign only exact values to magnitudes, the consequence is that many magnitudes pertaining to system are indefinite. In the case of the spin $\frac{1}{2}$-particle this implies that the magnitudes of 'spin in $\tilde{v}$ direction' are indefinite-valued for all $\tilde{v} \neq \tilde{x}$. If, on the other hand, one chooses to assign values to all magnitudes, then a magnitude A^α is generally assigned a set of values, denoted by $[A^\alpha] \in^* \Gamma$, where the $*$ accompanying the $\in$ sign is a reminder that this assignment should be taken as that A^α has an inexact value restricted to the set Γ. In the case of the spin $\frac{1}{2}$-particle it follows that for all $\tilde{v} \neq \tilde{x}$ the value of the spin $S_{\tilde{v}}^\sigma$ is given by the trivial assignment $[S_{\tilde{v}}^\sigma] \in^* \{-\frac{1}{2}\hbar, +\frac{1}{2}\hbar\}$ which does not imply that $S_{\tilde{v}}^\sigma$ has either the value $-\frac{1}{2}\hbar$ or the value $+\frac{1}{2}\hbar$.

The bi, spectral and atomic modal interpretations thus succeed in assigning at each instant exact values to some physical magnitudes but fail to assign such values to all magnitudes. Again I believe that this failure need not be problematic. It's strange, for sure, but maybe an experimentalist will demonstrate that sometimes magnitudes indeed do not have exact values (what ever that may mean), so that we can get used to this phenomenon.

14.2 Correlations and perspectivalism

Modal interpretations can with varying success correlate the properties they ascribe. Consider, firstly, the bi modal interpretation. This interpretation ascribes properties to all the subsystems of composites ω which have a pure state $|\Psi^\omega\rangle\langle\Psi^\omega|$. And this interpretation gives correlations between the properties ascribed to two disjoint subsystems α and β if these two systems bisect ω, that is, if $\alpha\beta$ is equal to ω. These correlations are captured by the joint probabilities

$$p(P_a^\alpha, P_b^\beta) = \mathrm{Tr}^\omega(|\Psi^\omega\rangle\langle\Psi^\omega|\,[P_a^\alpha \otimes P_b^\beta]). \tag{14.1}$$

However, the bi modal interpretation is silent about the correlations between the properties ascribed to two non-disjoint subsystems of ω or to more general sets of subsystems.

If one accepts perspectivalism as discussed in Section 4.3 and as possibly embraced by Kochen (1985), the joint probabilities (14.1) give the correlations between all the properties one can consider simultaneously. For, according to perspectivalism, one can only simultaneously consider the properties of subsystems if these systems can be considered from one and the same perspective. And a perspective in the bi modal interpretation is given by a bisection of the composite ω into two disjoint subsystems. Hence, from the perspective 'ω bisected into α and β' one can simultaneously consider the properties of the subsystems α and β, and from the perspective 'ω bisected into γ and δ' (with $\omega = \gamma\delta$) one can simultaneously consider the properties of the subsystems γ and δ. However, one cannot, in general, simultaneously consider the properties of all these subsystems α, β, γ and δ. For if $\gamma \neq \alpha$ and $\gamma \neq \beta$, then the bisection of ω into γ and δ serves as a perspective which is incompatible with the perspective 'ω bisected into α and β.' Hence, with perspectivalism the bi modal interpretation satisfies the criterion of Internal Completeness with regard to the correlations, because it correlates by means of (14.1) all the properties of the systems which can be considered simultaneously.

This completeness is, however, obtained at the expense of the content of the bi modal interpretation; by adopting perspectivalism most of the interesting questions in quantum mechanics are simply evaded. The perspectival bi modal interpretation is, for instance, silent about questions on the correlations between outcomes of two or more measurements as was illustrated by the example of the Einstein–Podolsky–Rosen experiment in Section 4.3. Let's assume therefore along the lines of Dieks (1994a) that the bi modal interpretation should correlate the properties of all the subsystems of a composite ω with a pure state. If one accepts the criterion of Empirical Adequacy and the assumptions of Instantaneous Autonomy and Dynamical Autonomy for measurements (see Section 3.3), it can be proved that the bi modal interpretation is equivalent to the spectral modal interpretation (see Appendix A). And, as is discussed in the next paragraph, the spectral modal interpretation is also in the end condemned to perspectivalism. So, if one wants to add correlations to the bi modal interpretation but avoid perspectivalism, one should reject one of the mentioned assumptions and arrive at a modal interpretation different to the spectral modal interpretation.[151]

---

[151] Papers by Dieks (1998a) and Hemmo (1998) can be seen as attempts to develop the bi modal interpretation differently compared to how it is developed in Vermaas and Dieks (1995).

Consider, secondly, the spectral modal interpretation. This interpretation ascribes properties to any system and gives correlations between the properties of disjoint systems $\{\alpha, \beta, \ldots\}$. Let ω be the composite $\alpha\beta\cdots$. The correlations are then given by the joint probabilities

$$p(P_a^\alpha, P_b^\beta, \ldots) = \text{Tr}^\omega(W^\omega \, [P_a^\alpha \otimes P_b^\beta \otimes \cdots]). \tag{14.2}$$

In Section 6.3 it has been proved that there cannot exist classical joint probabilities for collections of non-disjoint systems which are compatible with the above joint probabilities (14.2) for disjoint systems. Hence, if one assumes that the spectral modal interpretation ascribes properties simultaneously to all possible systems and that it gives correlations between these properties in terms of classical joint probabilities, then one should judge that the spectral modal interpretation is internally incomplete.

A possible way out of this predicament can be found in Svetlichny, Redhead, Brown and Butterfield (1988). They proved the following. Consider a (sufficiently random) sequence of three simultaneously occurring events denoted by A, A' and B. Then, if one construes probabilities as Von Mises–Church relative frequencies, it is possible that: (A) the sequence is such that one can define probabilities (as limiting relative frequencies) for the occurrence of the events A, A' and B separately, (B) one can define joint probabilities for the simultaneous occurrence of A and B as well as for the simultaneous occurrence of A' and B, but (C) one cannot define joint probabilities (as limiting relative frequencies) for the simultaneous occurrence of A and A'. This result shows that the non-existence of the joint probabilities $p(A, A')$ need not be inconsistent with the assumption that the events A, A' and B occur simultaneously.

If this result generalises to the property ascription to systems in the spectral modal interpretation, the following position with regard to correlations becomes possible. Firstly, disjoint and non-disjoint systems possess their properties simultaneously. Secondly, for every separate system there exist probabilities that it possesses its properties, and for specific collections of systems there exist joint probabilities that they simultaneously possess their properties (for collections of disjoint systems these joint probabilities are equal to (14.2)). And, thirdly, there are collections of (non-disjoint) systems for which there do not exist joint probabilities that they simultaneously possess their properties, and this means that the limiting relative frequencies corresponding to these joint probabilities are not defined. The advantage of such a position is that one can account for the non-existence of joint probabilities for collections of non-disjoint systems and still maintain that these systems possess their properties simultaneously. Whether this position is in the end tenable (mathematically or otherwise) is not yet clear.

A second way out of this predicament is by again introducing perspectivalism by proclaiming that one can only simultaneously consider the properties of a collection of systems $\{\alpha, \beta, \ldots\}$ if these systems are disjoint and divide a composite ω. It then no longer makes sense to try to find correlations between the properties of non-disjoint systems because these properties cannot be viewed from one mutual perspective (see Section 4.4).

If one accepts this perspectivalism, it still is possible to claim that the properties ascribed to a system are objective in the sense that they do not depend on the specific perspective from which they are considered. So, the core properties of a system α part of a composite $\alpha\beta\gamma\delta$, for instance, are represented by the projections $\{P_j^\alpha\}_j$ irrespectively of whether one considers α from, say, the perspective '$\alpha\beta\gamma\delta$ divided by α, β, γ and δ' or from the perspective '$\alpha\beta\gamma\delta$ divided by α, β and $\gamma\delta$.' Also the probabilities $\{p(P_j^\alpha)\}_j$ with which these core properties are ascribed are objective in this sense. Finally, the correlations have this objectivity as well since the joint probabilities $p(P_j^\alpha, P_k^\beta)$ are given by $\mathrm{Tr}^{\alpha\beta}(W^{\alpha\beta}[P_j^\alpha \otimes P_k^\beta])$ in all the perspectives from which one can simultaneously consider the properties of α and β.

A drawback of perspectivalism is, however, that the objectivity of the transition probabilities $p(P_k^\alpha(t)/P_j^\alpha(s))$ which describe the dynamics of the core properties of a system α is not easily achieved. More precisely, in Section 9.2 it has been proved that if one accepts (A) perspectivalism, (B) that the core properties of freely evolving systems evolve deterministically (that is, if α evolves freely, then $p(P_k^\alpha(t)/P_j^\alpha(s)) = \delta_{jk}$, where $P_k^\alpha(t)$ is equal to $U^\alpha(t,s)\,P_k^\alpha(s)\,U^\alpha(s,t)$), and (C) that there exist joint transition probabilities $p(P_c^\alpha(t), P_d^\beta(t)/P_a^\alpha(s), P_b^\beta(s))$ for two disjoint systems α and β, then the transition probabilities of a system as determined from one perspective can differ from the transitions probabilities of that same system as determined from another perspective. Hence, given the assumptions (A), (B) and (C), it is impossible to claim that transition probabilities are objective.

The reader might wonder whether this perspective dependency of the transition probabilities $p(P_k^\alpha(t)/P_j^\alpha(s))$ is consistent with the objectivity of the single time probabilities $p(P_j^\alpha(s))$ and $p(P_k^\alpha(t))$. A provisional answer is that it need not be inconsistent: since the single time probabilities $p(P_j^\alpha(s))$ and $p(P_k^\alpha(t))$ do not uniquely fix the transition probabilities $p(P_k^\alpha(t)/P_j^\alpha(s))$ (see Section 8.1), it is possible that these transition probabilities are different in the relevant perspectives whereas the single time probabilities are always the same.

Finally, the atomic modal interpretation correlates without difficulty the properties of atomic and molecular systems. Let $\{\beta, \gamma, \ldots\}$ be any set of molecules, let ω be any system which has these molecules β, γ, ... as subsystems and let $\{\alpha_j\}_{j=1}$ be the collection of atoms in ω. The correlations between these systems are then given by

$$p(P^{\beta}_{b_k \cdots b_m}, P^{\gamma}_{c_q \cdots c_s}, P^{\alpha_1}_{a_1}, P^{\alpha_2}_{a_2}, \ldots) =$$
$$\mathrm{Tr}^{\omega}(W^{\omega} \, [P^{\beta}_{b_k \cdots b_m} \otimes \mathbb{I}^{\omega/\beta}][P^{\gamma}_{c_q \cdots c_s} \otimes \mathbb{I}^{\omega/\gamma}][P^{\alpha_1}_{a_1} \otimes P^{\alpha_2}_{a_2} \otimes \cdots]). \tag{14.3}$$

14.3 Discontinuities and instabilities

In the analysis of the evolution of the core properties I have distinguished two types of evolution. On the one hand one has the evolution of the actually possessed core property of a system and on the other hand there is the evolution of the set of possibly possessed core properties of a system. Both types of evolution give rise to discontinuities.

Consider an atom α with a state for which the spectral resolution is given by $W^{\alpha}(t) = \sum_j w_j(t) P_j^{\alpha}(t)$. The set of core properties ascribed to the atom is then in the bi, spectral and atomic modal interpretations given by the set of eigenprojections $\{P_j^{\alpha}(t)\}_j$. Assume now that the actually possessed core property of α is given by $P_k^{\alpha}(s)$ at instant s. Then, in the case that the atom evolves freely, this actual core property evolves continuously to the actual core property $P_k^{\alpha}(t) = U^{\alpha}(t,s) P_k^{\alpha}(s) U^{\alpha}(s,t)$ for every instant t, where $U^{\alpha}(x,y)$ is equal to $\exp([(x-y)/i\hbar] H^{\alpha})$ (see Section 8.2). However, in the case that the atom interacts with other systems, the actually possessed core property can evolve discontinuously. Consider, for instance, the state evolution $W^{\sigma}(t) = \cos^2 t \, |u_{\hat{z}}^{\sigma}\rangle\langle u_{\hat{z}}^{\sigma}| + \sin^2 t \, |d_{\hat{z}}^{\sigma}\rangle\langle d_{\hat{z}}^{\sigma}|$ of a spin $\frac{1}{2}$-particle σ. The actual core property of this particle jumps discontinuously from $|u_{\hat{z}}^{\sigma}\rangle\langle u_{\hat{z}}^{\sigma}|$ to $|d_{\hat{z}}^{\sigma}\rangle\langle d_{\hat{z}}^{\sigma}|$ at some instant between $s = 0$ and $t = \pi/2$.

The set of possible core properties of a system evolves too: for an atom α with a state with a spectral resolution $W^{\alpha}(t) = \sum_j w_j(t) P_j^{\alpha}(t)$, this set evolves from $\{P_j^{\alpha}(s)\}_j$ at instant s to $\{P_j^{\alpha}(t)\}_j$ at instant t. This evolution can now be continuous if, for instance, the eigenvalues $\{w_j(t)\}_j$ are always different and if the eigenprojections $\{P_j^{\alpha}(t)\}_j$ are time-independent. But this evolution can also be discontinuous. If, for instance, the eigenvalues $\{w_j(t)\}_j$ are not always different, it can happen that $W^{\alpha}(t)$ is non-degenerate at one instant s but degenerate at another instant t. The set of eigenprojections of W^{α} then evolves discontinuously from a set containing only one-dimensional projections to a set containing (also) higher-dimensional eigenprojections. Consequently, the set of possible core properties of α evolves discontinuously.

In Chapter 7 it has been assessed whether one can remove the discontinuities in the evolution of the set of the possible core properties which arise due to a passing degeneracy of a state (a passing degeneracy is a degeneracy which occurs at an isolated instant; consider, for instance, the degeneracy in the above given state $W^\sigma(t)$ of the spin $\frac{1}{2}$-particle σ at $t_0 = \pi/4$). The idea was that one should identify the core properties of a system α at time t_0 not with the eigenprojections $\{P_j^\alpha(t_0)\}_j$ given by the instantaneous spectral resolution of the state $W^\alpha(t_0)$, but with the projections $\{T_q^\alpha(t_0)\}_q$ which lie on continuous trajectories of eigenprojections of the state. These continuous trajectories $\{T_q^\alpha(t)\}_q$ are defined by means of the spectral resolutions of the state $W^\alpha(t)$ in a time interval around t_0. With this modified property ascription one obtains the extended modal interpretation (Section 7.1).

Unfortunately, the state dynamics can be such that continuous trajectories of eigenprojections do not exist. More precisely, continuous trajectories $\{T_q^\alpha(t)\}_q$ exist if and only if the state satisfies the Dynamical Decomposition condition (see page 105). If the evolution of the state is analytic, then this condition is satisfied. But Example 5.6 of Bacciagaluppi, Donald and Vermaas (1995) (see (7.34) on page 126) proves that there also exist states which evolve by the Schrödinger equation and which violate the Dynamical Decomposition condition. So, it is impossible, in general, to use continuous trajectories of eigenprojections to remove discontinuities in the evolution of the set of the possible core properties ascribed by the bi, spectral and atomic modal interpretations.

In order to save the extended modal interpretation as a generally applicable interpretation, one can try to argue that not every theoretically imaginable evolution is physically possible and that physically possible evolutions do not violate the Dynamical Decomposition condition. The idea that not every imaginable function $W^\alpha(t)$ gives a physically possible evolution is in itself not new. For example, evolutions $W^\alpha(t)$, where the expectation value for the energy becomes infinite, are not physically allowed. Elaborating on this, one may try to argue that physically possible state evolutions are analytic evolutions, which, incidentally, is the case if the Hamiltonian of the whole universe is bounded with regard to the operator norm. To my knowledge such an argument has not yet been given, so the question of whether the extended modal interpretation is generally applicable cannot yet be answered positively.

The analysis of the evolution of the set of the possible core properties also yielded that this evolution need not harmonise with the evolution of the state and that this evolution can be highly unstable. The first phenomenon was illustrated by Example 5.1 of Bacciagaluppi, Donald and Vermaas (1995) (see page 129), where the state of a system evolves periodically but the set of the possible core properties does not. The second phenomenon is discussed in Section 7.4, where it was proved that if a state has a spectral resolution which is nearly degenerate, then an arbitrarily small change of that state (by an interaction with the environment or by internal dynamics) can maximally change the set of the possible core properties.

Perhaps this instability is one of the more serious defects of modal interpretations because, firstly, it can have consequences for their ability to solve the measurement problem. For even when a modal interpretation ascribes readings to a pointer at a specific instant, a small fluctuation of the state may mean that at the next instant the pointer possesses properties which are radically different to readings (see Section 14.5 below). Secondly, instability can have epistemological consequences. For if one wants to determine the state of a system and ascribe by this state properties to that system, then a small error in the determination of the state can make that the property ascription is substantially wrong (see Section 14.7 below). On the other hand, instability is not a defect which specifically haunts modal interpretations, but is instead a more generic problem also present in classical physics (consider chaotic systems, for instance).

14.4 Determinism and the lack of Dynamical Autonomy and of Locality

The results about the evolution of the actually possessed core properties of systems were both limited and worrying. A first result concerns freely evolving systems (Section 8.2). Namely, if one accepts the criterion of Empirical Adequacy and the assumption of Dynamical Autonomy for whole systems, one can derive that the core properties of freely evolving systems evolve deterministically in the bi and spectral modal interpretations. And if one accepts Empirical Adequacy and Dynamical Autonomy for atomic systems, one can derive that the core properties of freely evolving atoms evolve deterministically in the atomic modal interpretation. That is, if a system α (an atom in the case of the atomic modal interpretation) evolves freely and possesses at an instant s the core property $P_j^\alpha(s)$, then it possesses with probability 1 at any second instant t the core property $P_k^\alpha(t) = U^\alpha(t,s) P_k^\alpha(s) U^\alpha(s,t)$. And this result is rather attractive since it supports the idea that the properties of a system evolve undisturbed and stably when outside influences on the system are absent.

A second result is that, given this deterministic evolution of the properties of freely evolving systems, one can determine in special cases the evolution of the core properties of interacting systems in the spectral and atomic modal interpretations. This evolution, captured by transition probabilities $p(P_k^\alpha(t)/P_j^\alpha(s))$, can be calculated for an interacting system α provided that the properties of α are at time s or at time t one-to-one correlated with the properties of a freely evolving snooper system (see Section 8.3). It can be proved that these transition probabilities are truly stochastic, non-Markovian and different to the probabilities calculated with the Born rule.

Thirdly, Bacciagaluppi and Dickson (1997) have given for the bi and spectral modal interpretations (both with perspectivalism) and for the atomic modal interpretation a framework for choosing transition probabilities for all freely evolving and interacting systems (see Sections 8.4 and 8.5). The existence of this framework illustrates that such general transition probabilities do exist but that we presently do not have the right arguments to fix them uniquely. However, future research may reveal that a specific choice for general transition probabilities by Bacciagaluppi and Dickson (the choice by means of the generalised Schrödinger current, see Section 8.5) can be backed up by sufficient argumentation.

Further and less attractive results are that the spectral and atomic modal interpretations violate the assumptions of Dynamical Autonomy (Section 9.2) in a number of ways. This means that it is, in general, impossible to give a unique functional relationship between the evolution of the state of a freely evolving composite and the transition probabilities which govern the evolution of the properties possessed by (parts of) that composite. For the spectral modal interpretation the results are that if one accepts Empirical Adequacy and Dynamical Autonomy for whole systems such that the properties of freely evolving systems evolve deterministically, then it is sometimes impossible to give the joint transition probabilities $p(P_c^\alpha(t), P_d^\beta(t)/P_a^\alpha(s), P_b^\beta(s))$ for two disjoint systems. When adopting perspectivalism, these joint transition probabilities can again exist but then the spectral modal interpretation violates the assumption of Dynamical Autonomy for composite systems. Hence, the transition probabilities for an interacting system α part of a freely evolving composite $\alpha\beta$ are not uniquely fixed by the evolution of the state of $\alpha\beta$. It is furthermore impossible to maintain that the transition probabilities of a system α are objective in the sense of being independent of the perspective from which one considers α (see also Section 14.2).

For the atomic modal interpretation the results are that if one accepts Empirical Adequacy and Dynamical Autonomy for atomic systems such that the properties of freely evolving atoms evolve deterministically, then its property ascription violates both Dynamical Autonomy for composite systems and Dynamical Autonomy for whole systems. Hence, in the atomic modal interpretation it is even impossible to uniquely fix the transition probabilities for an freely evolving molecule by means of the state of that molecule.

These violations of the assumptions of Dynamical Autonomy are in my opinion worrying for they imply that in modal interpretations states of systems codify far less information about the properties of systems than states usually do in physics. In deterministic theories, for instance, the state of a system at time t uniquely fixes the actual properties of the system at t. And the dynamics of the state of a system therefore uniquely fixes the evolution of the properties of that system. In statistical theories the states of systems no longer uniquely fix the actual properties of systems, but in such theories the state of a system at time t still gives unique statistical information about the actual properties of the system. My expectation is that the dynamics of the state of a freely evolving composite system also gives unique statistical information about the evolution of the properties of the composite itself and about the evolution of the properties of the systems which are part of that composite. This expectation is not now met by modal interpretations of quantum mechanics. The state of a system at time t still uniquely determines the statistics of the actually possessed properties of the system at t (the assumption of Instantaneous Autonomy is not violated by modal interpretations). However, since Dynamical Autonomy is violated, the dynamics of the state of a freely evolving composite need not uniquely determine the statistics of the evolution of the properties of the composite, and need not uniquely determine the statistics of the evolution of the properties of the systems in that composite. Hence, modal interpretations break with the idea that states of freely evolving composite systems encode all the information necessary to (statistically) describe the behaviour of the system and its parts.

An unpleasant consequence of the fact that modal interpretations break with this idea is that the results already derived by means of the Dynamical Autonomy assumptions become questionable in retrospect. Take, for instance, the deterministic evolution for freely evolving systems. This evolution was derived in the bi and spectral modal interpretations by assuming that these interpretations satisfy Dynamical Autonomy for whole systems. And in the atomic modal interpretation this deterministic evolution was derived by assuming that it satisfies Dynamical Autonomy for atomic systems. But how can one justify these assumptions now that one knows that the state of a freely evolving system does not need to codify the information necessary to describe the behaviour of that system? Instead, it seems much more natural to take the violations of Dynamical Autonomy in general as a strong indication for that Dynamical Autonomy is also not satisfied in the special cases of a freely evolving whole system and a freely evolving atomic system. If one indeed concludes that Dynamical Autonomy is also not satisfied in these special cases, one is (or, more precisely, I am) empty-handed when trying to derive deterministic evolution for freely evolving systems or when trying to derive any other result presented in this book which makes use of deterministic evolution. Hence, the violations of Dynamical Autonomy cast doubts on many of the results obtained for modal interpretations.

A further and more practical consequence of the violations of Dynamical Autonomy is that it is, in general, impossible in experimental setups to control the evolution of the properties ascribed by modal interpretations. Consider an experiment in which a composite system is prepared in some initial state and in which the evolution of this state is controlled in the sense that the Hamilitonian of the composite is known. The experimenter can then exactly fix at all times the state of the composite and the states of the subsystems of that composite. It follows that the experimenter can also exactly determine at all times the possible possessed properties of the composite and of the subsystems, as well as the probabilities with which these possible properties are actually possessed. However, the experimenter cannot deduce the transition probabilities which govern the evolution of these properties. In the atomic modal interpretation the transition probabilities for the properties of the composite are not uniquely fixed by the dynamics of its state since the atomic modal interpretation violates Dynamical Autonomy for whole molecular systems. And the transition probabilities for the properties of the subsystems are not uniquely fixed by the dynamics of the states since all modal interpretations violate Dynamical Autonomy for composite systems. Hence, even if the state of the composite is precisely controlled in the experiment, the evolution of the properties of at least the subsystems cannot be controlled. It may thus happen that this evolution changes, in the sense that the transition probabilities change, while the state of the composite remains the same (see Section 14.7 for a more extensive discussion of the elusiveness of the properties ascribed by modal interpretations).

Finally, it has been proved that the properties ascribed by modal interpretations are non-local (Chapter 9). Specifically, it is possible that the evolution of the properties of the subsystems of a freely evolving composite ω, confined to a space-time region R, changes if the evolution of the state of another composite ω', confined to a second space-time region R', space-like separated from R, changes (see Section 9.3). This explicit non-locality of the modal transition probabilities is a special case of the phenomenon, sketched in the last paragraph, that the transition probabilities for the subsystems of a freely evolving composite can change even though the dynamics of the state of the composite is fixed.

This non-locality of modal interpretations is worrying as well. And since one is always in some way or another confronted with non-locality in quantum mechanics (in its standard formulation or in any interpretation), one has to conclude that modal interpretations certainly do not have the advantage of in some way resolving this quantum mechanical non-locality.

14.5 The measurement problem and empirical adequacy

Measurements have definite outcomes and it is impossible to confirm this fact with modal interpretations for every measurement interaction one can theoretically imagine. Hence, modal interpretations do not, in general, solve the measurement problem. The question addressed in Chapter 10 was now whether modal interpretations still satisfy the criterion of Empirical Adequacy by solving the measurement problem for *physically realistic* measurement interactions.

Bacciagaluppi and Hemmo (1994, 1996) have proved that if pointers of measurement devices interact with an environment and if pointers are (effectively) described by few-dimensional Hilbert spaces, then, by decoherence, the bi and spectral modal interpretations approximately ascribe pointer readings after measurements (see Section 10.2). It is clear that every pointer of a realistic measurement device interacts with its environment but it is also clear that there exist pointers which are described by high- or infinite-dimensional Hilbert spaces. For instance, a pointer which can assume a continuous set of readings is modelled by an infinite-dimensional Hilbert space. So, in spite of the fact that Bacciagaluppi and Hemmo considered measurement interactions which comprise many types of measurements (perfect measurements, measurements of the second kind, unsharp measurements), their results fail to prove that the bi and spectral modal interpretations ascribe outcomes after every physically realistic measurement interaction.

In Sections 10.3 and 10.4 I have defined sufficient conditions for exactly ascribing after measurements the pointer readings with modal interpretations. These sufficient conditions constrain the internal structure of measurement devices but are compatible with many types of measurements (perfect measurements, erroneous measurements, measurements of the second kind, measurements perturbed by the environment). It is, however, an open question whether all physically realistic measurements do indeed obey these conditions.

A vulnerable aspect of the sufficient conditions (10.40), (10.41) and (10.42)

(see page 190) for ascribing outcomes with the bi or spectral modal interpretation is that they imply that there exist strict one-to-one correlations
between the properties $\{D_j^{\bar{\mu}}\}_j$ ascribed to the mechanism $\bar{\mu}$ of the measurement device and the readings $\{R_j^{\pi}\}_j$ ascribed to the pointer of the device (that
is, $[D_j^{\bar{\mu}}] = 1$ if and only if $[R_j^{\pi}] = 1$). Whenever it can be proved that realistic
measurements fail to comply with these strict correlations, or whenever it
can be argued that environmental perturbations are bound to disrupt such
correlations, my results also fail to prove that the bi and spectral modal
interpretations solve the measurement problem for realistic measurements.

The sufficient conditions (10.15) and (10.19) (see pages 183 and 184) for
ascribing outcomes with the atomic modal interpretation are vulnerable as
well. Take two readings represented by

$$\left. \begin{aligned}
R_1^{\pi} &= |D_1^{\alpha}\rangle\langle D_1^{\alpha}| \otimes |D_1^{\beta}\rangle\langle D_1^{\beta}| \otimes |D_3^{\gamma}\rangle\langle D_3^{\gamma}| \otimes \cdots, \\
R_2^{\pi} &= |D_2^{\alpha}\rangle\langle D_2^{\alpha}| \otimes |D_2^{\beta}\rangle\langle D_2^{\beta}| \otimes |D_3^{\gamma}\rangle\langle D_3^{\gamma}| \otimes \cdots.
\end{aligned} \right\} \quad (14.4)$$

Condition (10.15) then demands that for every atom X in the pointer
the projections $|D_1^X\rangle\langle D_1^X|$ and $|D_2^X\rangle\langle D_2^X|$ are either orthogonal or equal.[152]
However, it is quite conceivable that the readings (14.4) do not meet this
demand. Assume that there indeed exists a number of atoms X for which
$|D_1^X\rangle\langle D_1^X|$ and $|D_2^X\rangle\langle D_2^X|$ are orthogonal such that R_1^{π} and R_2^{π} are orthogonal.
Then it may very well be the case that there exist other atoms Y for
which $|D_1^Y\rangle\langle D_1^Y|$ and $|D_2^Y\rangle\langle D_2^Y|$ are neither orthogonal nor equal. One can, for
instance, think in terms of wave functions and conclude that there surely
exist some atoms for which the tails of the wave functions that correspond
to the states $|D_1^Y\rangle$ and $|D_2^Y\rangle$ extend to a common region in space. It then
follows that the inner product $\langle D_1^Y|D_2^Y\rangle$ has a value between 0 and 1 such
that $|D_1^Y\rangle\langle D_1^Y|$ and $|D_2^Y\rangle\langle D_2^Y|$ are neither orthogonal nor equal. The readings
R_1^{π} and R_2^{π} then do not satisfy condition (10.15).

If there indeed exist physically realistic measurements for which modal
interpretations do not solve the measurement problem exactly, one still can
hope that modal interpretations solve this problem in good approximation.
So, following Bacciagaluppi and Hemmo, one can consider the interaction
between the pointer and the environment and try to prove that decoherence
effects ensure that the properties of the pointer converge to the readings.
This approach has two disadvantages. Firstly, it is debatable whether the
ascription of properties to a pointer which, taken as projections in Hilbert

---

[152] Condition (10.15) says that there exist sets of pair-wise orthogonal atomic projections $\{P_a^{\alpha}\}_a$, $\{P_b^{\beta}\}_b$,
... such that $R_j^{\pi} = \sum_{\langle a,b,\dots\rangle \in I_j} P_a^{\alpha} \otimes P_b^{\beta} \otimes \cdots + \tilde{R}_j^{\pi}$. This implies that if $R_1^{\pi} = |D_1^{\alpha}\rangle\langle D_1^{\alpha}| \otimes \cdots$ and
$R_2^{\pi} = |D_2^{\alpha}\rangle\langle D_2^{\alpha}| \otimes \cdots$, then $|D_1^{\alpha}\rangle\langle D_1^{\alpha}|$ and $|D_2^{\alpha}\rangle\langle D_2^{\alpha}|$ are orthogonal or equal.

space, are close to pointer readings, is a proper solution to the measurement problem.[153] Secondly, if interactions with the environment mean that the pointer acquires its readings, then the solution of the measurement problem can become unstable. For if interactions with the environment can change the pointer properties to properties which are close to readings, these interactions can also change the pointer properties within a small time interval from readings to properties which are not close to readings (see also Section 14.3).

Measurements not only have definite outcomes but these outcomes also occur with certain frequencies and with certain correlations. The standard formulation of quantum mechanics predicts these frequencies and correlations by means of the Born rule, and these predictions are empirically correct. If one now assumes that the bi, spectral and atomic modal interpretations indeed ascribe outcomes after measurements, it is possible to prove that these interpretations reproduce these empirically correct predictions with regard to the occurrence of outcomes of single measurements (see Section 11.1). The spectral and atomic modal interpretations also reproduce these predictions with regard to the correlations between the outcomes of simultaneously performed measurements, and, in so far as they can be calculated, with regard to the correlations between the outcomes of sequentially performed measurements (see Section 11.2).

So, if physically realistic measurement interactions satisfy the above mentioned sufficient conditions, modal interpretations satisfy Empirical Adequacy. For, if these conditions hold, then modal interpretations ascribe exactly definite outcomes after measurements. These outcomes are then, according to the last paragraph, automatically ascribed with the empirically correct probabilities and correlations. If, in contrast, there exist realistic measurement interactions which do not satisfy these conditions and decoherence should make pointers possess their outcomes after measurements, then the fact that the modal property ascription can be highly unstable prevents a proper solution of the measurement problem.

14.6 The lack of reductionism and of holism

The quantum mechanical properties of a composite system ω can be divided into reducible and holistic properties. Examples of reducible properties are the ones represented by $Q^\omega = \sum_j Q_j^\alpha \otimes Q_j^\beta \otimes \cdots$ and examples of holistic properties are the ones represented by $|\Psi^\omega\rangle\langle\Psi^\omega|$, where $|\Psi^{\alpha\beta}\rangle$ is a non-trivially entangled vector. An interpretation is holistic if it can ascribe every

---

[153] See Bacciagaluppi and Hemmo (1994, 1996), Dieks (1994a,b), Arntzenius (1998) and Ruetsche (1998).

holistic property of a composite to that composite. And an interpretation is reductionistic if it respects the relations between any reducible property and the properties to which this reducible property is reduced. That is, an interpretation is reductionistic if it yields that ω possesses $Q^{\omega} = \sum_j Q_j^{\alpha} \otimes Q_j^{\beta} \otimes \cdots$ if and only if the subsystems α, β, etc., possess, respectively, the properties $\langle Q^{\alpha}, Q^{\beta}, \ldots \rangle \in \{\langle Q_j^{\alpha}, Q_j^{\beta}, \ldots \rangle\}_j$ (see Section 13.1).

In Section 13.2 it has been proved that the bi and spectral modal interpretations are holistic but not reductionistic, and that the atomic modal interpretation is not holistic but reductionistic with regard to the reducible properties $Q^{\omega} = \sum_j Q_j^{\alpha} \otimes Q_j^{\beta} \otimes \cdots$ for which it holds that the sets $\{Q_j^{\alpha}\}_j$, $\{Q_j^{\beta}\}_j$, etc., are sets of pair-wise orthogonal projections. Moreover, it was argued that it is impossible to merge the bi, spectral and atomic modal interpretations into a super modal interpretation which is both reductionistic and holistic. Finally, it has been proved that a modal interpretation is reductionistic with regard to the reducible properties $Q^{\omega} = Q^{\alpha} \otimes Q^{\beta} \otimes \cdots$ if and only if its property ascription satisfies the conditions of Property Composition ($[Q^{\alpha}] = x$ implies $[Q^{\alpha} \otimes \mathbb{I}^{\beta}] = x$, for $x = 0, 1$) and of Property Decomposition ($[Q^{\alpha} \otimes \mathbb{I}^{\beta}] = x$ implies $[Q^{\alpha}] = x$).

The discussion in Section 13.3 of whether the lack of reductionism harms the bi and spectral modal interpretations focused on the related violations of Property Composition and Decomposition. Firstly, it has been shown that these violations lead to a contradiction except if one denies (A) that Q^{α} and $Q^{\alpha} \otimes \mathbb{I}^{\beta}$ represent the same magnitude, or if one accepts perspectivalism and denies (B) that the ascription of Q^{α} to α should be re-endorsed by the ascription of $Q^{\alpha} \otimes \mathbb{I}^{\beta}$ to $\alpha\beta$ and *vice versa*. Both options lead to bizarre consequences. If one denies (A), it can simultaneously be the case that the left-hand side of a table has the property 'green' and that the property 'greenness of the left-hand side' is indefinite for the table as a whole. And if one denies (B), it can be the case that from one perspective a system has the property 'green' and that from a second (incompatible) perspective the property 'green' is indefinite for that same system. However, these bizarre consequences do not imply that the bi and spectral modal interpretations violate the criteria of Consistency and Internal Completeness.

Secondly, it has been proved that the spectral modal interpretation satisfies Property Composition and Decomposition at the level of observation. That is, successive measurements of Q^{α} and $Q^{\alpha} \otimes \mathbb{I}^{\beta}$ have with probability 1 either two positive or two negative outcomes (Proposition 13.1 on page 239). Hence, the spectral modal interpretation satisfies Property Composition and Decomposition with respect to the observations. Therefore one can say that the spectral modal interpretation is observationally reductionistic with regard to the reducible properties $Q^\omega = Q^\alpha \otimes Q^\beta \otimes \cdots$. This result shows that the bizarre consequences of the violations of Property Composition and Decomposition are restricted to only unobservable or noumenal states of affairs. And this result shows that if one takes Property Composition and Decomposition as empirical conditions, then the spectral modal interpretation is in this respect satisfying Empirical Adequacy.

The impossibility of ascribing holistic properties to composites by means of the atomic modal interpretation does not also lead to violations of the criteria of Consistency and Internal Completeness. On the other hand, if observations yield that holistic properties are sometimes actually possessed properties, for instance, if measurement outcomes are holistic properties, then the atomic modal interpretation fails to be empirically adequate since it then cannot ascribe properties that systems actually possess. The proposals by Clifton (1996) and Dieks (1998b) to take holistic properties in the atomic modal interpretation as dispositions or as collective dynamical effects makes sense only if the atomic modal interpretation solves the measurement problem by ascribing outcomes as real (non-dispositional) properties (see Section 13.4).

This leaves us with the fact that the bi, spectral and atomic modal interpretations are not reductionistic with regard to every reducible property $Q^\omega = \sum_j Q_j^\alpha \otimes Q_j^\beta \otimes \cdots$. A first remark is that at the level of observation modal interpretations are reductionistic with regard to every property $Q^\omega = \sum_j Q_j^\alpha \otimes Q_j^\beta \otimes \cdots$. That is, if a series of measurements of the properties $\langle Q^\alpha, Q^\beta, \ldots \rangle \in \{\langle Q_j^\alpha, Q_j^\beta, \ldots \rangle\}_j$ all yield positive outcomes, then a subsequent measurement of Q^ω yields with probability 1 a positive outcome as well. And if a measurement of Q^ω yields a positive outcome, then there is a non-zero probability that subsequent measurements of the properties $\langle Q^\alpha, Q^\beta, \ldots \rangle \in \{\langle Q_j^\alpha, Q_j^\beta, \ldots \rangle\}_j$ all yield positive outcomes (the proofs are left to the reader). Hence, if Empirical Adequacy implies that interpretations are reductionistic at the level of observation, then one can make a case that the bi, spectral and atomic modal interpretations satisfy this criterion.

The discussion in Section 13.3 did not touch on the question of whether modal interpretations violate Consistency and Internal Completeness by not being reductionistic with regard to every reducible property. My conjecture is (but I confess that a thorough discussion has not yet been given) that one can address this question in a similar way to the question of whether the violations of Property Composition and Decomposition violate Consistency and Internal Completeness. That is, my conjecture is that the fact that modal interpretations do not respect the relations between the reducible property $Q^\omega = \sum_j Q_j^\alpha \otimes Q_j^\beta \otimes \cdots$ and the properties $\{\langle Q_j^\alpha, Q_j^\beta, \ldots \rangle\}_j$, leads to a contradiction. This contradiction can then be avoided by denying (A) that $Q^\omega = \sum_j Q_j^\alpha \otimes Q_j^\beta \otimes \cdots$ should be reducible to the properties $\{\langle Q_j^\alpha, Q_j^\beta, \ldots \rangle\}_j$, or by accepting perspectivalism and by denying (B) that the ascription of the property $Q^\omega = \sum_j Q_j^\alpha \otimes Q_j^\beta \otimes \cdots$ to the composite ω should be re-endorsed by the ascription of properties $\langle Q^\alpha, Q^\beta, \ldots \rangle \in \{\langle Q_j^\alpha, Q_j^\beta, \ldots \rangle\}_j$ to the subsystems α, β, etc., and *vice versa*. Both options again lead to bizarre consequences, but these consequences do not imply that the bi, spectral or atomic modal interpretations violate Consistency and Internal Completeness.

14.7 An elusive ontology

The bi, spectral and atomic modal interpretations provide descriptions of what reality could be like. And these descriptions comprise descriptions of phenomenal states of affairs, which we can observe, but also descriptions of noumenal states of affairs, which we can never observe (see Section 12.1).

An unattractive feature of these descriptions of reality is now that it is quite hard to acquire knowledge of them. If one knows in advance the exact evolution of the state of the whole universe, there is, of course, no problem: one can then determine for every system at every time the set of possible possessed properties and calculate the correlations between these properties. However, in practice we do not know in advance the state of the universe, or of any other system. So, our knowledge of the descriptions of reality can only come from direct observations of the properties actually possessed by systems, from the outcomes of measurements and from theoretical considerations.

The difficulty is now that the results of Section 12.2 prove that the actually possessed properties of a system α, as well as the outcomes of measurements performed on that system, are poor sources of information about the state of α. If one, for instance, directly observes the actually possessed properties of α, one can determine merely one eigenprojection of W^α according to the bi and spectral modal interpretations, and merely one eigenprojection of the state of each atom in α according to the atomic modal interpretation. Only in a few special cases can one fix (approximately) the state of a system by means of observations or measurements, namely if one has an ensemble of N systems α which are all in the same (unknown) state W^α. If one observes that $\#(C_j^\alpha)$ systems in such an ensemble actually possess the core property C_j^α ($j = 1, 2, \cdots$), then the state $\widetilde{W}^\alpha = \sum_j (\#(C_j^\alpha)/[N \dim(C_j^\alpha)]) C_j^\alpha$ is according to the bi and spectral modal interpretations a good approximation of the true state W^α of the systems (if N is large). And if one performs a measurement on each member, where the measurements are part of the special class of measurements defined in Section 12.2, the state $\widetilde{W}^\alpha = \sum_j (\#(|R_j^\pi\rangle\langle R_j^\pi|)/N)\, |\tilde{a}_j^\alpha\rangle\langle\tilde{a}_j^\alpha|$ is according to all modal interpretations a good approximation of the true state W^α of the systems after the measurements. Hence, although we are used to the fact that in classical physics knowledge of the actually possessed properties of a system or knowledge of the outcome of a measurement performed on that system generally gives immediate knowledge of the state of the system, this is usually not the case in the bi, spectral and atomic modal interpretations. And, as a consequence, it is usually impossible to determine by means of observations or by means of measurements the full description of reality as put forward by these interpretations.

However, maybe this is all too much to expect. Maybe one should already be happy if one can predict on the basis of observations or of measurements, the future properties ascribed by modal interpretations to systems. Now, in the bi and spectral modal interpretations it is in a number of cases indeed possible to predict the future actually possessed properties of systems but in the atomic modal interpretation the results are disappointing.[154] To see this, assume that one directly observes the actually possessed properties of a system α at time s or that one performs a measurement on α which yields an outcome at s. Assume furthermore that after s the system α evolves freely by means of some Hamiltonian H^α which is known to a good approximation. In classical physics such cases are straightforwardly described. On the basis of the observation or on the basis of the outcome of the measurement one determines the state of α at s. And because one knows the Hamiltonian of α to a good approximation, one also knows the evolution of the state of α after s to a good approximation. Any prediction about the future properties of α can then be determined by means of this approximate state evolution.

In the bi and spectral modal interpretations one can also give such predictions. Consider, firstly, the case of a direct observation of α and, say, the observation yields that α actually possesses the core property $P_j^\alpha(s)$. Then, because the core properties of freely evolving systems evolve deterministically, it follows that one can predict that α actually possesses with probability 1 at all times $t > s$ the core property $P_j^\alpha(t)$ given by $U^\alpha(t,s)\, P_j^\alpha(s)\, U^\alpha(s,t)$. Consider, secondly, the case of a measurement on α. The outcome of such a

---

[154] Note that the issue is not whether one can generate on the basis of actual measurement outcomes predictions about the outcomes of future measurements (one can use the effective state defined in Section 12.3 to arrive at such predictions); the issue is whether one can generate predictions about the actual *properties* of systems, independently of whether or not measurements are involved.

measurement does not, in general, determine the actually possessed properties of α after the measurement; only for the above mentioned special class of measurements one can uniquely infer from the outcome the final actually possessed core property of α. And in this special case, the above results for predictions again hold: if α actually possesses core property $P_j^\alpha(s)$ after the measurement, then, with probability 1, it actually possesses the core property $U^\alpha(t,s) P_j^\alpha(s) U^\alpha(s,t)$ at $t > s$.

In the atomic modal interpretation these nice results do not, however, hold. The properties we can directly observe are usually properties of huge molecular systems and for a molecule α it generally is not the case that if it possesses the core property $P_{jk\cdots}^\alpha(s)$ at s and α evolves freely, it then actually possesses the core property $U^\alpha(t,s) P_{jk\cdots}^\alpha(s) U^\alpha(s,t)$ at t (see Section 8.2). Instead, if one wants to predict how the core properties of such a molecule α evolve, one needs to know the state of α but that is, as I said, usually impossible. So, an observation of the actually possessed properties of a freely evolving molecule cannot be used to predict the future possessed properties of that molecule. Consider, secondly, a measurement on α. In the atomic modal interpretation it also holds that the actual outcome of a measurement does not, in general, determine the actually possessed core property of α after the measurement. Again there exists a special class of measurements for which one can uniquely infer from the outcome the final actually possessed core property of α. However, if α is a molecule, this determination of the final possessed core property in this case still does not give the means to predict the future actual core properties. Only if α is atomic, does it hold that if the actual core property after the measurement is $P_j^\alpha(s)$, then, with probability 1, α actually possesses the core property $U^\alpha(t,s) P_j^\alpha(s) U^\alpha(s,t)$ at $t > s$.

Moreover, even if one can precisely or approximately fix the state of α at time s, say, on theoretical grounds or by means of the above described ensembles, one still runs into trouble when trying to predict the future possible possessed properties of α. Echoing a point made by Donald (1998), one usually makes approximations in physics. One neglects, for instance, interactions between distant systems when determining the Hamiltonian of a system. The Hamiltonian is thus only approximately correct. So, even if one knows the exact state of α at s, one ends up with only an approximately correct state of α at $t > s$ when one applies the Schrödinger evolution to this exact state of α at s. Now, if one applies the bi, spectral and atomic modal interpretations to this approximately correct state, the resulting description of reality need no longer be approximately correct. The property ascriptions of the bi, spectral and atomic modal interpretations are unstable when states of systems are close to being degenerate (see Section 14.3). Hence, a small error in the determination of the state of α at t can, near a degeneracy, induce huge errors in the determination of the set of the possible possessed properties of α at t. (This problem that small errors in the determination of the state of a system can induce large errors in the future description of the system is, as I said before, not a problem which haunts only modal interpretations.)

So, the descriptions of reality provided by the bi, spectral and atomic modal interpretations are from an epistemological point of view often rather intractable and elusive.

15

Conclusions

Let's return to the three demands I have imposed on the bi, spectral and atomic modal interpretations. The first was the demand that these interpretations should provide well-developed descriptions of reality. And a first conclusion is that the bi, spectral and atomic modal interpretations indeed provide the necessary starting points to develop them. That is, they all describe reality by means of well-defined property ascriptions: for all it is clear on which points these descriptions of reality need improvement, and for all one has the means to make a start with these improvements.

A second conclusion is, however, that the success with which the bi, spectral and atomic modal interpretations can be developed varies. The results with regard to the full property ascription to a single system are in my opinion satisfactory for all three interpretations (Chapter 5). But with regard to the correlations between the properties ascribed to different systems the results start to diverge. For the atomic modal interpretation such correlations can be given (Section 6.4). But for the spectral modal interpretation it is proved that such correlations do not always exist for the properties ascribed to non-disjoint systems (Section 6.3). And for the bi modal interpretation correlations are unknown or, when one accepts Instantaneous Autonomy, Dynamical Autonomy for measurements and Empirical Adequacy (Section 3.3), do not exist as well (Section 6.3). Only if one adopts perspectivalism, that is, if one assumes that one can simultaneously only consider the properties of two disjoint systems in the bi modal interpretation and of n disjoint systems in the spectral modal interpretation, these two interpretations give all the defined correlations.

The results with regard to the dynamics of the ascribed properties also diverge. The bi, spectral and atomic modal interpretations all allow the derivation of transition probabilities which govern the dynamics of the properties of freely evolving systems (Section 8.2). However, for the case of

273

interacting systems it is generally impossible in the bi modal interpretation without perspectivalism to derive transition probabilities. And for the spectral modal interpretation without perspectivalism the results even turn out to be contradictory (Section 8.3). Only for the atomic modal interpretation and for the bi and spectral modal interpretations with perspectivalism do there exist general expressions for the transition probabilities for the ascribed properties (Sections 8.4 and 8.5).

From this, I conclude that only the atomic modal interpretation and the bi and spectral modal interpretations with perspectivalism can be taken as interpretations which allow well-developed descriptions of reality: these interpretations correlate the properties they simultaneously ascribe to system, and they allow candidate expressions for the transition probabilities for the dynamics of these properties. For the bi and spectral modal interpretations without perspectivalism these correlations and transition probabilities cannot be given.

However, this conclusion does not imply that all the results obtained for the atomic modal interpretation and the bi and spectral modal interpretations with perspectivalism are favourable. The facts that the set of properties possibly possessed by a system evolves discontinuously (Section 7.1) and exhibits instabilities (Section 7.4) are worrisome. These facts mean, for instance, that the problem of the classical limit of quantum mechanics becomes even more difficult than it already was: we now not only have to retrieve the phase space description of classical mechanics from quantum mechanics, we also have to prove the relative continuity and stability of the properties in our classical world from a quantum mechanical reality which is fundamentally discontinuous and unstable. Furthermore, the facts that the dynamics of the actually possessed properties of systems violates Dynamical Autonomy (Sections 8.3 and 9.2) and that this dynamics is explicitly non-local (Section 9.3) are unattractive as well. The violations of Dynamical Autonomy prove that the states of systems do not uniquely determine the statistics of the dynamics of the properties of those systems. And, even though quantum mechanics in the standard formulation is also non-local, the non-locality which is exhibited by our modal interpretations is much more manifest than the usual quantum non-localities. Now, in my opinion the instability of the property ascriptions and the violations of Dynamical Autonomy especially threaten the tenability of modal interpretations. Below I return to these threats.

The second demand on modal interpretations is that their descriptions of reality are empirically adequate in the sense that they solve the measurement problem. That is, the descriptions of reality must yield that measurements have outcomes and that these outcomes appear with the probablities and correlations that are predicted by the standard formulation of quantum mechanics. A first conclusion is now that the bi modal interpretation with perspectivalism cannot meet the second part of this demand. With perspectivalism the bi modal interpretation can only give correlations between at most two measurement outcomes, whereas the standard formulation of quantum mechanics predicts that the outcomes of three or more measurements are also correlated. A second conclusion is that it is beyond doubt that the atomic modal interpretation and the spectral modal interpretation with perspectivalism do not solve the measurement problem for every theoretically imaginable measurement: for both interpretations one can give measurement interactions for which they do not ascribe outcomes after the interaction (Section 10.1). So, if these interpretations are to be empirically adequate, they are so by solving the measurement problem for only the class of physically realistic measurements.

I have defined schemes for measurements for which, due to the internal structure of the measurement device, the spectral and atomic modal interpretations exactly ascribe the outcomes after measurements (Sections 10.3 and 10.4). But I believe it is fair to say that it is conceivable that there exist realistic measurements which do not fit this scheme (Section 14.5). So, one probably has to hope that the spectral and atomic modal interpretations solve the measurement problem by ascribing approximately the outcomes after measurements. And for the spectral modal interpretation there indeed exist measurements where the interactions between the device and the environment (decoherence) make that after the measurement the device is ascribed properties which are close to the outcomes (Section 10.2). This solution to the measurement problem has, however, a serious drawback. And this is that the interactions with the environment are not only able to change the properties of a measurement device to properties that are approximately outcomes, but that they sometimes also make the properties of the device change from outcomes to properties which are not outcomes. And, due to the mentioned instability of the property ascription, this latter change may be possible within a small time interval (Sections 10.2 and 14.5). It is here that the instability of the modal property ascription seriously threatens the spectral and atomic modal interpretations: this instability may make these interpretations empirically inadequate in the sense that they do not yield a stable solution to the measurement problem.

A positive result is that if the spectral and atomic modal interpretations solve the measurement problem, then they automatically ascribe the outcomes with the correct Born probabilities and with the correct correlations (Chapter 11).

With regard to the third demand, I have taken an interpretation of quantum mechanics as metaphysically tenable if it satisfies the criteria of Consistency and Internal Completeness. And since modal interpretations give descriptions of states of affairs which are noumenal in the sense that these states of affairs cannot, in principle, be observed, I have defended the position that one should not impose further criteria on metaphysically tenable interpretations (Section 12.1). Given this position, my conclusion is that on the basis of the problems and results presented in this book, the atomic modal interpretation and the spectral modal interpretation with perspectivalism are metaphysically tenable.

This conclusion again does not imply that these interpretations yield descriptions which are philosophically or physically attractive, or that they yield descriptions which are in full harmony with our every-day notions about reality. It rather means that they are consistent with our observations and that the most poignant violations of every-day notions do not show up at the level of the phenomena, but are restricted to the descriptions of the noumenal states of affairs only. For instance, the facts that the atomic modal interpretation is non-holistic and that the spectral modal interpretation is non-reductionistic (Section 13.2) do not harm their metaphysical tenability. And this means that these facts do not violate Consistency and Internal Completeness, and that at the level of observation these facts need not bother us. The descriptions of our observations and of the outcomes of measurement by the spectral modal interpretation do satisfy reductionism (Section 13.3). And if the atomic modal interpretation solves the measurement problem, then this interpretation can assign in a sensible way holistic properties as dispositions to systems (Section 13.4).

I have also argued that it is often impossible to obtain knowledge about the descriptions of reality by the spectral and atomic modal interpretations (Section 14.7). Hence, from an epistemological point of view one has to hold against these interpretations that they put forward descriptions of reality which are often unknowable to us.

Hence, to summerise, I take the bi and spectral modal interpretations without perspectivalism as interpretations which cannot be developed to full descriptions of reality. And I take the bi modal interpretation with perspectivalism as empirically inadequate. The atomic modal interpretation and the spectral modal interpretation with perspectivalism are metaphysically tenable. However, it is questionable whether they are empirically adequate for it may very well be the case that there exist measurement interactions for which the spectral and atomic modal interpretations do not manage to solve the measurement problem in a stable way.

I have become rather hesitant about the modal interpretations considered in this book. My hesitance has not only to do with the somewhat disappointing results about the well-developedness, empirical adequacy and metaphysical tenability of these interpretations. It also has to do with the starting points of these interpretations. The bi, spectral and atomic modal interpretations belong, as I said when I introduced the different modal interpretations, to the same programme towards an interpretation of quantum mechanics (Section 4.1). Methodologically, this programme is characterised by the common feature that the property ascriptions to systems are implicitly definable from the states of the systems. This implicit definability is achieved because the bi, spectral and atomic modal interpretations all employ the spectral resolutions of states to define the properties of systems. And this made these modal interpretations, when they were proposed, promising interpretations of quantum mechanics: they not only rejected the projection postulate but also advanced descriptions of reality which kept close to the quantum formalism because it is the quantum states themselves which fix the properties of systems. And that all seems natural.

However, in retrospect one can detect a certain incoherence in the spectral and atomic modal interpretations which has to do with the violations of Dynamical Autonomy (it is unclear whether the following applies to the bi modal interpretation because this interpretation is, in general, silent about the tenability of Dynamical Autonomy). The spectral and atomic modal interpretations define, as I have said, the properties of a system by means of the state of the system and the property ascription therefore satisfies Instantaneous Autonomy. That is, there exists a unique relation between a state of the system and the properties possessed by that system. However, since the spectral and atomic modal interpretations violate Dynamical Autonomy for composite systems, there does not exist a unique relation between the evolution of the state of a freely evolving composite system and the evolution of the properties of the subsystems of that composite. And since the atomic modal interpretation violates Dynamical Autonomy for whole systems as well, there also does not exist a unique relation in that interpretation between the evolution of the state of a freely evolving molecule and the evolution of the properties of that molecule.

There is, of course, no logical contradiction involved in this difference between the way in which states and properties are related and the way in which the evolution of states and properties is related. However, it is at least slightly incoherent and a possible ground to question the spectral and atomic modal interpretations. For if it is the case that the evolution of the properties of systems is not uniquely related to the evolution of the states of those systems, why should it then be the case that the properties themselves are uniquely related to those states? By their violations of Dynamical Autonomy, the spectral and atomic modal interpretations force us to understand the physics of the properties of a quantum system not as being uniquely fixed by the state of that system, but as being dependent on the state of the whole universe. But if this is the case, why then not be brave and also take the properties of a quantum system itself as being dependent on the state of the whole universe? So, the inability of these modal interpretations to satisfy Dynamical Autonomy undermines, in my mind, their very starting point that the properties of a system should be defined from the state of the system.

To conclude, even if in the end the spectral and atomic modal interpretations are rejected because they are proved to be empirically inadequate or because they are generally seen as being incoherent, I believe that all the work done on these interpretations will prove its worth. The development of their property ascriptions to fully-developed descriptions of reality, the results obtained when checking their empirical adequacy and the analysis of their metaphysical tenability have yielded fertile techniques for interpreting quantum mechanics without being committed to the projection postulate. That is, many of the results obtained for the spectral and atomic modal interpretations can be applied directly to other (modal) interpretations. Examples of such easily transposable results are the full property ascription (Chapter 5) and the proofs about the probabilities and the correlations with which modal interpretations ascribe outcomes after measurements (Chapter 11). So, even though I expect that the ultimate interpretation will not ascribe properties to quantum systems by means of the spectral resolutions of the states of systems, I surely believe that modal interpretations represent a general and powerful approach to the interpretation of quantum mechanics.

Possibly Bub's fixed modal interpretation (Section 4.6) may eventually be recognised as a modal interpretation which yields a full, empirically adequate and tenable description of reality. And the reason for this is that the fixed modal interpretation does not suffer from the problems that the spectral and atomic modal interpretations have. Its property ascription is by definition fixed and thus stable. And since the core property ascription to a system is only loosely connected with the state of the system, the fixed modal interpretation violates Instantaneous Autonomy equally as it violates Dynamical Autonomy. The only problem I see for the fixed modal interpretation is the question of how to define the core property ascription (or, equivalently, of how to define the preferred magnitude). As I said before, the attractive feature of the bi, spectral and atomic modal interpretations is that their property ascriptions stay close to quantum mechanics. In the fixed modal interpretation it now seems difficult to define a property ascription which shares this feature. Just to assume, for instance, that for pointers of measurement devices the core properties are given by the pointer readings, seems like putting the desired description of reality in 'by hand' and does not provide the attractive feature of being close to quantum mechanics. Instead, what seems to be required is an argument which starts from the quantum formalism and which yields a natural core property ascription for the fixed modal interpretation. And then, if this core property ascription is applied to measurements, it should be a consequence (and not an assumption) that pointers assume their readings. (An obvious route to go is to follow the example of Bohmian mechanics and to argue that the quantum formalism singles out a core property ascription which ascribes positions to systems.) And if it is indeed possible to define a property ascription which follows naturally from quantum mechanics and which makes the fixed modal interpretation solve the measurement problem, then the framework advanced in this book turns this fixed modal interpretation into a well-developed interpretation of quantum mechanics.

Appendix A

From the bi to the spectral modal interpretation

The bi modal interpretation generalises uniquely to the spectral modal interpretation if one assumes that (A) all systems possess properties, that (B) there exist correlations between the properties of mutually disjoint systems, and that the property ascription meets (C) Instantaneous Autonomy, (D) Dynamical Autonomy for measurements and (E) Empirical Adequacy (see Section 3.3).[155]

Firstly, I prove by these assumptions that the bi modal interpretation ascribes core properties to any system independently of whether or not systems are part of a composite with a pure state. Secondly, I show that there exists for any system at any time a measurement which reveals with probability 1 the actually possessed core property of that system. Thirdly, I derive for arbitrary sets of disjoint systems the correlations between the core properties of those systems.

A.1 The general core property ascription

Consider a system α with a state W^α. If α is part of a composite ω with a pure state $|\Psi^\omega\rangle$, then the core property ascription to α is in the bi modal interpretation given by

$$[P_a^\alpha] = 1 \quad \text{with probability} \quad p(P_a^\alpha) = \text{Tr}^\alpha(W^\alpha P_a^\alpha). \tag{A.1}$$

Here, P_a^α is an eigenprojection of W^α corresponding to the non-zero eigenvalue w_a^α.

This property ascription is only valid in the special case that α is part of a composite ω with a pure state. However, by invoking Instantaneous Autonomy, one can turn this property ascription into a generally valid one (I again copy the line of reasoning followed at the end of Section 6.2). Consider any system α with a state W^α. If α is part of a composite ω with a pure state, the core property ascription to α is in the bi modal interpretation given by (A.1). If α is not a part of a composite with a pure state, α still possesses properties according to my assumption (A), but the core property of α is unknown. Instantaneous Autonomy now demands that the property ascription to α is equal in both cases since the state of α is in both cases equal.[156] Hence, also if α is not a part of a composite with a pure state, the core property ascription to α is in the bi modal interpretation given by (A.1).

---

[155] The proof is based on the one given in Vermaas and Dieks (1995).

[156] Instantaneous Autonomy can be used because the property ascription (A.1) to α satisfies the necessary condition that it is a function of the state of α only.

A.2 Core property revealing measurements

Consider the magnitude represented by the state $W^\alpha = \sum_j w_j^\alpha P_j^\alpha$ of a system α. If $\{|p_{jk}^\alpha\rangle\}_{j,k}$ is a set of pair-wise orthogonal vectors which expands the eigenprojections of W^α as $\sum_k |p_{jk}^\alpha\rangle\langle p_{jk}^\alpha| = P_j^\alpha$, a measurement interaction of this magnitude is given by

$$U^{\alpha\pi}(t_2, t_1)|p_{jk}^\alpha\rangle \otimes |\mathrm{R}_0^\pi\rangle = |p_{jk}^\alpha\rangle \otimes |\mathrm{R}_{jk}^\pi\rangle, \qquad (A.2)$$

where π is a pointer and $\{|\mathrm{R}_{jk}^\pi\rangle\}_{j,k}$ are pair-wise orthogonal vectors which are related to the eigenprojections $\{R_j^\pi\}_j$ of the pointer reading magnitude $M^\pi = \sum_j m_j R_j^\pi$ as $R_j^\pi = \sum_k |\mathrm{R}_{jk}^\pi\rangle\langle \mathrm{R}_{jk}^\pi|$, for all j.

According to the bi modal interpretation this measurement reveals with probability 1 the core property that α actually possesses before the measurement. To prove this, I consider a special model in which two of these measurements (A.2) are embedded. The model comprises α and two pointers π and π'. The initial state of $\alpha\pi'\pi$ at time t_0 is $\sum_{jk} \sqrt{w_j^\alpha} |p_{jk}^\alpha\rangle \otimes |\mathrm{R}_0^{\pi'}\rangle \otimes |\mathrm{R}_0^\pi\rangle$ where the values $\{w_j^\alpha\}_j$ are assumed to be the eigenvalues of the state W^α at t_1. Then, from t_0 to t_1 a first measurement (A.2) is performed on α with pointer π' and from t_1 to t_2 a second measurement (A.2) is performed on α, now with the pointer π. The states of the model at t_1 and t_2 are then

$$\left.\begin{aligned}
|\Psi^{\alpha\pi'\pi}(t_1)\rangle &= \sum_{j,k} \sqrt{w_j^\alpha}\, |p_{jk}^\alpha\rangle \otimes |\mathrm{R}_{jk}^{\pi'}\rangle \otimes |\mathrm{R}_0^\pi\rangle, \\
|\Psi^{\alpha\pi'\pi}(t_2)\rangle &= \sum_{j,k} \sqrt{w_j^\alpha}\, |p_{jk}^\alpha\rangle \otimes |\mathrm{R}_{jk}^{\pi'}\rangle \otimes |\mathrm{R}_{jk}^\pi\rangle.
\end{aligned}\right\} \qquad (A.3)$$

The second measurement is the one I am interested in. (Note that the model poses no restrictions on the state of α before this second measurement; by tuning the coefficients $\{w_j^\alpha\}_j$ and the vectors $\{|p_{jk}^\alpha\rangle\}_{j,k}$, one can obtain every possible state $W^\alpha(t_1)$.)

Now apply the bi modal interpretation to the systems α and π' at t_1 and to π at t_2. At t_1 the state of α is $W^\alpha(t_1) = \sum_j w_j^\alpha P_j^\alpha$, so its core properties are $\{P_j^\alpha\}_j$. The state of π' is $W^{\pi'}(t_1) = \sum_j w_j^\alpha R_j^{\pi'}$, so its core properties are the readings $\{R_j^{\pi'}\}_j$. And the state of π at t_2 is $W^\pi(t_2) = \sum_j w_j^\alpha R_j^\pi$, so its core properties are $\{R_j^\pi\}_j$. Since all these properties are by assumption (B) correlated, one can introduce the joint probability $p(P_a^\alpha(t_1), R_b^{\pi'}(t_1), R_c^\pi(t_2))$ that α actually possesses P_a^α at t_1, that π' actually possesses $R_b^{\pi'}$ at t_1, and that π actually possesses R_c^π at t_2. The state of the composite $\alpha\pi'$ is pure at t_1, so one can derive with the bi modal interpretation the joint probabilities $p(P_a^\alpha(t_1), R_b^{\pi'}(t_1))$. By (4.10) these joint probabilities are zero if $a \neq b$. It follows that $p(P_a^\alpha(t_1), R_b^{\pi'}(t_1), R_c^\pi(t_2))$ is zero if $a \neq b$ and one can conclude that

$$p(R_c^\pi(t_2)/P_a^\alpha(t_1)) = p(R_c^\pi(t_2)/R_a^{\pi'}(t_1)). \qquad (A.4)$$

Hence, the left-hand conditional probability, which correlates the actually possessed property of α before the second measurement with the final reading of that measurement, is equal to the conditional probability for the two readings possessed by π' at t_1 and by π at t_2. This latter conditional probability (the right-hand side of (A.4)) yields predictions about measurement outcomes and should, according to Empirical Adequacy, be equal to the conditional probabilities generated by the standard formulation of quantum mechanics.

Apply therefore the standard formulation to the model. If at t_1 the pointer π' assumes the outcome $R_a^{\pi'}$, the state of $\alpha\pi'\pi$ collapses in the standard formulation to

$$\widehat{W}^{\alpha\pi'\pi}(t_1) = \frac{[\mathbb{I}^{\alpha\pi} \otimes R_a^{\pi'}] \, W^{\alpha\pi'\pi}(t_1) \, [\mathbb{I}^{\alpha\pi} \otimes R_a^{\pi'}]}{\mathrm{Tr}^{\alpha\pi'\pi}(W^{\alpha\pi'\pi}(t_1) \, [\mathbb{I}^{\alpha\pi} \otimes R_a^{\pi'}])}$$

$$= \frac{1}{\mathrm{Tr}^{\alpha}(P_a^{\alpha})} \sum_{q,r} |p_{aq}^{\alpha}\rangle\langle p_{ar}^{\alpha}| \otimes |R_{aq}^{\pi'}\rangle\langle R_{ar}^{\pi'}| \otimes |R_0^{\pi}\rangle\langle R_0^{\pi}|. \qquad (A.5)$$

This collapsed state evolves with (A.2) to

$$\widehat{W}^{\alpha\pi'\pi}(t_2) = \frac{1}{\mathrm{Tr}^{\alpha}(P_a^{\alpha})} \sum_{q,r} |p_{aq}^{\alpha}\rangle\langle p_{ar}^{\alpha}| \otimes |R_{aq}^{\pi'}\rangle\langle R_{ar}^{\pi'}| \otimes |R_{aq}^{\pi}\rangle\langle R_{ar}^{\pi}|. \qquad (A.6)$$

From the Born rule it follows that

$$p_{\mathrm{Born}}(R_c^{\pi}(t_2)/R_a^{\pi'}(t_1)) = \mathrm{Tr}^{\alpha\pi'\pi}(\widehat{W}^{\alpha\pi'\pi}(t_2) \, [\mathbb{I}^{\alpha\pi'} \otimes R_c^{\pi}]) = \delta_{ac}. \qquad (A.7)$$

So, by using Empirical Adequacy, one can fix the right-hand side of (A.4) and conclude that

$$p(R_c^{\pi}(t_2)/P_a^{\alpha}(t_1)) = \delta_{ac}. \qquad (A.8)$$

This result can be made generally valid with Dynamical Autonomy for measurements. Take any measurement (A.2) performed on a system α with an initial state $W^{\alpha} = \sum_j w_j^{\alpha} P_j^{\alpha}$. In the case that this measurement is embedded in the above model, the conditional probabilities $p(R_c^{\pi}(t_2)/P_a^{\alpha}(t_1))$ are equal to δ_{ac}. In any other case, these conditional probabilities are not known. Dynamical Autonomy for measurements demands that these conditional probabilities are the same in both cases.[157] Hence, in all cases it holds that $p(R_c^{\pi}(t_2)/P_a^{\alpha}(t_1)) = \delta_{ac}$.

So, every measurement (A.2) on a system α with the state $W^{\alpha} = \sum_j w_j^{\alpha} P_j^{\alpha}$ reveals with probability 1 the actual core property of α before the measurement.

A.3 The correlations between disjoint systems

Consider finally an arbitrary set of disjoint systems α, β, γ, ... with a composite state W^{ω}, where $\omega = \alpha\beta\cdots$. The core properties of these systems are correlated by assumption (B). Suppose now that one measures one by one the core properties of these systems with a series of the property revealing measurements which I defined above. So, a first pointer π_1 interacts with α, a second π_2 interacts with β, and so on. They all reveal with certainty the initial actually possessed core property. So, the probability that these pointers π_1, π_2, etc., possess simultaneously the readings $R_a^{\pi_1}$, $R_b^{\pi_2}$, etc., after the measurements, is equal to the probability $p(P_a^{\alpha}, P_b^{\beta}, \ldots)$ that α, β, ... possess simultaneously their core properties before the measurements. Hence,

$$p(P_a^{\alpha}, P_b^{\beta}, \ldots) = p(R_a^{\pi_1}, R_b^{\pi_2}, \ldots). \qquad (A.9)$$

This latter probability is again a joint probability for measurement outcomes so must by Empirical Adequacy be equal to the joint probability for measurement outcomes predicted by the standard formulation. A straightforward calculation yields that the

---

[157] Dynamical Autonomy for measurements can be used here because the conditional probabilities $p(R_c^{\pi}(t_2)/P_a^{\alpha}(t_1)) = \delta_{ac}$ derived with the model do not depend on the specific state of the pointer π'.

standard formulation predicts that the joint probability to obtain the outcomes $R_a^{\pi_1}$, $R_b^{\pi_2}$, etc., after the measurements is given by $\mathrm{Tr}^\omega(W^\omega [P_a^\alpha \otimes P_b^\beta \otimes \cdots])$. Hence,

$$p(P_a^\alpha, P_b^\beta, \ldots) = \mathrm{Tr}^\omega(W^\omega [P_a^\alpha \otimes P_b^\beta \otimes \cdots]). \tag{A.10}$$

By invoking Instantaneous Autonomy[158] one can transpose this result, derived for this special case of a series of property revealing measurements, to the general case when no measurements are performed on the systems α, β, etc. (The argument is a variation to the arguments given at the end of the two previous sections.)

The conclusion is thus that if one generalises the property ascription of the bi modal interpretation to a property ascription which is applicable to any system, which establishes correlations between the properties of sets of disjoint systems and which satisfies Instantaneous Autonomy, Dynamical Autonomy for measurements and Empirical Adequacy, one uniquely arrives at the property ascriptions (A.1) and (A.10) of the spectral modal interpretation. Conversely, Lemma 6.2 on page 96 proves that the property ascription of the bi modal interpretation can be regained as a special case of the property ascription of the spectral modal interpretation.

---

[158] The joint probabilities (A.10) for $\{\alpha, \beta, \ldots\}$ are only a function of the state of ω as demanded by the necessary condition of Instantaneous Autonomy. One can thus use this assumption.

Glossary

$\alpha, \beta, \gamma, \ldots$	physical systems
$\{\alpha_q\}_q$	atomic systems
$\alpha\beta \cdots$	composite system containing the disjoint systems $\alpha, \beta, \ldots$
λ_d	spectral value of a biorthogonal decomposition of $\|\Psi^{\alpha\beta}\rangle$
μ, ν	measurement devices
$\bar{\mu}$	mechanism of the measurement device μ equal to μ/π
π, ρ	pointers of measurement devices
$\|\psi^{\alpha}\rangle, \|\phi^{\alpha}\rangle, \ldots$	vectors in the Hilbert space $\mathscr{H}^{\alpha}$
$\|\Psi^{\omega}\rangle$	vector in the Hilbert space $\mathscr{H}^{\omega}$ with ω a composite system
ω	composite system
ω/α	the system equal to the composite ω minus its subsystem α
a_j	eigenvalue of the operator A^{α}
$\|a_{jk}^{\alpha}\rangle$	eigenvector of A^{α} corresponding to the eigenvalue a_j
$\sum_k \|a_{jk}^{\alpha}\rangle\langle a_{jk}^{\alpha}\|$	eigenprojection of A^{α} corresponding to the eigenvalue a_j
A^{α}, B^{α}	operators on the Hilbert space $\mathscr{H}^{\alpha}$ (usually self-adjoint)
$\mathscr{B}(\{S_j^{\alpha}\}_j)$	Boolean algebra generated by the projections $\{S_j^{\alpha}\}_j$
C_j^{α}	projection on $\mathscr{H}^{\alpha}$ representing the core property of the system α
$\|d_{\hat{v}}^{\sigma}\rangle$	vector representing spin down in $\hat{v}$ direction of spin $\frac{1}{2}$-particle σ
$\|D^{\mu}\rangle$	vector in the Hilbert space $\mathscr{H}^{\mu}$ of the measurement device μ
$\|D^{\bar{\mu}}\rangle$	vector in the Hilbert space $\mathscr{H}^{\bar{\mu}}$ of the mechanism $\bar{\mu}$

$D^{\bar{\mu}}$	projection on $\mathcal{H}^{\bar{\mu}}$ representing a property of the mechanism $\bar{\mu}$
$\mathcal{DP}_j$	set of definite-valued projections
$\{|e_j^\alpha\rangle\}_j,\ \{|f_k^\beta\rangle\}_k$	sets of normalised and pair-wise orthogonal vectors
E_j	eigenvalue of the Hamiltonian H^α
$|E_{jk}^\alpha\rangle$	eigenvector of the Hamiltonian H^α
$E_A^\alpha(\Gamma)$	spectral projection of the operator A^α corresponding to the set Γ
$|\mathscr{E}^\omega\rangle$	vector in the Hilbert space $\mathcal{H}^\omega$ with ω the environment
$f_q^\alpha(t)$	continuous function of eigenvalues of the state $W^\alpha(t)$
F^α	preferred magnitude (in the fixed modal interpretation)
$\mathscr{F}(\{S_j^\alpha\}_j)$	faux-Boolean algebra generated by the projections $\{S_j^\alpha\}_j$
H^α	Hamiltonian of the system α
$\mathcal{H}^\alpha$	Hilbert space associated with the system α
$J(t)$	probability current
m_j	eigenvalue of the pointer reading magnitude M^π
$M^\pi = \sum_j m_j R_j^\pi$	reading magnitude of the pointer π
$\mathcal{N}(W^\alpha)$	the set of projections onto subspaces of the null space of W^α
$p_j,\ p(\cdots)$	probabilities
P_j^α	eigenprojection of the state W^α
$P_{abc\cdots}^\beta$	projection $P_a^{\alpha_1} \otimes P_b^{\alpha_2} \otimes \cdots$ on $\mathcal{H}^\beta$ with $\{\alpha_q\}_q$ the atoms in β
$P^\alpha(\lambda_d)$	projection on $\mathcal{H}^\alpha$ defined by a biorthogonal decomposition
$P_{(a,b)}^\alpha(t)$	the projection $\sum_{r_i^\alpha(t)\in(a,b)} |r_i^\alpha(t)\rangle\langle r_i^\alpha(t)|$
P_j	the set of projections $\{P_{j_\alpha}^\alpha, P_{j_\beta}^\beta, \ldots\}$
Q^α	projection on the Hilbert space $\mathcal{H}^\alpha$
$\mathcal{Q}^\alpha$	subspace of $\mathcal{H}^\alpha$ onto which the projection Q^α projects
$r_i^\alpha(t)$	ordered eigenvalue function of the state $W^\alpha(t)$
$|r_i^\alpha(t)\rangle\langle r_i^\alpha(t)|$	ordered one-dimensional eigenprojection function of $W^\alpha(t)$
$\tilde{r}_i^\alpha(t)$	eigenvalue function of the state $W^\alpha(t)$
$|\tilde{r}_i^\alpha(t)\rangle\langle\tilde{r}_i^\alpha(t)|$	one-dimensional eigenprojection function of $W^\alpha(t)$
$|\mathrm{R}_{jk}^\pi\rangle$	eigenvector of M^π corresponding to the eigenvalue m_j
$R_j^\pi = \sum_k |\mathrm{R}_{jk}^\pi\rangle\langle\mathrm{R}_{jk}^\pi|$	eigenprojection of M^π corresponding to the eigenvalue m_j

$R_j^{\pi;\omega}$	the projection $R_j^\pi \otimes \mathbb{I}^{\omega/\pi}$ on the Hilbert space $\mathcal{H}^\omega$
$S_{\vec{v}}^\sigma$	spin magnitude in the direction $\vec{v}$ of spin particle σ
$\{S_j^\alpha\}_j$	set of pair-wise orthogonal projections on $\mathcal{H}^\alpha$
$T_q^\alpha(t)$	continuous trajectory of eigenprojections of $W^\alpha(t)$
$T_{kj}(t)$	infinitesimal transition probabilities
$\lvert u_{\vec{v}}^\sigma \rangle$	vector representing spin up in $\vec{v}$ direction of spin $\frac{1}{2}$-particle σ
U^α	unitary operator on the Hilbert space $\mathcal{H}^\alpha$
$U^\alpha(x, y)$	unitary time evolution operator $\exp([(x - y)/i\hbar]\, H^\alpha)$ on $\mathcal{H}^\alpha$
w_j	eigenvalue of the state W^α
W^α	density operator representing the state of the system α
$W_{R_j^\pi}^\alpha$	effective state of α given preparational measurement outcome R_j^π
$[.]_j$	value assignment to properties and magnitudes
$\mathbb{I}^\alpha$	unit operator on the Hilbert space $\mathcal{H}^\alpha$
$\mathbb{O}^\alpha$	null operator on the Hilbert space $\mathcal{H}^\alpha$
$\neg$	negation
$\wedge$	conjunction
$\vee$	disjunction
$\lVert \lvert \psi^\alpha \rangle \rVert$	Hilbert space norm of the vector $\lvert \psi^\alpha \rangle$
$\lVert A^\alpha \rVert$	operator norm of the operator A^α
$\lVert A^\alpha \rVert_1$	trace norm of the operator A^α

Bibliography

Albert, D. Z (1992). *Quantum Mechanics and Experience.* Cambridge,
Massachusetts: Harvard University Press.
Albert, D. Z and B. Loewer (1990). Wanted Dead or Alive: Two Attempts to
Solve Schrödinger's Paradox. In Fine, Forbes, and Wessels (1990),
pp. 277–285.
Albert, D. Z and B. Loewer (1993). Non-Ideal Measurements. *Foundations of
Physics Letters* **6**, 297–305.
Arntzenius, F. (1990). Kochen's Interpretation of Quantum Mechanics. In Fine,
Forbes, and Wessels (1990), pp. 241–249.
Arntzenius, F. (1998). Curiouser and Curiouser: A Personal Evaluation of Modal
Interpretations. In Dieks and Vermaas (1998), pp. 337–377.
Bacciagaluppi, G. (1995). A Kochen–Specker Theorem in the Modal Interpretation
of Quantum Mechanics. *International Journal of Theoretical Physics* **34**,
1205–1216.
Bacciagaluppi, G. (1996a). Delocalised Properties in the Modal Interpretation
of a Continuous Model of Decoherence. University of Cambridge
preprint.
Bacciagaluppi, G. (1996b). Topics in the Modal Interpretation of Quantum
Mechanics. Ph. D. thesis, University of Cambridge. The reader may
alternatively consult Bacciagaluppi (2000).
Bacciagaluppi, G. (1998). Bohm–Bell Dynamics in the Modal Interpretation. In
Dieks and Vermaas (1998), pp. 177–211.
Bacciagaluppi, G. (2000). *Modal Interpretations of Quantum Mechanics.* Cambridge:
Cambridge University Press. Forthcoming.
Bacciagaluppi, G. and W. M. Dickson (1997). Dynamics for Density Operator
Interpretations of Quantum Theory. Los Alamos e-print archives,
quant-hp/9711048.
Bacciagaluppi, G., M. J. Donald, and P. E. Vermaas (1995). Continuity and
Discontinuity of Definite Properties in the Modal Interpretation. *Helvetica
Physica Acta* **68**, 679–704.
Bacciagaluppi, G. and M. Hemmo (1994). Making Sense of Approximate
Decoherence. In D. Hull, M. Forbes, and R. Burian (Eds.), *Proceedings of the
1994 Biennial Meeting of the Philosophy of Science Association*, Volume 1. East
Lansing: Philosophy of Science Association, pp. 345–354.
Bacciagaluppi, G. and M. Hemmo (1996). Modal Interpretations, Decoherence and

Measurements. *Studies in History and Philosophy of Modern Physics* **27**, 239–277.

Bacciagaluppi, G. and M. Hemmo (1998). State Preparation in the Modal Interpretation. In Healey and Hellman (1998), pp. 95–114.

Bacciagaluppi, G. and P. E. Vermaas (1999). Virtual Reality: Consequences of No-Go Theorems for the Modal Interpretation of Quantum Mechanics. In M. Dalla Chiara, F. Laudisa, and R. Giuntini (Eds.), *Language, Quantum, Music*. Dordrecht: Kluwer Academic Publishers. Forthcoming.

Bell, J. S. (1987). *Speakable and Unspeakable in Quantum Mechanics*. Cambridge: Cambridge University Press.

Beltrametti, E. G. and M. J. Maczynski (1993). On the Characterization of Probabilities: A Generalization of Bell's Inequalities. *Journal of Mathematical Physics* **34**, 4919–4929.

Bohm, D. (1952). A Suggested Interpretation of Quantum Theory in Terms of "Hidden Variables". *Physical Review* **85**, 166–193. Parts I and II.

Bohm, D. and B. J. Hiley (1993). *The Undivided Universe: An Ontological Interpretation of Quantum Theory*. London: Routledge.

Bohr, N. (1949). Discussion with Einstein on Epistemological Problems in Atomic Physics. In P. A. Schilpp (Ed.), *Albert Einstein: Philosopher-Scientist* (third ed.). La Salle, Illinois: Open Court, pp. 199–241.

Bub, J. (1992). Quantum Mechanics without the Projection Postulate. *Foundations of Physics* **22**, 737–754.

Bub, J. (1993). Measurement: It Ain't Over Till It's Over. *Foundations of Physics Letters* **6**, 21–35.

Bub, J. (1995). Maximal Structures of Determinate Propositions in Quantum Mechanics. *International Journal of Theoretical Physics* **34**, 1–10.

Bub, J. (1997). *Interpreting the Quantum World*. Cambridge: Cambridge University Press.

Bub, J. (1998a). Decoherence in Bohmian Modal Interpretations. In Dieks and Vermaas (1998), pp. 241–252.

Bub, J. (1998b). Schrödinger's Cat and Other Entanglements of Quantum Mechanics. In J. Earman and J. Norton (Eds.), *The Cosmos of Science*. Pittsburgh: University of Pittsburgh Press, pp. 274–298.

Bub, J. and R. K. Clifton (1996). A Uniqueness Theorem for 'No Collapse' Interpretations of Quantum Mechanics. *Studies in History and Philosophy of Modern Physics* **27**, 181–219.

Busch, P., P. J. Lahti, and P. Mittelsteadt (1991). *The Quantum Theory of Measurement*. Berlin: Springer.

Butler, S. (1872). *Erewhon*. London: Trübner & Co.

Clifton, R. K. (1994). The Triorthogonal Uniqueness Theorem and Its Irrelevance to the Modal Interpretation of Quantum Mechanics. In K. V. Laurikainen, C. Montonen, and K. Sunnarborg (Eds.), *Symposium on the Foundations of Modern Physics 1994: 70 Years of Matters Waves*. Gif-sur-Yvette: Editions Frontières, pp. 45–60.

Clifton, R. K. (1995a). Independent Motivation of the Kochen–Dieks Modal Interpretation of Quantum Mechanics. *British Journal for the Philosophy of Science* **46**, 33–57.

Clifton, R. K. (1995b). Making Sense of the Kochen–Dieks 'No-Collapse' Interpretation of Quantum Mechanics Independent of the Measurement Problem. In D. Greenberger and A. Zeilinger (Eds.), *Fundamental Problems in*

Quantum Theory, Volume 755. New York: Annals of the New York Academy of Science, pp. 570–578.

Clifton, R. K. (1995c). Why Modal Interpretations of Quantum Mechanics Must Abandon Classical Reasoning About Physical Properties. *International Journal of Theoretical Physics* **34**, 1302–1312.

Clifton, R. K. (1996). The Properties of the Modal Interpretations of Quantum Mechanics. *British Journal for the Philosophy of Science* **47**, 371–398.

de Saint-Exupéry, A. (1943). *Le Petit Prince*. New York: Reynal and Hitchcock.

d'Espagnat, B. (1966). An Elementary Note about "Mixtures". In A. De-Shalit, H. Feshbach, and L. van Hove (Eds.), *Preludes in Theoretical Physics*. Amsterdam: North–Holland, pp. 185–191.

d'Espagnat, B. (1971). *Conceptual Foundations of Quantum Mechanics*. Menlo Park, California: W. A. Benjamin.

Dickson, W. M. (1995a). Faux-Boolean Algebras and Classical Models. *Foundations of Physics Letters* **8**, 401–415.

Dickson, W. M. (1995b). Faux-Boolean Algebras, Classical Probabilities, and Determinism. *Foundations of Physics Letters* **8**, 231–242.

Dickson, W. M. (1998a). On the Plurality of Dynamics: Transition Probabilities and Modal Interpretations. In Healey and Hellman (1998), pp. 160–182.

Dickson, W. M. (1998b). *Quantum Chance and Non-Locality: Probability and Non-Locality in the Interpretations of Quantum Mechanics*. Cambridge: Cambridge University Press.

Dickson, W. M. and R. K. Clifton (1998). Lorentz-Invariance in Modal Interpretations. In Dieks and Vermaas (1998), pp. 9–47.

Dieks, D. (1988). The Formalism of Quantum Theory: An Objective Description of Reality? *Annalen der Physik* **7**, 174–190.

Dieks, D. (1989). Quantum Mechanics without the Projection Postulate and Its Realistic Interpretation. *Foundations of Physics* **19**, 1397–1423.

Dieks, D. (1993). The Modal Interpretation of Quantum Mechanics and Some of Its Relativistic Aspects. *International Journal of Theoretical Physics* **32**, 2363–2375.

Dieks, D. (1994a). Modal Interpretation of Quantum Mechanics, Measurements, and Macroscopic Behavior. *Physical Review A* **49**, 2290–2300.

Dieks, D. (1994b). Objectification, Measurement and Classical Limit According to the Modal Interpretation of Quantum Mechanics. In P. Busch, P. J. Lahti, and P. Mittelsteadt (Eds.), *Symposium on the Foundations of Modern Physics 1993: Quantum Measurement, Irreversibility and the Physics of Information*. Singapore: World Scientific, pp. 160–167.

Dieks, D. (1995). Physical Motivation of the Modal Interpretation of Quantum Mechanics. *Physics Letters A* **197**, 367–371.

Dieks, D. (1998a). Locality and Lorentz-Covariance in the Modal Interpretation of Quantum Mechanics. In Dieks and Vermaas (1998), pp. 49–67.

Dieks, D. (1998b). Preferred Factorizations and Consistent Property Attribution. In Healey and Hellman (1998), pp. 144–159.

Dieks, D. and P. E. Vermaas (Eds.) (1998). *The Modal Interpretation of Quantum Mechanics*, Volume 60 of *The Western Ontario Series in Philosophy of Science*. Dordrecht: Kluwer Academic Publishers.

Donald, M. J. (1998). Discontinuity and Continuity of Definite Properties in the Modal Interpretation. In Dieks and Vermaas (1998), pp. 213–222.

Doob, J. L. (1953). *Stochastic Processes*. New York: Wiley.

Einstein, A., B. Podolsky, and N. Rosen (1935). Can Quantum-Mechanical Description of Physical Reality be Considered Complete? *Physical Review* **47**, 777–780.

Elby, A. (1993). Why "Modal" Interpretations of Quantum Mechanics don't Solve the Measurement Problem. *Foundations of Physics Letters* **6**, 5–19.

Elby, A. and J. Bub (1994). Triorthogonal Uniqueness Theorem and Its Relevance to the Interpretation of Quantum Mechanics. *Physical Review A* **49**, 4213–4216.

Feller, W. (1950). *An Introduction to Probability Theory and Its Applications*, Volume I. New York: Wiley.

Feyerabend, P. (1975). *Against Method*. London: New Left Books.

Fine, A. (1973). Probabilities and the Interpretation of Quantum Mechanics. *British Journal for the Philosophy of Science* **24**, 1–37.

Fine, A. (1982). Joint Distributions, Quantum Correlations, and Commuting Observables. *Journal of Mathematical Physics* **23**, 1306–1310.

Fine, A., M. Forbes, and L. Wessels (Eds.) (1990). *Proceedings of the 1990 Biennial Meeting of the Philosophy of Science Association*, Volume 1. East Lansing: Philosophy of Science Association.

Healey, R. A. (1989). *The Philosophy of Quantum Mechanics: An Interactive Interpretation*. Cambridge: Cambridge University Press.

Healey, R. A. (1993a). Measurement and Quantum Indeterminateness. *Foundations of Physics Letters* **6**, 307–316.

Healey, R. A. (1993b). Why Error-Prone Quantum Measurements have Outcomes. *Foundations of Physics Letters* **6**, 37–54.

Healey, R. A. (1994). Nonseparable Processes and Causal Explanation. *Studies in History and Philosophy of Modern Physics* **25**, 337–374.

Healey, R. A. (1995). Dissipating the Quantum Measurement Problem. *Topoi* **14**, 55–65.

Healey, R. A. (1998). "Modal" Interpretations, Decoherence, and the Quantum Measurement Problem. In Healey and Hellman (1998), pp. 52–86.

Healey, R. A. and G. Hellman (Eds.) (1998). *Quantum Measurement: Beyond Paradox*, Volume 17 of *Minnesota Studies in the Philosophy of Science*. Minneapolis: University of Minnesota Press.

Hemmo, M. (1996). Quantum Mechanics Without Collapse: Modal Interpretation, Histories and Many Worlds. Ph. D. thesis, University of Cambridge.

Hemmo, M. (1998). Quantum Histories in the Modal Interpretation. In Dieks and Vermaas (1998), pp. 253–277.

Holland, P. R. (1993). *The Quantum Theory of Motion: An Account of the de Broglie–Bohm Causal Interpretation of Quantum Mechanics*. Cambridge: Cambridge University Press.

Honderich, T. (Ed.) (1995). *The Oxford Companion to Philosophy*. Oxford: Oxford University Press.

Jauch, J. M. (1968). *Foundations of Quantum Mechanics*. Reading, Massachusetts: Addison–Wesley Publishing Company.

Kant, I. (1781/1787). *Kritik der reinen Vernunft*. Riga: Johann Friederich Hartknoch.

Kato, T. (1976). *Perturbation Theory for Linear Operators* (second ed.). Berlin: Springer.

Kawata, T. (1972). *Fourier Analysis in Probability Theory*. New York: Academic Press.

Kernaghan, M. (1994). Bell–Kochen–Specker Theorem for 20 Vectors. *Journal of Physics A* **27**, L829–L830.

Kochen, S. (1985). A New Interpretation of Quantum Mechanics. In P. J. Lahti and P. Mittelsteadt (Eds.), *Symposium on the Foundations of Modern Physics*. Singapore: World Scientific, pp. 151–169.

Kochen, S. and E. Specker (1967). The Problem of Hidden Variables in Quantum Mechanics. *Journal of Mathematics and Mechanics* **17**, 59–87.

Krips, H. (1987). *The Metaphysics of Quantum Theory*. Oxford: Clarendon.

Lacey, A. R. (1976). *A Dictionary of Philosophy*. London: Routledge & Kegan Paul.

Leeds, S. and R. A. Healey (1996). A Note on Van Fraassen's Modal Interpretation of Quantum Mechanics. *Philosophy of Science* **63**, 91–104.

Malament, D. (1977). Causal Theories of Time and the Conventionality of Simultaneity. *Noûs* **11**, 293–300.

More, T. (1516). *Utopia*. Antwerp: Thierry Maartens.

Peyo (1967). *De Zwarte Smurfen*. Den Haag: Dupuis.

Pitowsky, I. (1989). *Quantum Probability — Quantum Logic*. Berlin: Springer.

Redhead, M. L. G. (1987). *Incompleteness, Nonlocality and Realism*. Oxford: Clarendon.

Reed, M. and B. Simon (1972). *Methods of Modern Mathematical Physics*, Volume 1. New York: Academic Press.

Reed, M. and B. Simon (1975). *Methods of Modern Mathematical Physics*, Volume 2. New York: Academic Press.

Reeder, N. (1995). Property Composition in Healey's Interpretation of Quantum Mechanics. *Foundations of Physics Letters* **8**, 497–522.

Reeder, N. (1998). Projection Operators, Properties, and Idempotent Variables in the Modal Interpretations. In Dieks and Vermaas (1998), pp. 149–176.

Reeder, N. and R. K. Clifton (1995). Uniqueness of Prime Factorizations of Linear Operators in Quantum Mechanics. *Physics Letters A* **204**, 198–204.

Rellich, F. (1969). *Perturbation Theory of Eigenvalue Problems*. New York: Gordon and Breach.

Ruetsche, L. (1995). Measurement Error and the Albert–Loewer Problem. *Foundations of Physics Letters* **8**, 327–344.

Ruetsche, L. (1998). How Close is "Close Enough"? In Dieks and Vermaas (1998), pp. 223–239.

Schmidt, E. (1907). Zur Theorie der linearen und nichtlinearen Integralgleichungen. I. Teil: Entwicklung willkürlicher Funktionen nach Systemen vorgeschriebener. *Mathematische Annalen* **63**, 433–476.

Schrödinger, E. (1935). Discussion of Probability Relations between Separated Systems. *Proceedings of the Cambridge Philosophical Society* **31**, 555–563.

Sudbery, A. (1986). *Quantum Mechanics and the Particles of Nature: An Outline for Mathematicians*. Cambridge: Cambridge University Press.

Svetlichny, G., M. L. G. Redhead, H. Brown, and J. N. Butterfield (1988). Do the Bell Inequalities Require the Existence of Joint Probability Distributions? *Philosophy of Science* **55**, 387–401.

Swift, J. (1726). *Gulliver's Travels*. London.

Van Fraassen, B. C. (1972). A Formal Approach to the Philosophy of Science. In R. Colodny (Ed.), *Paradigms and Paradoxes: Philosophical Challenges of the Quantum Domain*. Pittsburgh: University of Pittsburgh Press, pp. 303–366.

Van Fraassen, B. C. (1973). Semantic Analysis of Quantum Logic. In C. A. Hooker (Ed.), *Contemporary Research in the Foundations and Philosophy of Quantum*

Theory. Dordrecht: Reidel, pp. 80–113..

Van Fraassen, B. C. (1976). The Einstein–Podolsky–Rosen Paradox. In P. Suppes (Ed.), *Logic and Probability in Quantum Mechanics.* Dordrecht: Reidel, pp. 283–301.

Van Fraassen, B. C. (1980). *The Scientific Image.* Oxford: Clarendon.

Van Fraassen, B. C. (1981). A Modal Interpretation of Quantum Mechanics. In E. G. Beltrametti and B. C. Van Fraassen (Eds.), *Current Issues in Quantum Logic.* New York: Plenum, pp. 229–258.

Van Fraassen, B. C. (1990). The Modal Interpretation of Quantum Mechanics. In P. J. Lahti and P. Mittelsteadt (Eds.), *Symposium on the Foundations of Modern Physics.* Singapore: World Scientific, pp. 440–460.

Van Fraassen, B. C. (1991). *Quantum Mechanics: An Empiricist View.* Oxford: Clarendon.

Van Fraassen, B. C. (1997). Modal Interpretation of Repeated Measurement: A Rejoinder to Leeds and Healey. *Philosophy of Science* **64**, 669–676.

Vermaas, P. E. (1996). Unique Transition Probabilities in the Modal Interpretation. *Studies in History and Philosophy of Modern Physics* **27**, 133–159.

Vermaas, P. E. (1997). A No-Go Theorem for Joint Property Ascriptions in Modal Interpretations of Quantum Mechanics. *Physical Review Letters* **78**, 2033–2037.

Vermaas, P. E. (1998a). Expanding the Property Ascriptions in the Modal Interpretation of Quantum Theory. In Healey and Hellman (1998), pp. 115–143.

Vermaas, P. E. (1998b). Possibilities and Impossibilities of Modal Interpretations of Quantum Mechanics. Ph. D. thesis, Utrecht University.

Vermaas, P. E. (1998c). The Pros and Cons of the Kochen–Dieks and the Atomic Modal Interpretation. In Dieks and Vermaas (1998), pp. 103–148.

Vermaas, P. E. (1999). Two No-Go Theorems for Modal Interpretations of Quantum Mechanics. *Studies in History and Philosophy of Modern Physics* **30**(3). Forthcoming.

Vermaas, P. E. and D. Dieks (1995). The Modal Interpretation of Quantum Mechanics and Its Generalization to Density Operators. *Foundations of Physics* **25**, 145–158.

Vink, J. (1993). Quantum Mechanics in Terms of Discrete Beables. *Physical Review A* **48**, 1808–1818.

Von Neumann, J. (1932). *Mathematische Grundlagen der Quantenmechanik.* Berlin: Springer Verlag. I refer to the English translation Von Neumann (1955).

Von Neumann, J. (1955). *Mathematical Foundations of Quantum Mechanics.* Princeton: Princeton University Press.

Zeh, H. D. (1970). On the Interpretation of Measurements in Quantum Theory. *Foundations of Physics* **1**, 69–76.

Zeh, H. D. (1973). Towards a Quantum Theory of Observation. *Foundations of Physics* **3**, 109–116.

Zimba, J. and R. K. Clifton (1998). Valuations on Functionally Closed Sets of Quantum Mechanical Observables and Von Neumann's 'No-Hidden-Variables' Theorem. In Dieks and Vermaas (1998), pp. 69–101.

Zurek, W. H. (1981). Pointer Basis of Quantum Apparatus: Into What Mixture Does the Wave Packet Collapse? *Physical Review D* **24**, 1516–1525.

Zurek, W. H. (1982). Environment-Induced Superselection Rules. *Physical Review D* **26**, 1862–1880.

Index

The bold numbers refer to definitions and more explicit descriptions

Printed in the United States
By Bookmasters